■ **TABLE 1.4**

Conversion Factors from SI Units to BG and EE Units[a]

	To Convert from	to	Multiply by
Acceleration	m/s^2	ft/s^2	3.281
Area	m^2	ft^2	1.076 E + 1
Density	kg/m^3	lbm/ft^3	6.243 E − 2
	kg/m^3	$slugs/ft^3$	1.940 E − 3
Energy	J	Btu	9.478 E − 4
	J	ft·lb	7.376 E − 1
Force	N	lb	2.248 E − 1
Length	m	ft	3.281
	m	in.	3.937 E + 1
	m	mile	6.214 E − 4
Mass	kg	lbm	2.205
	kg	slug	6.852 E − 2
Power	W	ft·lb/s	7.376 E − 1
	W	hp	1.341 E − 3
Pressure	N/m^2	in. Hg (60 °F)	2.961 E − 4
	N/m^2	lb/ft^2 (psf)	2.089 E − 2
	N/m^2	$lb/in.^2$ (psi)	1.450 E − 4
Specific weight	N/m^3	lb/ft^3	6.366 E − 3
Temperature	°C	°F	$T_F = 1.8\,T_C + 32°$
	K	°R	1.800
Velocity	m/s	ft/s	3.281
	m/s	mi/hr (mph)	2.237
Viscosity (dynamic)	$N{\cdot}s/m^2$	$lb{\cdot}s/ft^2$	2.089 E − 2
Viscosity (kinematic)	m^2/s	ft^2/s	1.076 E + 1
Volume flowrate	m^3/s	ft^3/s	3.531 E + 1
	m^3/s	gal/min (gpm)	1.585 E + 4

[a]If more than four-place accuracy is desired, refer to Appendix A.

Student Solutions Manual & Study Guide
To Accompany

Fundamentals of Fluid Mechanics
Fifth Edition

Bruce R. Munson
Iowa State University

Donald F. Young
Iowa State University

Theodore H. Okiishi
Iowa State University

WILEY
JOHN WILEY & SONS, INC.

Cover Photo Credit: ©Herman Eisenbeiss/Photo Researchers, Inc.

To order books or for customer service call 1-800-CALL-WILEY (225-5945).

ISBN 0-471-71896-3

Printed in the United States of America

10 9 8 7 6 5 4 3 2 1

Printed and bound by Malloy, Inc.

Preface

This *Student Solution Manual and Study Guide* has been developed as a supplement to the *Fundamentals of Fluid Mechanics*, by Munson, Young, and Okiishi (www.wiley.com/college/munson). A set of 456 completely worked-out examples is available to the user: 109 examples in the *Study Guide* section of this manual, 183 examples in the *Student Solutions Manual* section, and 164 examples in the textbook.

The *Study Guide* portion of this manual contains a concise overview of the essential points of most of the main sections in the text, along with appropriate equations and illustrations. A series of bulleted items forms this overview. Material from all of the chapters in the text is included. For most of the sections outlined in the *Study Guide*, a *Caution* is added to alert users to possible difficulties associated with the concepts and equations discussed. These "cautions" are based on many years of classroom experience dealing with students' questions and common mistakes.

At the end of most of the sections in the *Study Guide*, an example problem is given along with a worked-out solution. These examples are very brief and are presented to emphasize the concepts and equations noted in the section. In the solutions to the examples, intermediate calculations and answers are given to three significant figures. Unless otherwise indicated in the example statement, values of fluid properties used are those given in the properties tables found on the inside of the front cover of the *Study Guide*.

The *Study Guide* can serve as a brief, but informative, outline of the essentials in the various chapters and sections of the text and can be useful when reviewing class notes and preparing for exams. It should be emphasized that the material contained in the Study Guide is intended as a supplement to the text. It does not contain the necessary detailed explanations and examples found in the ext. To gain the greatest benefit from the Study guide, it is important that the full text be used.

The book website (www.wiley.com/college/munson) contains a series of review problems. These problems are representative of the types of problems that students should be able to solve after completing the chapter, and the *Student Solution Manual* portion of this supplement contains the detailed solutions to these review problems. We believe that as students prepare for an examination, or feel the need for some additional work on a particular topic, it will be helpful to have available such a set of review problems with their corresponding solutions.

Each review problem is preceded by brief phrases which give an indication of the main topics to be used in solving the problem. Thus, the student can conveniently select those topics, and the corresponding review problems, of interest. This information is also presented in the table of contents. The solutions contained in this manual are worked in a logical, systematic way with sufficient detail so that they can be readily followed. Except where a greater accuracy is warranted, all intermediate calculations and answers are given to three significant figures. For convenience, some tables of properties and tables of unit conversion factors taken from *Fundamentals of Fluid Mechanics* are included on the inside of the front and back covers of this manual. Unless otherwise indicated in the problem statement, values of fluid properties used in the solutions are those given in the properties tables found on the inside of the front cover. Some occasional references to equations made in the solutions refer back to *Fundamentals of Fluid Mechanics*.

The authors hope that the *Student Solution Manual and Study Guide* will prove to be a useful additional tool and aid to the student in acquiring knowledge of the fundamental concepts of fluid mechanics. We believe that practice through solving a variety of problems, with immediately available feedback by way of a *Student Solution Manual and Study Guide*, can be a valuable component in the spectrum of teaching tools needed in the study of fluid mechanics. Any suggestions and comments from you, the user, are certainly welcome and appreciated.

Bruce R. Munson
Donald F. Young
Theodore H. Okiishi

Table of Contents

Chapter 3
Elementary Fluid Dynamics – The Bernoulli Equation

Section	Page	Review Problems Topics	
3.1 Newton's Second Law	3-1	F=ma along streamline	3.1R
3.2 F=ma Along a Streamline	3-2	F=ma normal to streamline	3.2R
3.3 F=ma Normal to a Streamline	3-4	Stagnation pressure	3.3R
3.5 Static, Stagnation, Dynamic, and Total Pressure	3-5	Bernoulli equation	3.4R
3.6 Examples of Use of the Bernoulli Equation	3-6	Heads	3.5R
3.7 The Energy Line and the Hydraulic Grade Line	3-10	Free jet	3.6R
		Bernoulli/ continuity	3.7R, 3.8R, 3.9R, 3.10R, 3.12R
		Bernoulli/ continuity/ Pitot tube	3.11R
		Cavitation	3.13R
		Flowrate	3.14R
		Channel flow	3.15R, 3.16R
		Energy line/ hydraulic grade line	3.17R
		Restrictions on Bernoulli equation	3.18R, 3.19R

Chapter 4
Fluid Kinematics

Section	Page	Review Problems Topics	
4.1 The Velocity Field	4-1	Streamlines	4.1R, 4.2R
4.2 The Acceleration Field	4-3	Material derivative	4.3R
4.3 Control Volume and System Representation	4-5	Acceleration	4.4R, 4.5R, 4.6R, 4.7R
4.4 The Reynolds Transport Theorem	4-5	Reynolds transport theorem	4.8R
		Flowrate	4.9R, 4.10R
		Control volume/ system	4.11R

Chapter 5
Finite Control Volume Analysis

		Review Problems Topics	
5.1 Conservation of Mass – The Continuity Equation	5-1	Continuity equation	5.1R, 5.2R, 5.3R
5.2 Newton's Second Law – The Linear Momentum and Moment-of-Momentum Equations	5-3	Average velocity	5.4R
5.3 First Law of Thermodynamics – The Energy Equation	5-6	Linear momentum	5.5R, 5.6R, 5.7R, 5.8R, 5.9R, 5.10R, 5.11R, 5.12R
		Moment-of-momentum	5.13R, 5.14R, 5.15R, 5.16R
		Energy equation	5.17R, 5.18R, 5.20R, 5.21R, 5.22R
		Linear momentum/ energy	5.19R

Chapter 6
Differential Analysis of Fluid Flow

		Review Problems Topics	
6.1 Fluid Element Kinematics	6-1	Acceleration	6.1R
6.2 Conservation of Mass	6-4	Vorticity	6.2R
6.3 Conservation of Linear Momentum	6-7	Conservation of mass	6.3R, 6.4R
6.4 Inviscid Flow	6-8	Stream function	6.5R
6.5 Some Basic, Plane Potential Flows	6-9	Velocity potential	6.6R
6.6 Superposition of Basic, Plane Potential Flows	6-11	Potential flow	6.7R, 6.10R, 6.11R
6.8 Viscous Flow	6-13	Inviscid flow	6.8R, 6.9R
6.9 Some Simple Solutions for Viscous, Incompressible Fluids	6-14	Viscous flow	6.12R, 6.13R, 6.14R, 6.15R, 6.16R

Chapter 7
Similitude, Dimensional Analysis, and Modeling

Review Problems Topics

Chapter 8
Viscous Flow in Pipes

Review Problems Topics

Chapter 9 Flow Over Immersed Bodies

Chapter 10 Open-Channel Flow

Chapter 11
Compressible Flow

Chapter 12
Turbomachines

1
Introduction

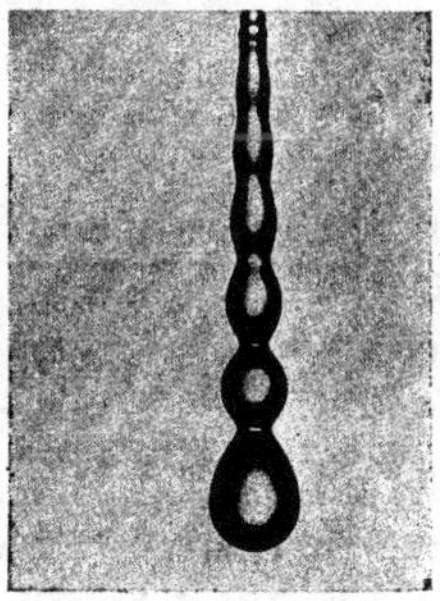

1.2 Dimensions, Dimensional Homogeneity, and Units

- Fluid characteristics can be described both *qualitatively* and *quantitatively*. The qualitative aspect serves to identify the type of characteristic, whereas the quantitative aspect provides a numerical measure of the characteristic through the use of a unit system.
- The qualitative description of fluid characteristics is expressed in terms of primary quantities (basic dimensions) such as length, L, time, T, and mass, M. Secondary quantities such as velocity, V, can be expressed in terms of the basic dimensions, e.g., $V \doteq L\,T^{-1}$, where the symbol $\doteq$ denotes "dimensions of".
- Newton's second law states that force is equal to mass times acceleration so that $F \doteq MLT^{-2}$ and therefore $M \doteq FL^{-1}T^{2}$. Thus, secondary quantities expressed in terms of L, T, and M can also be expressed in terms of L, T, and F.
- All theoretically derived equations are *dimensionally homogeneous* so that the dimensions of the left side of an equation must be the same as the dimensions on the right side of the equation.
- *Caution:* It would be incorrect to use L, T, F, and M as basic dimensions.

EXAMPLE: A common fluid property is the fluid density, ρ, defined as the fluid mass per unit volume. **(a)** Express the dimensions of ρ in terms of L, T, and F. **(b)** The Froude number is a common parameter used in the study of river flow and is defined as $V/\sqrt{g\,d}$ where V is a velocity, g the acceleration of gravity, and d the fluid depth. Determine the dimensions of the Froude number.
SOLUTION:

(a) $\rho \doteq ML^{-3} \doteq (FL^{-1}T^{2})L^{-3} = \underline{\underline{FL^{-4}T^{2}}}$

(b) Froude number $= V/\sqrt{gd} \doteq (LT^{-1})[(LT^{-2})L]^{-1/2} = \underline{\underline{L^{0}T^{0}}}$

Note that the Froude number is dimensionless.

Systems of Units

- For a quantitative description of fluid characteristics *units* are required. In this text two systems of units are used, the British Gravitational (BG) System (feet, seconds, pounds, and slugs) and the International (SI) System (meters, seconds, newtons, and kilograms).
- When solving problems it is important to use a consistent system of units; e.g., don't mix BG and SI units.
- *Caution:* In mechanics it is necessary to distinguish between weight, $\mathcal{W}$, and mass, m. The relationship between the weight (which is the force due to gravity) and the corresponding mass is $\mathcal{W} = m\,g$.

EXAMPLE: (a) Make use of Table 1.3 to express a velocity of 16 ft/s in SI units. **(b)** Make use of Table 1.4 to express a pressure of 200 kN/m^2 in BG units. **(c)** A tank of liquid weighs 150 lb. Determine its mass in slugs.

SOLUTION:

(a) From Table 1.3, $16\ \text{ft/s} = 16\ \text{ft/s}\,(3.048\times10^{-1}) = \underline{\underline{4.88\ \text{m/s}}}$

(b) From Table 1.4, $200\times10^3\ \text{N/m}^2 = 200\times10^3\ \text{N/m}^2\,(2.089\times10^{-2})$

$= \underline{\underline{4180\ \text{lb/ft}^2}}$

(c) $\mathcal{W} = mg$ or $m = \mathcal{W}/g = 150\ \text{lb}/(32.2\ \text{ft/s}^2)$

$= 4.66\ \text{lb}\cdot\text{s}^2/\text{ft} = 4.66\ (\text{slug}\cdot\text{ft/s}^2)\text{s}^2/\text{ft}$

$= \underline{\underline{4.66\ \text{slugs}}}$

Recall that $1\ \text{lb} = 1\ \text{slug}\cdot\text{ft/s}^2$

1.4 Measures of Fluid Mass and Weight

Density

- The density, ρ, of a fluid is defined as its *mass* per unit volume.

Specific Weight

$$\gamma = \rho g \qquad \textbf{(1.6)}$$

- The specific weight, γ, of a fluid is defined as its *weight* per unit volume.
- The specific weight is related to density through the acceleration of gravity as shown by Eq. 1.6.

Specific Gravity

- The *specific gravity,* SG, of a fluid is defined as the ratio of the density of the fluid to the density of water at some specified temperature (usually at 4° C).

EXAMPLE: The specific gravity of a certain liquid is 2.6. Determine the density and specific weight of the liquid in SI units.
SOLUTION:

$SG = \dfrac{\rho}{\rho_{H_2O}}$ so that $\rho = SG\,\rho_{H_2O} = 2.6\,(1000\ kg/m^3) = \underline{\underline{2.6\times10^3\ kg/m^3}}$

From Eq. 1.6, $\gamma = \rho g = 2.6\times10^3\ kg/m^3\,(9.81\ m/s^2) = 25.5\times10^3\ kg/m^2 s^2$
$= \underline{\underline{25.5\ kN/m^3}}$

Recall that $1N = 1\ kg\cdot m/s^2$

1.5 Ideal Gas Law

$$p = \rho RT \qquad \textbf{(1.8)}$$

- The relationship between pressure, density, and temperature for an ideal (or perfect) gas is given by Eq. 1.8, where R is the gas constant.
- *Caution: Absolute* pressures and temperatures must be used in Eq. 1.8.

EXAMPLE: A pressure gage connected to a tank of oxygen reads 50 psi, and the temperature of the oxygen is 80° F. If the local atmospheric pressure is 14.7 psia determine the density of the oxygen in BG units.

SOLUTION:

From Eq. 1.8, $\rho = p/RT$ where $T = (80 + 460)°R = 540°R$ and

$p = 50\,psi + 14.7\,psi = 64.7\,lb/in.^2\,(abs)\,(144\,in.^2/ft^2) = 9320\,lb/ft^2\,(abs)$

Also, for oxygen $R = 1.554 \times 10^3\,ft \cdot lb/slug \cdot °R$ Thus,

$\rho = 9320\,lb/ft^2 / [(1.554 \times 10^3\,ft \cdot lb/slug \cdot °R)(540°R)] = \underline{\underline{0.0111\,slugs/ft^3}}$

1.6 Viscosity

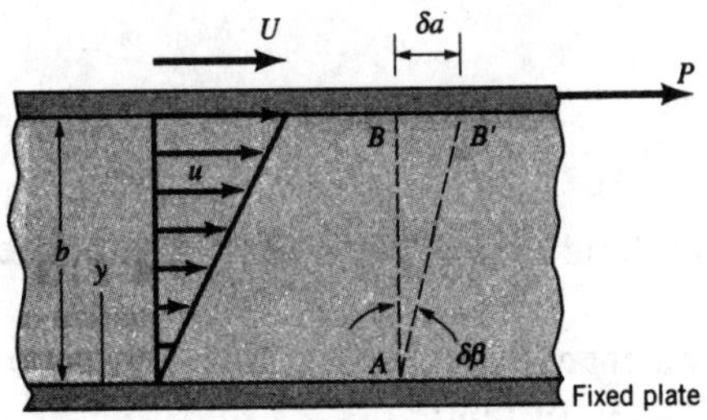

■ FIGURE 1.3 **Behavior of a fluid placed between two parallel plates.**

$$\tau = \mu \frac{du}{dy} \qquad \textbf{(1.9)}$$

- Real fluids, even though they are moving, always "stick" to the solid boundaries that contain them (the *no-slip condition*).
- As illustrated in Fig. 1.3, when a fluid is placed between parallel plates with the bottom plate fixed and the upper plate moving, a velocity gradient, du/dy, develops between the plates due to the no-slip condition.
- As shown in Eq. 1.9, the shearing stress, τ, created by the moving plate is directly related to the velocity gradient, du/dy. The constant of proportionality, μ, is called the *viscosity* of the fluid.
- Fluids for which the shearing stress is directly proportional to the velocity gradient, as in Eq. 1.9, are called *Newtonian* fluids.
- In many situations the viscosity is combined with the fluid density, and the ratio, $\nu = \mu/\rho$, is called the *kinematic* viscosity.
- *Caution:* If the velocity gradient is not constant Eq. 1.9 still applies, but the velocity distribution $u = u(y)$ must be known to determine the du/dy.

EXAMPLE: A Newtonian liquid is placed between parallel plates as shown in Fig. 1.3 with $b = 0.5$ in. When a force $P = 0.2$ lb is applied to the upper plate it moves with a velocity of 0.1 ft/s. If the area of the upper plate in contact with the liquid is 2.0 ft^2, determine the viscosity of the liquid.

SOLUTION:

From Eq. 1.9, $\tau = \mu \frac{du}{dy}$ where $\frac{du}{dy} = \frac{U}{b} = \frac{0.1\ \text{ft/s}}{(0.5/12)\ \text{ft}} = 2.4\ \frac{1}{s}$

Also, $\tau = P/A = 0.2\ \text{lb}/(2\ \text{ft}^2) = 0.1\ \text{lb/ft}^2$

Thus, $\mu = \tau/(du/dy) = 0.1\ \text{lb/ft}^2/(2.4\ \frac{1}{s}) = \underline{\underline{0.0417\ \text{lb}\cdot\text{s/ft}^2}}$

1.7 Compressibility of Fluids

Bulk Modulus

$$E_v = -\frac{dp}{d\forall/\forall} \qquad \textbf{(1.12)}$$

- A property used to characterize the compressibility of a fluid is the *bulk modulus, E_v,* defined in Eq. 1.12.
- As shown in Eq. 1.12, the bulk modulus indicates how a change in pressure, *dp,* is related to a corresponding change in volume, $d\forall$, for a given volume, $\forall$, of fluid.
- *Caution:* A positive change in pressure ($dp > 0$) produces a negative change in volume ($d\forall < 0$). Thus, the bulk modulus is a positive number.

EXAMPLE: How much pressure is required to compress a certain volume of water by 1 %?
SOLUTION:

From Eq. 1.12, $dp = -E_v \frac{d\forall}{\forall}$, where $\frac{d\forall}{\forall} = -0.01$

Thus, $dp = -(3.12\times10^5 \text{ lb/in.}^2)(-0.01) = \underline{\underline{3120 \text{ psi}}}$

Note that it takes a large pressure change to compress water.

Compression and Expansion of Gases

$$\frac{p}{\rho} = \text{constant} \qquad \textbf{(1.14)}$$

$$\frac{p}{\rho^k} = \text{constant} \qquad \textbf{(1.15)}$$

- When gases are compressed or expanded, the relationship between pressure and density depends on the nature of the process.
- For compression or expansion of an ideal gas under constant temperature conditions (an isothermal process), Eq. 1.14 applies.
- For compression or expansion that is frictionless and with no heat exchanged with the surroundings (an isentropic process), Eq. 1.15 applies.
- *Caution:* Absolute pressure must be used in Eqs. 1.14 and 1.15.

Speed of Sound

$$c = \sqrt{\frac{E_v}{\rho}} \qquad \textbf{(1.19)}$$

$$c = \sqrt{kRT} \qquad \textbf{(1.20)}$$

- **As shown in Eq. 1.19, the speed of sound, *c*, in a fluid medium depends on the bulk modulus and density of the fluid.**
- **As shown in Eq. 1.20, the speed of sound in an ideal gas depends on the gas properties, *k* and *R*, for the particular gas of interest and is proportional to the square root of the absolute temperature.**
- *Caution:* Absolute temperature must be used in Eq. 1.20.

EXAMPLE: Determine the values for the speed of sound in water and in standard air.

SOLUTION:

From Eq. 1.19, $c = \sqrt{E_v/\rho}$

For water, $c = \left[3.12 \times 10^5 \text{ lb/in.}^2 (144 \text{ in.}^2/\text{ft}^2)/(1.94 \text{ slugs/ft}^3)\right]^{1/2}$

$= \underline{\underline{4810 \text{ ft/s}}}$

For air at $T = 59°\text{F} = (59 + 460)°\text{R} = 519°\text{R}$, from Eq. 1.20

$c = \sqrt{kRT} = \left[1.4 (1716 \text{ ft}\cdot\text{lb/slug}\cdot°\text{R})(519°\text{R})\right]^{1/2} = \underline{\underline{1120 \text{ ft/s}}}$

1.8 Vapor Pressure

- Vapor pressure is closely associated with molecular activity and its value for a particular liquid depends on temperature.
- *Boiling*, which is the formation of vapor bubbles within a fluid mass, is initiated when the absolute pressure in the fluid reaches the vapor pressure.
- In flowing fluids it is possible to reach very low pressures and if the pressure is lowered to the vapor pressure, boiling will occur. Vapor bubbles formed in the flowing fluid are swept into regions of higher pressure where the bubbles may collapse with sufficient intensity to cause structural damage. This phenomenon is called *cavitation*.

EXAMPLE: At a certain altitude in the mountains the atmospheric pressure is found to be 11.5 psia. Determine the temperature at which you would expect water to boil at this altitude.

SOLUTION:

The water will boil when the temperature is such that the vapor pressure reaches 11.5 psia. From Table B.1 in Appendix B of the text $p_v = 11.5$ psia when $T = 200°F$. Thus, the water will boil at 200°F.

1.9 Surface Tension

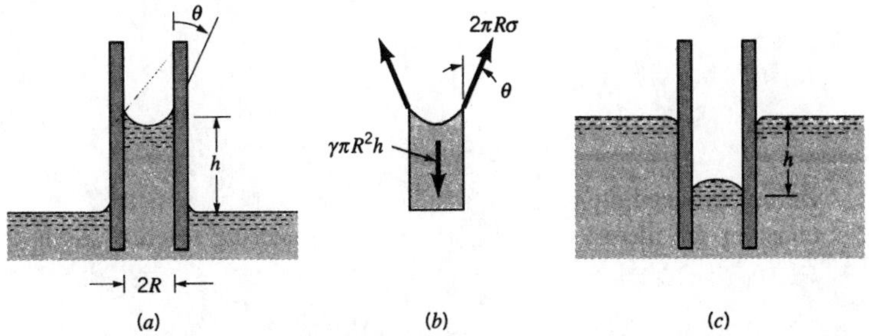

■ **FIGURE 1.8** **Effect of capillary action in small tubes. (*a*) Rise of column for a liquid that wets the tube. (*b*) Free-body diagram for calculating column height. (*c*) Depression of column for a nonwetting liquid.**

$$h = \frac{2\sigma \cos \theta}{\gamma R} \qquad \textbf{(1.22)}$$

- Due to unbalanced cohesive forces acting on liquid molecules at a fluid surface, a hypothetical "skin or membrane" is created at the surface. The intensity of the molecular attraction per unit length along any line in the surface is called the *surface tension*., σ.
- As illustrated in Fig. 1.8 and in Eq. 1.22, one important consequence of surface tension is the rise (or fall) of a liquid column in a small vertical tube.
- *Caution:* For nonwetting liquids, such as mercury, the column of liquid will be depressed as shown in Fig. 1.8*c*.

EXAMPLE: Determine the height that a column of water would rise in a clean 5-mm-diameter glass tube ($\theta = 0°$) inserted into water as shown in Fig. 1.8*a*.
SOLUTION:

From Eq. 1.22, $h = 2\sigma \cos\theta / \gamma R$ where $R = 2.5 \times 10^{-3} m$

or

$h = 2(7.34 \times 10^{-2} N/m)(\cos 0)/(9.80 \times 10^{3} N/m^3)(2.5 \times 10^{-3} m)$

$= 0.00599 \ m = \underline{\underline{5.99 \ mm}}$

Solutions

to Review Problems

Chapter 1

1.1R

1.1R (Dimensions) During a study of a certain flow system the following equation relating the pressures p_1 and p_2 at two points was developed:

$$p_2 = p_1 + \frac{f\ell V}{Dg}$$

In this equation V is a velocity, ℓ the distance between the two points, D a diameter, g the acceleration of gravity, and f a dimensionless coefficient. Is the equation dimensionally consistent?

(ANS: No)

$$p_2 = p_1 + \frac{f\ell V}{Dg}$$

$$[FL^{-2}] \doteq [FL^{-2}] + \left[\frac{(F^0L^0T^0)(L)(LT^{-1})}{(L)(LT^{-2})}\right]$$

$$[FL^{-2}] \doteq [FL^{-2}] + [T]$$

Since each term in the equation does not have the same dimensions, the equation is not dimensionally consistent. No.

1.2R

1.2R (Dimensions) If V is a velocity, ℓ a length, $\mathcal{W}$ a weight, and μ a fluid property having dimensions of $FL^{-2}T$, determine the dimensions of: **(a)** $V\ell\mathcal{W}/\mu$, **(b)** $\mathcal{W}\mu\ell$, **(c)** $V\mu/\ell$, and **(d)** $V\ell^2\mu/\mathcal{W}$.

(ANS: L^4T^{-2}; $F^2L^{-1}T$; FL^{-2}; L)

(a) $$\frac{V\ell\mathcal{W}}{\mu} \doteq \frac{(LT^{-1})(L)(F)}{(FL^{-2}T)} \doteq L^4T^{-2}$$

(b) $$\mathcal{W}\mu\ell \doteq (F)(FL^{-2}T)(L) \doteq F^2L^{-1}T$$

(c) $$\frac{V\mu}{\ell} \doteq \frac{(LT^{-1})(FL^{-2}T)}{(L)} \doteq FL^{-2}$$

(d) $$\frac{V\ell^2\mu}{\mathcal{W}} \doteq \frac{(LT^{-1})(L^2)(FL^{-2}T)}{(F)} \doteq L$$

1.3R

1.3R (Units) Make use of Table 1.4 to express the following quantities in BG units: **(a)** 465 W, **(b)** 92.1 J, **(c)** 536 N/m^2, **(d)** 85.9 mm^3, **(e)** 386 kg/m^2.

(ANS: 3.43×10^2 ft·lb/s; 67.9 ft·lb; 11.2 lb/ft^2; 3.03×10^{-6} ft^3; 2.46 $slugs/ft^2$)

$$(a)\quad 465\ W = (465\ W)\left(7.376\times10^{-1}\ \frac{\frac{ft\cdot lb}{s}}{W}\right) = \underline{\underline{3.43\times10^{2}\ \frac{ft\cdot lb}{s}}}$$

$$(b)\quad 92.1\ J = (92.1\ J)\left(7.376\times10^{-1}\ \frac{ft\cdot lb}{J}\right) = \underline{\underline{67.9\ ft\cdot lb}}$$

$$(c)\quad 536\ \frac{N}{m^2} = \left(536\ \frac{N}{m^2}\right)\left(2.089\times10^{-2}\ \frac{\frac{lb}{ft^2}}{\frac{N}{m^2}}\right) = \underline{\underline{11.2\ \frac{lb}{ft^2}}}$$

$$(d)\quad 85.9\ mm^3 = (85.9\times10^{-9}\ m^3)\left[(3.281)^3\ \frac{ft^3}{m^3}\right] = \underline{\underline{3.03\times10^{-6}\ ft^3}}$$

$$(e)\quad 386\ \frac{kg}{m^2} = \left(386\ \frac{kg}{m^2}\right)\left(6.852\times10^{-2}\ \frac{slugs}{kg}\right)\left(\frac{1}{1.076\times10\ \frac{ft^2}{m^2}}\right)$$

$$= \underline{\underline{2.46\ \frac{slugs}{ft^2}}}$$

1.4R

1.4R (Units) A person weighs 165 lb at the earth's surface. Determine the person's mass in slugs, kilograms, and pounds mass.

(ANS: 5.12 slugs; 74.8 kg; 165 lbm)

$$mass = \frac{weight}{g}$$

In slugs,

$$mass = \frac{165\ lb}{32.2\ ft/s^2} = \underline{\underline{5.12\ slugs}}$$

In kg,

$$mass = \frac{(165\ lb)\left(4.448\ \frac{N}{lb}\right)}{9.81\ m/s^2} = \underline{\underline{74.8\ kg}}$$

In lbm,

$$mass = \underline{\underline{165\ lbm}}$$

1.5R

1.5R (Specific gravity) Make use of Fig. 1.1 to determine the specific gravity of water at 22 and 89 °C. What is the specific volume of water at these two temperatures?

(ANS: 0.998; 0.966; $1.002 \times 10^{-3} m^3/kg$; $1.035 \times 10^{-3} m^3/kg$)

From Fig. 1.1:

at 22°C $\rho_{H_2O} = 998 \frac{kg}{m^3}$

at 89°C $\rho_{H_2O} = 966 \frac{kg}{m^3}$

Thus,

$$SG = \frac{\rho}{\rho_{H_2O} @ 4°C} = \frac{998 \frac{kg}{m^3}}{1000 \frac{kg}{m^3}} = \underline{\underline{0.998 \ @ \ 22°C}}$$

and

$$SG = \frac{966 \frac{kg}{m^3}}{1000 \frac{kg}{m^3}} = \underline{\underline{0.966 \ @ \ 89°C}}$$

Since specific volume $v = \frac{1}{\rho}$

$$v = \frac{1}{998 \frac{kg}{m^3}} = \underline{\underline{1.002 \times 10^{-3} \frac{m^3}{kg} \ @ \ 22°C}}$$

and

$$v = \frac{1}{966 \frac{kg}{m^3}} = \underline{\underline{1.035 \times 10^{-3} \frac{m^3}{kg} \ @ \ 89°C}}$$

1.6R

1.6R (Specific weight) A 1-ft-diameter cylindrical tank that is 5 ft long weighs 125 lb and is filled with a liquid having a specific weight of 69.6 lb/ft³. Determine the vertical force required to give the tank an upward acceleration of 9 ft/s².

(ANS: 509 lb up)

weight of tank $= W_t = 125\,\text{lb}$

weight of liquid $= W_\ell = \gamma_\ell \times \forall$

$$W_\ell = \left(69.6\ \frac{\text{lb}}{\text{ft}^3}\right)\left[\frac{\pi}{4}(1\ \text{ft}^2)(5\ \text{ft})\right]$$

$$= 273\ \text{lb}$$

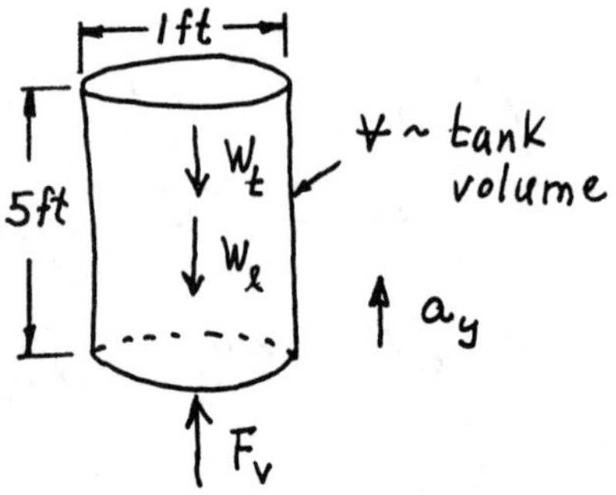

$(\uparrow +)$ $$\Sigma F_y = m\,a_y$$

$$F_v - W_t - W_\ell = m\,a_y$$

$$F_v - 125\,\text{lb} - 273\,\text{lb} = \left(\frac{125\,\text{lb} + 273\,\text{lb}}{32.2\ \frac{\text{ft}}{\text{s}^2}}\right) \times 9\ \frac{\text{ft}}{\text{s}^2}$$

$$F_v = 509\ \text{lb}\ \uparrow$$

1.7R

1.7R (Ideal gas law) Calculate the density and specific weight of air at a gage pressure of 100 psi and a temperature of 100 °F. Assume standard atmospheric pressure.

(ANS: 1.72×10^{-2} slugs/ft³; 0.554 lb/ft³)

From the ideal gas law:

$$\rho = \frac{p}{RT} = \frac{\left(100\ \frac{\text{lb}}{\text{in.}^2} + 14.7\ \frac{\text{lb}}{\text{in.}^2}\right)\left(\frac{144\ \text{in.}^2}{\text{ft}^2}\right)}{\left(1.716 \times 10^3\ \frac{\text{ft}\cdot\text{lb}}{\text{slug}\cdot{}^\circ\text{R}}\right)\left[(100\,^\circ\text{F} + 460)\,^\circ\text{R}\right]}$$

$$= 1.72 \times 10^{-2}\ \frac{\text{slugs}}{\text{ft}^3}$$

$$\gamma = \rho g = \left(1.72 \times 10^{-2}\ \frac{\text{slugs}}{\text{ft}^3}\right)\left(32.2\ \frac{\text{ft}}{\text{s}^2}\right) = 0.554\ \frac{\text{lb}}{\text{ft}^3}$$

1.8R

1.8R (Ideal gas law) A large dirigible having a volume of 90,000 m^3 contains helium under standard atmospheric conditions [pressure = 101 kPa (abs) and temperature = 15 °C]. Determine the density and total weight of the helium.
(ANS: 0.169 kg/m^3; 1.49×10^5N)

volume = 90,000 m^3

From the ideal gas law,

$$\rho = \frac{p}{RT} = \frac{101 \times 10^3 \frac{N}{m^2}}{\left(2077 \frac{J}{kg \cdot K}\right)\left[(15°C + 273)K\right]} = \underline{\underline{0.169 \frac{kg}{m^3}}}$$

$$\text{weight} = \rho g \times \text{volume} = \left(0.169 \frac{kg}{m^3}\right)\left(9.81 \frac{m}{s^2}\right)\left(9 \times 10^4 m^3\right)$$

$$= \underline{\underline{1.49 \times 10^5 N}}$$

1.9R

1.9R (Viscosity) A Newtonian fluid having a specific gravity of 0.92 and a kinematic viscosity of 4×10^{-4} m^2/s flows past a fixed surface. The velocity profile near the surface is shown in Fig. P1.9R. Determine the magnitude and direction of the shearing stress developed on the plate. Express your answer in terms of U and δ, with U and δ expressed in units of meters per second and meters, respectively.
(ANS: 0.578 U/ δ N/m^2 acting to right on plate)

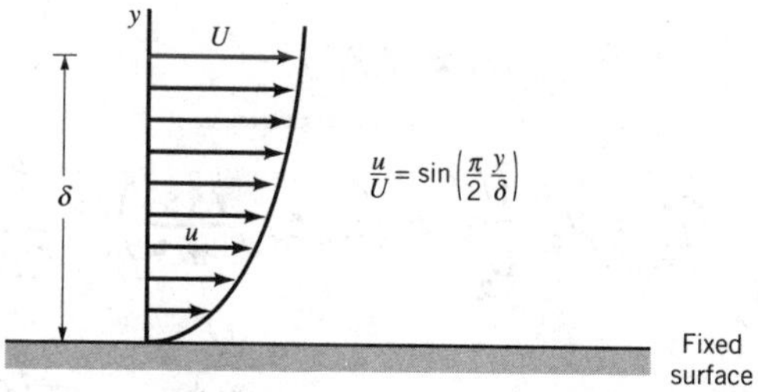

■ FIGURE P1.9R

$$\tau_{\substack{surface \\ (y=0)}} = \mu\left(\frac{du}{dy}\right)_{y=0} \quad \text{where } \mu = \nu\rho$$

$$\frac{du}{dy} = \frac{\pi}{2}\frac{U}{\delta}\cos\left(\frac{\pi}{2}\frac{y}{\delta}\right)$$

At $y=0$, $\frac{du}{dy} = \frac{\pi}{2}\frac{U}{\delta}(1)$

Since, $\mu = \nu\rho$ where $\rho = SG\,\rho_{H_2O} = 0.92\left(1000 \frac{kg}{m^3}\right)$

$$\tau_{surface} = \nu\rho\left(\frac{\pi}{2}\frac{U}{\delta}\right)$$

$$= \left(4 \times 10^{-4} \frac{m^2}{s}\right)\left(0.92 \times 10^3 \frac{kg}{m^3}\right)\left(\frac{\pi}{2}\right)\frac{U}{\delta}$$

$$= \underline{\underline{0.578 \frac{U}{\delta} \; N/m^2 \text{ acting to right on plate}}}$$

1.10R

1.10R (Viscosity) A large movable plate is located between two large fixed plates as shown in Fig. P1.10R. Two Newtonian fluids having the viscosities indicated are contained between the plates. Determine the magnitude and direction of the shearing stresses that act on the fixed walls when the moving plate has a velocity of 4 m/s as shown. Assume that the velocity distribution between the plates is linear.

(ANS: 13.3 N/m^2 in direction of moving plate)

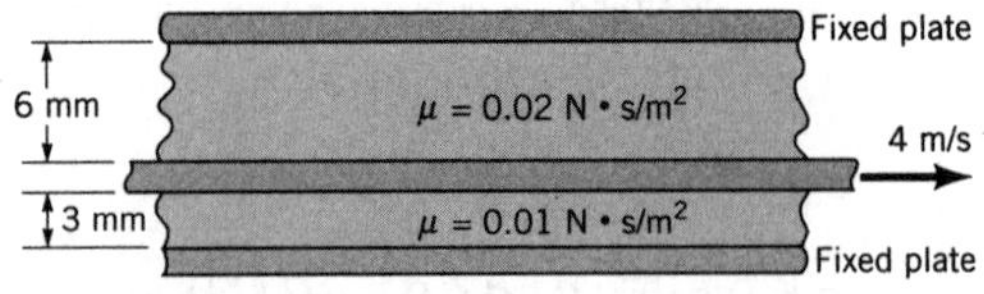

■ FIGURE P1.10R

$$\tau = \mu \frac{du}{dy} = \mu \frac{U}{b} \text{ so that}$$

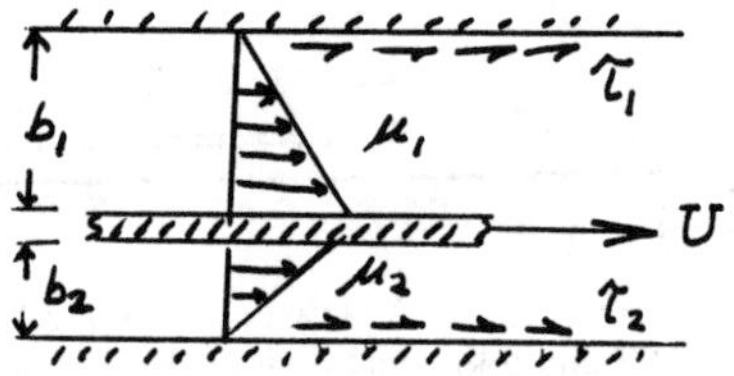

$$\tau_1 = \mu_1 \frac{U}{b_1} = \left(0.02 \frac{N \cdot s}{m^2}\right)\left(\frac{4 \frac{m}{s}}{0.006 m}\right)$$

$$= 13.3 \frac{N}{m^2}$$

$$\tau_2 = \mu_2 \frac{U}{b_2} = \left(0.01 \frac{N \cdot s}{m^2}\right)\left(\frac{4 \frac{m}{s}}{0.003 m}\right)$$

$$= 13.3 \frac{N}{m^2}$$

Stresses act on fixed walls in direction of moving plate.

1.11R

1.11R (Viscosity) Determine the torque required to rotate a 50-mm-diameter vertical cylinder at a constant angular velocity of 30 rad/s inside a fixed outer cylinder that has a diameter of 50.2 mm. The gap between the cylinders is filled with SAE 10 oil at 20 °C. The length of the inner cylinder is 200 mm. Neglect bottom effects and assume the velocity distribution in the gap is linear. If the temperature of the oil increases to 80 °C, what will be the percentage change in the torque?

(ANS: 0.589 N m; 92.0 percent)

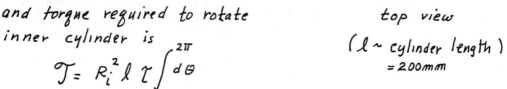

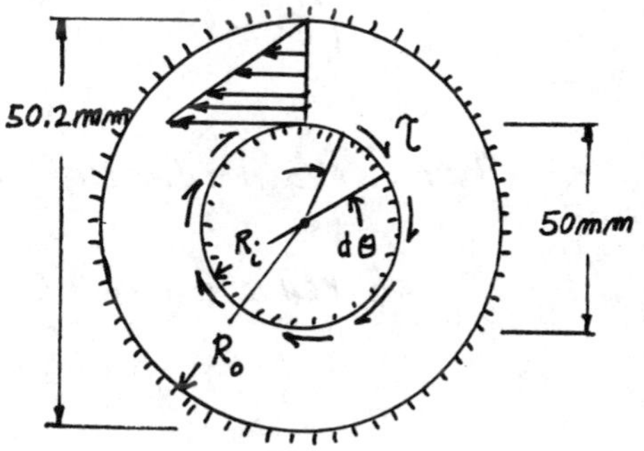

top view

(ℓ ~ cylinder length) = 200mm

Torque, $d\mathcal{T}$, due to shearing stress on inner cylinder is equal to

$$d\mathcal{T} = R_i \, \tau \, dA$$

where $dA = (R_i \, d\theta)\ell$. Thus,

$$d\mathcal{T} = R_i^2 \ell \, \tau \, d\theta$$

and torque required to rotate inner cylinder is

$$\mathcal{T} = R_i^2 \ell \, \tau \int_0^{2\pi} d\theta$$

$$= 2\pi R_i^2 \ell \, \tau$$

For a linear velocity distribution in the gap,

$$\tau = \mu \frac{R_i \omega}{R_o - R_i}$$

so that

$$\mathcal{T} = \frac{2\pi R_i^3 \ell \mu \omega}{R_o - R_i}$$

From Fig. B.1 in Appendix B:

(for SAE 10 oil at 20°C) $\mu = 1.0 \times 10^{-1} \ \frac{N \cdot s}{m^2}$

(for SAE 10 oil at 80°C) $\mu = 8.0 \times 10^{-3} \ \frac{N \cdot s}{m^2}$

(contined)

1.11R continued

Thus, at 20°C

$$\mathcal{T} = \frac{2\pi \left(\frac{0.050}{2} m\right)^3 (0.2 m)(1.0 \times 10^{-1} \frac{N \cdot s}{m^2})(30 \frac{rad}{s})}{\left(\frac{0.0502 m - 0.0500 m}{2}\right)}$$

$$= \underline{\underline{0.589 \ N \cdot m}}$$

At 80°C,

$$\mathcal{T} = (0.589 \ N \cdot m) \frac{\mu_{80°}}{\mu_{20°}}$$

so that % reduction in torque is equal to:

$$\% \text{ reduction in } \mathcal{T} = \left[\frac{0.589 - 0.589 \frac{\mu_{80°}}{\mu_{20°}}}{0.589}\right] \times 100$$

$$= \left[1 - \frac{8.0 \times 10^{-3}}{1.0 \times 10^{-1}}\right] \times 100 = \underline{\underline{92.0\%}}$$

1.12R

1.12R (Bulk modulus) Estimate the increase in pressure (in psi) required to decrease a unit volume of mercury by 0.1%.
(ANS: 4.14×10^3 psi)

$$E_v = -\frac{dp}{d\forall/\forall}, \text{ where from Table 1.5, } E_v = 4.14 \times 10^6 \frac{lb}{in.^2}$$

Thus,

$$\Delta p \approx -\frac{E_v \Delta\forall}{\forall} = -(4.14 \times 10^6 \frac{lb}{in.^2})(-0.001)$$

$$\Delta p \approx \underline{\underline{4.14 \times 10^3 \ psi}}$$

1.13R

1.13R (Bulk modulus) What is the isothermal bulk modulus of nitrogen at a temperature of 90 °F and an absolute pressure of 5600 lb/ft^2?

(ANS: 5600 lb/ft^2)

For isothermal bulk modulus,

$$E_v = p \qquad \text{(Eq. 1.16)}$$

so that

$$E_v = 5600 \ \frac{lb}{ft^2}$$

1.14R

1.14R (Speed of sound) Compare the speed of sound in mercury and oxygen at 20 °C.

(ANS: $c_{Hg}/c_{O2} = 4.45$)

For mercury,

$$c = \sqrt{\frac{E_v}{\rho}} \qquad \text{(Eq. 1.19)}$$

so that

$$c = \sqrt{\frac{2.85 \times 10^{10} \ \frac{N}{m^2}}{1.36 \times 10^4 \ \frac{kg}{m^3}}} = 1.45 \times 10^3 \ \frac{m}{s}$$

For oxygen,

$$c = \sqrt{kRT} \qquad \text{(Eq. 1.20)}$$

so that

$$c = \sqrt{(1.40)\left(259.8 \ \frac{J}{kg \cdot K}\right)\left[(20°C + 273)K\right]} = 326 \ \frac{m}{s}$$

Thus,

$$\frac{c\,(\text{mercury})}{c\,(\text{oxygen})} = \frac{1.45 \times 10^3 \ \frac{m}{s}}{326 \ \frac{m}{s}} = 4.45$$

(See Tables 1.6 and 1.8 for values of E_v, k, and R.)

1.15R

1.15R (Vapor pressure) At a certain altitude it was found that water boils at 90 °C. What is the atmospheric pressure at this altitude?

(ANS: 70.1 kPa (abs))

The vapor pressure of water at 90°C is 7.01×10^4 Pa (abs) (from Table B.2 in Appendix B). Thus, if water boils at this temperature, the atmospheric pressure must be equal to

$$p_{atm} = 7.01 \times 10^4 \text{ Pa} = \underline{\underline{70.1 \text{ kPa (abs)}}}$$

2
Fluid Statics

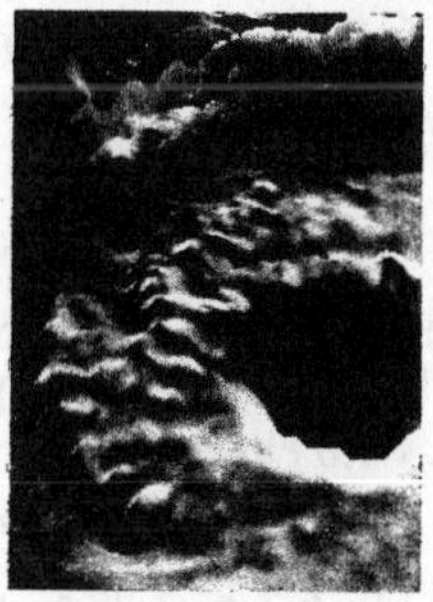

2.1 Pressure at a Point

- The pressure at a point in a fluid at rest, or in motion, is independent of direction as long as there are no shearing stresses present (Pascal's law).
- *Caution:* For moving fluids in which there are shearing stresses the pressure at a point may not be the same in all directions. (See Chapter 6.)

2.2 Basic Equation for Pressure Field

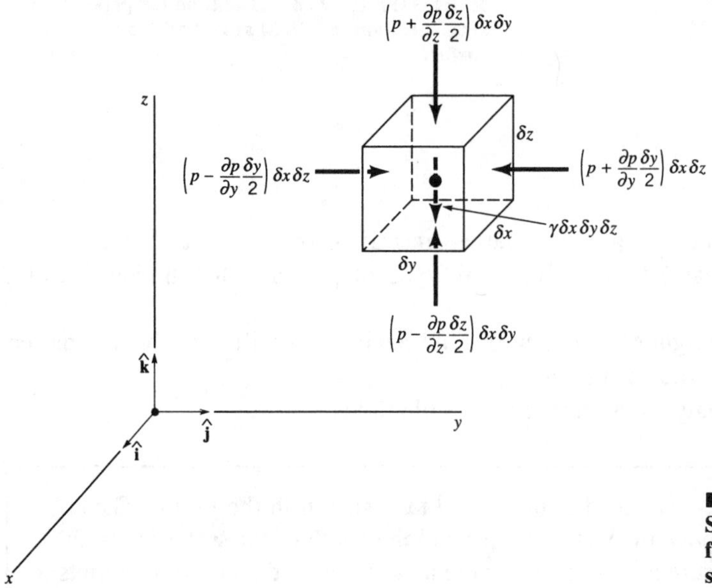

■ **FIGURE 2.2** **Surface and body forces acting on small fluid element.**

$$-\nabla p - \gamma\hat{\mathbf{k}} = \rho\mathbf{a} \qquad (2.2)$$

- For a small rectangular fluid element in a fluid mass in which there are no shearing stresses there are two types of forces acting on the element -- *surface* forces due to the pressure and a *body* force due to the weight of the element. (See Fig. 2.2.)
- Application of Newton's second law to the fluid element shown in Fig. 2.2 yields Eq. 2.2, which is the general equation of motion for a fluid in which there are no shearing stresses.

2.3 Pressure Variation in a Fluid at Rest

$$\frac{dp}{dz} = -\gamma \qquad \textbf{(2.4)}$$

- For a fluid at rest there is no pressure variation from point to point in a *horizontal* plane.
- Equation 2.4 is the fundamental equation for fluids at rest and is used to determine how pressure changes with elevation.
- Since the pressure gradient, *dp/dz,* is negative, the pressure decreases as we move upward.
- *Caution:* To use Eq. 2.4 it is necessary to stipulate how the specific weight, γ, varies with the vertical distance, *z*.

Incompressible Fluid

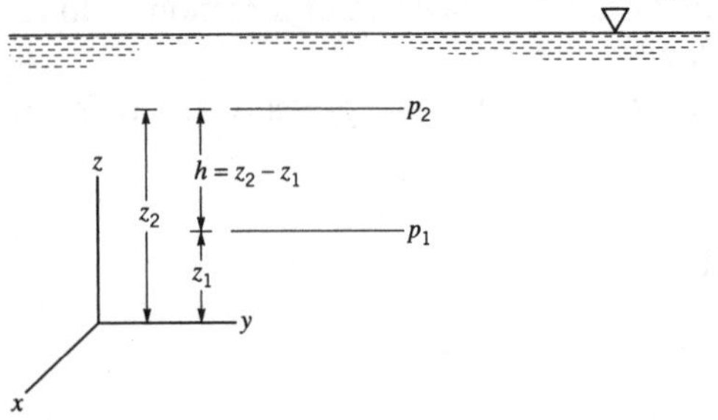

■ **FIGURE 2.3 Notation for pressure variation in a fluid at rest with a free surface.**

$$p_1 = \gamma h + p_2 \qquad \textbf{(2.7)}$$

- Equation 2.7 indicates that the pressure varies linearly with the fluid depth, *h*, for an incompressible (γ = constant) fluid at rest. This type of pressure distribution is called a *hydrostatic* distribution.
- The pressure *head* is the height of a column of fluid having a specific weight γ required to give a specified pressure, *p*, i.e., head = p/γ.
- *Caution:* Equation 2.7 is only valid for incompressible fluids.

EXAMPLE: Water is contained in the closed tank shown in the figure. The air pressure in the tank above the water is 5 psi and the depth of the water is h = 20 ft. Determine the pressure at the bottom of the tank. Express the pressure in units of psi, and as a pressure head in feet of water.

SOLUTION:

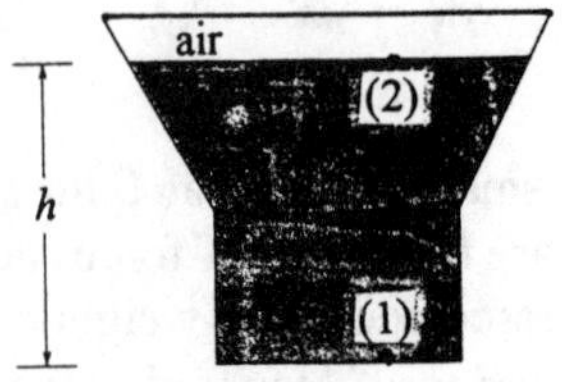

From Eq. 2.7 with $p_2 = 5\ psi$,

$$p_1 = \gamma h + p_2 = (62.4\,lb/ft^3)(20\,ft)(1\,ft^2/144\,in.^2) + 5\,lb/in^2$$
$$= 13.7\ psi$$

Thus, pressure head $= p_1/\gamma$

$$= (13.7\,lb/in.^2)(144\,in.^2/ft^2)/(62.4\,lb/ft^3)$$
$$= 31.6\ ft\ of\ water$$

Compressible Fluid

- The specific weight of gases such as air, oxygen, nitrogen, etc. can change significantly with changes in pressure and temperature. Such fluids are considered to be *compressible* fluids.
- In general, for compressible fluids at rest, the pressure distribution will not be hydrostatic. However, for small changes in elevation typically found in pipes and tanks containing gases at rest, the gas pressure remains essentially constant for most practical purposes.

2.4 Measurement of Pressure

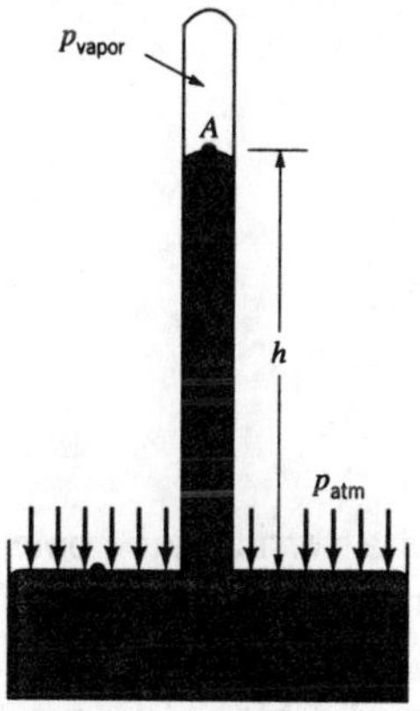

■ FIGURE 2.8 **Mercury barometer.**

- Pressure at a point in a fluid is designated as either an *absolute* pressure or a *gage* pressure. Absolute pressure is measured relative to a perfect vacuum (absolute zero). Gage pressure is measured relative to the local atmospheric pressure.
- Atmospheric pressure can be measured with a barometer of the type shown in Fig. 2.8.

EXAMPLE: A simple mercury barometer consists of a glass tube closed at one end with the open end immersed in a container of mercury as shown in Fig.2.8. **(a)** For a mercury column height $h = 30.1$ in. determine the atmospheric pressure (neglect p_{vapor}). **(b)** A gage attached to a pressurized air tank located near the barometer reads 18 psi. What is the absolute pressure of the air in the tank?

SOLUTION:

(a) For the static mercury in the barometer, $p_B = p_A + \gamma_{Hg} h$,
where $p_A = p_{vapor} \approx 0$ (absolute) and $p_B = p_{atm}$.

Thus, $p_{atm} = \gamma_{Hg} h = (847 \text{ lb/ft}^3)(30.1 \text{ in.})(1 \text{ ft}/12 \text{ in.})^3 = \underline{\underline{14.8 \text{ psi (absolute)}}}$

(b) $p_{tank}\text{(absolute)} = p_{tank}\text{(gage)} + p_{atm} = 18 \text{ psi} + 14.8 \text{ psi} = \underline{\underline{32.8 \text{ psia}}}$

2.5 Manometry

- A standard technique for measuring pressure involves the use of liquid columns in vertical or inclined tubes. Pressure measuring devices based on this technique are called *manometers.*

Piezometer Tube

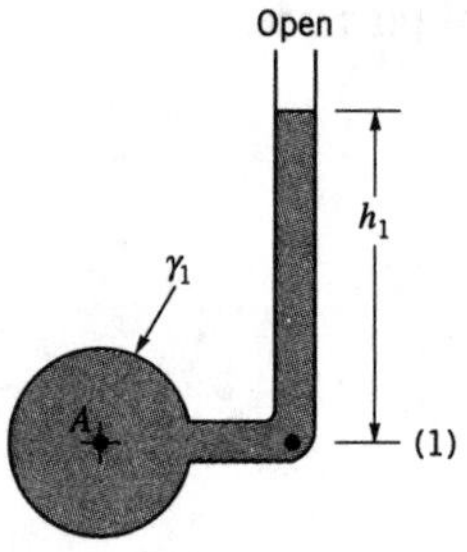

■ FIGURE 2.9 **Piezometer tube.**

$$p = \gamma h + p_0 \quad \text{(2.8)}$$

- The simplest type of manometer, called a *piezometer tube*, consists of a vertical tube open at the top and attached to the container for which the pressure is desired as illustrated in Fig. 2.9.
- For the piezometer tube the pressure is simply determined from Eq. 2.8 by measuring the column height, h_1, and setting $p_0 = 0$ (open column).
- *Caution:* Although the piezometer tube is a simple and accurate pressure measuring device, it is not suitable for measuring moderate or high pressures because of the impractical column heights required.

EXAMPLE: A piezometer tube is connected to a pipe containing water. If the water rises to a height of 4 ft in the tube, what is the pressure (in psi) in the pipe?
SOLUTION:

From Eq. 2.8, $p_{pipe} = \gamma h + p_0 = \gamma h + 0$

$= (62.4 \text{ lb/ft}^3)(4 \text{ ft})(1 \text{ ft}^2/144 \text{ in.}^2) = 1.73 \text{ psi}$

Note that the 1.73 psi is a gage pressure since we have set $p_0 = 0$, which is a gage pressure.

U-Tube Manometer

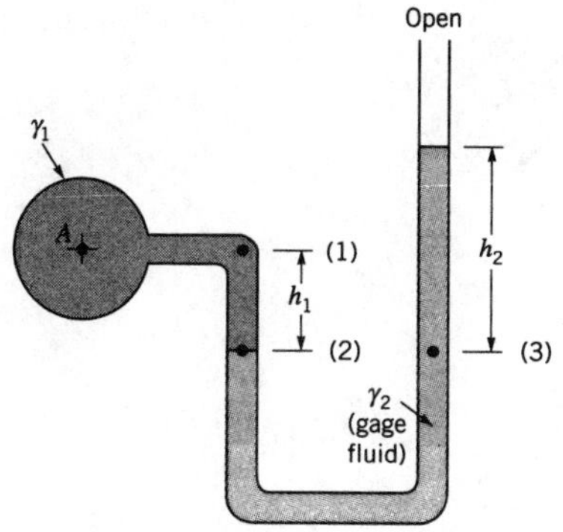

■ FIGURE 2.10 **Simple U-tube manometer.**

- A tube formed into the shape of a U as shown in Fig. 2.10 is a common type of manometer called a *U-tube manometer*. The liquid in the manometer, such as mercury, is called the *gage fluid.*
- Since any liquids in the manometer are incompressible and at rest, the pressure in the liquids varies in accordance with the hydrostatic pressure distribution, $p = \gamma h + p_0$. The effect of the weight of any gas columns in the manometer can be neglected.
- *Caution:* When solving manometer problems, remember that the pressure in the manometer tube *increases* with decreasing elevation and *decreases* with increasing elevation.

EXAMPLE: For the U-tube manometer shown in Fig. 2.10, pipe A contains water and the gage fluid is mercury. Measurements indicate that $h_1 = 6$ in. and $h_2 = 12$ in. Determine the pressure in pipe A.

SOLUTION:

Start at level (A) and work around to the open end using the fact that the pressure distribution is hydrostatic. Also, $p_2 = p_3$ since these two points are at the same elevation in a homogeneous fluid at rest. Thus,

$$p_A + \gamma_1 h_1 - \gamma_2 h_2 = 0, \text{ or}$$

$$p_A = \gamma_2 h_2 - \gamma_1 h_1 = SG_{Hg}\,\gamma_{H_2O}\,h_2 - \gamma_{H_2O}\,\gamma_1 = 62.4\,\text{lb/ft}^3[13.6\,(1\,\text{ft}) - (6/12)\,\text{ft}]$$

$$= \underline{\underline{817\ \text{lb/ft}^2}}$$

Inclined-Tube Manometer

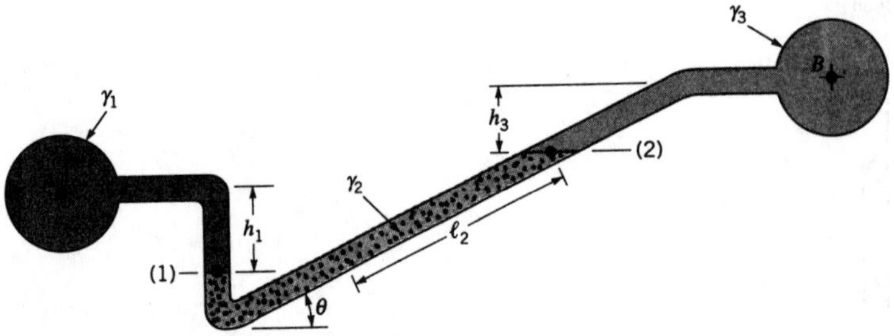

■ FIGURE 2.12 **Inclined-tube manometer.**

- For the measurement of small pressures or pressure changes an inclined-tube manometer of the type shown in Fig. 2.12 is frequently used. The differential reading is measured along the inclined tube.
- *Caution:* The change in pressure between points (1) and (2) is due to the *vertical* distance between the points.

EXAMPLE: The inclined-tube manometer shown in Fig. 2.12 is used to measure the air pressure in pipe *A*. The gage fluid is water, $\theta = 30°$, and pipe *B* contains air at atmospheric pressure. Determine the air pressure in pipe *A* when the differential reading is $\ell_2 = 9$ in.

SOLUTION:

Application of the hydrostatic pressure-depth relationship from point (A) to point (B) gives

$p_A + \gamma_1 h_1 - \gamma_2 \ell_2 \sin\theta - \gamma_3 h_3 = p_B$

where $p_B = 0$ and $\gamma_1 = \gamma_3 = \gamma_{air} << \gamma_2 = \gamma_{H_2O}$. Thus,

$p_A = \gamma_{H_2O} \ell_2 \sin\theta = (62.4\ lb/ft^3)(9/12\ ft) \sin 30° = \underline{\underline{23.4\ lb/ft^2}}$

2.8 Hydrostatic Force on a Plane Surface

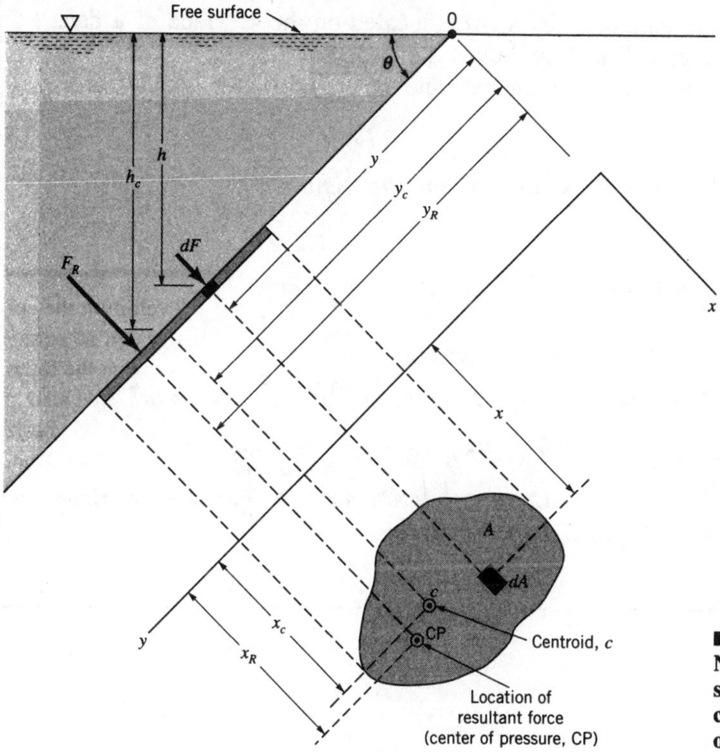

■ **FIGURE 2.17 Notation for hydrostatic force on an inclined plane surface of arbitrary shape.**

$$F_R = \gamma h_c A \quad \textbf{(2.18)}$$

$$y_R = \frac{I_{xc}}{y_c A} + y_c \quad \textbf{(2.19)}$$

- When an inclined plane surface, such as a tank wall or the face of a dam, is in contact with a fluid at rest, a hydrostatic force develops on the surface as illustrated in Fig. 2.17.
- The magnitude of the resultant fluid force, F_R, acting on a plane area is equal to pressure, $p_c = \gamma h_c$, acting at the centroid of the area multiplied by the area, A, as shown in Eq. 2.18.
- The resultant fluid force is perpendicular to the surface since there are no shearing stresses present.
- The location of the resultant fluid force is given by Eq. 2.19. The point through which the resultant acts is called the *center of pressure.*
- *Caution:* As shown by Eq. 2.19, the center of pressure is *not* located at the centroid of the area but, for non-horizontal surfaces, is always below it.

EXAMPLE: A square gate (4 m by 4 m) is located on the 45° face of a dam. The top edge of the gate lies 1 m below the water surface. Determine the resultant force of the water acting on the gate and the point through which it acts.

SOLUTION:

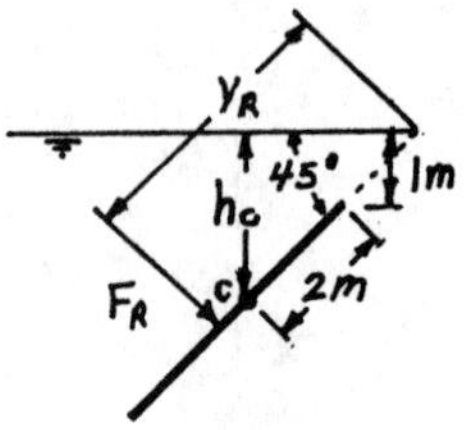

From Eq. 2.18, $F_R = \gamma h_c A$, where from the figure

$h_c = 1m + 2\sin 45° m = 2.41m$

Thus, $F_R = (9.80\ kN/m^3)(2.41m)[(4m)(4m)] = \underline{\underline{378\ kN}}$

From Eq. 2.19, $y_R = (I_{xc}/y_c A) + y_c$ where

$y_c = (1m/\sin 45°) + 2m = 3.41m$ and $I_{xc} = \frac{1}{12}(4m)(4m)^3 = 21.3\ m^4$

Thus, $y_R = (21.3m^4)/[3.41m(4m)(4m)] + 3.41m = \underline{\underline{3.80m}}$

The 378 kN resultant force acts normal to the gate at $y_R = 3.80m$ along the axis of symmetry of the gate.

2.9 Pressure Prism

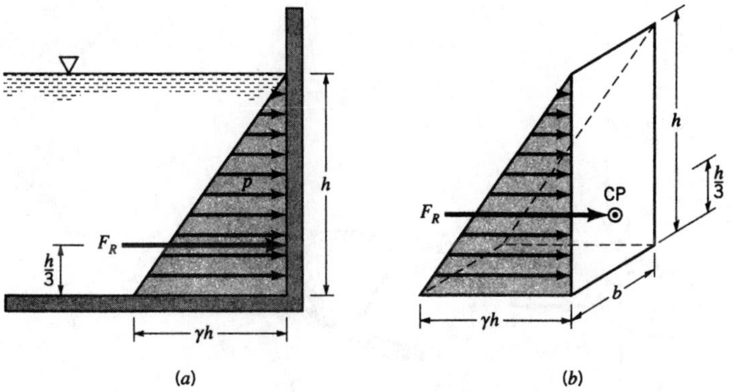

■ **FIGURE 2.19**
Pressure prism for vertical rectangular area.

- As illustrated in Fig. 2.19, a plot of the pressure acting on a vertical rectangular wall yields a three-dimensional representation of the pressure distribution. This 'volume" is called a *pressure prism.*
- The magnitude of the resultant fluid force acting on the rectangular wall is equal to the volume of the pressure prism, and the force passes though the centroid of the volume.

EXAMPLE: The vertical wall forming one end of a swimming pool is 30 ft wide and the depth of water at the wall is 10 ft. Use the concept of the pressure prism to determine the resultant force of the water on the wall and its location.

SOLUTION:

The pressure prism is shown in Fig. 2.19 with $b = 30\text{ft}$ and $h = 10\text{ft}$.

Thus, $F_R = \text{volume} = \frac{1}{2}(\gamma h)(bh) = \frac{1}{2}(62.4\text{lb/ft}^3)(10\text{ft})(30\text{ft})(10\text{ft})$

or $F_R = \underline{\underline{93{,}600\text{ lb}}}$

The centroid of the volume corresponds to the center of pressure and is located along the vertical axis of symmetry at $h/3 = 10\text{ft}/3 = \underline{\underline{3.33\text{ ft}}}$ above the bottom of the pool.

2.10 Hydrostatic Force on a Curved Surface

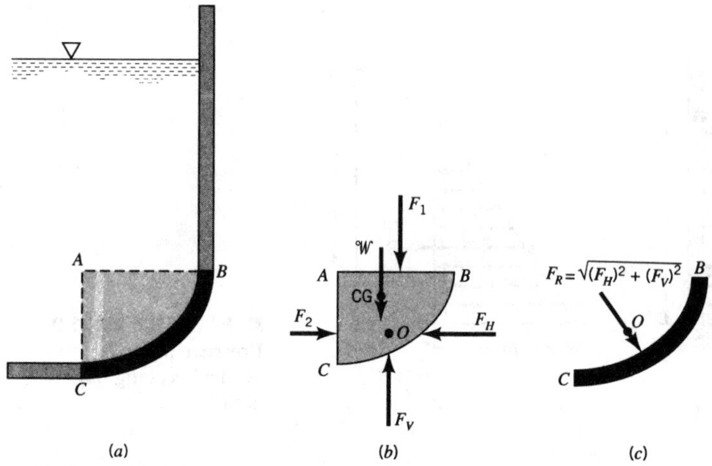

■ FIGURE 2.23 **Hydrostatic force on a curved surface.**

- For fluids at rest in contact with a curved surface, the pressure distribution remains hydrostatic, but the direction of the pressure varies over the surface so that the equations developed for a plane surface do not apply.
- A relatively simple approach to determining the resultant fluid force acting on a curved surface is to consider the equilibrium of a fluid volume enclosed by the curved surface of interest and the horizontal and vertical projections of the surface as shown in Fig. 2.23.

EXAMPLE: An open tank having the cross section shown in the figure contains a liquid (specific weight = 9 kN/m^3) at a depth of 10 m. The weight of the liquid in a 1-m length of the tank is 250 kN. Determine the magnitude of the force per 1-m length of the tank that the liquid exerts on the curved section *BC*.

SOLUTION:

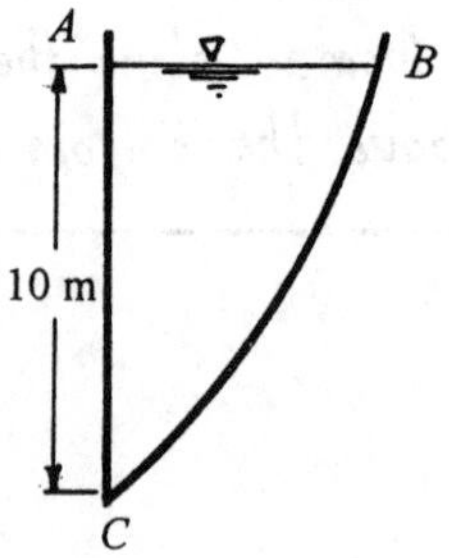

For equilibrium of the liquid in the tank, the free body diagram is as shown in Fig. 2.23(b) with $F_1 = 0$ since the top of the tank is at atmospheric pressure ($p = 0$).

Thus, $F_V = \mathcal{W} = 250\text{ kN}$ and

$F_2 = F_H = \gamma h_c A = (9\text{ kN/m}^3)(5\text{ m})[(10\text{ m})(1\text{ m})] = 450\text{ kN}$

Thus, $F_R = [F_H^{\,2} + F_V^{\,2}]^{1/2} = [(250\text{ kN})^2 + (450\text{ kN})^2]^{1/2}$

$= \underline{\underline{515\text{ kN}}}$

2.11 Buoyancy, Flotation, and Stability

$$F_B = \gamma \forall \qquad (2.22)$$

- The resultant fluid force acting on a stationary body that is completely submerged or floating in a static fluid is called the *buoyant force.*
- As shown in Eq. 2.22 , the magnitude of the buoyant force is equal to the weight of the fluid displaced by the body (Archimedes' principle).
- The buoyant force is directed vertically upward and passes through the centroid of the displaced volume.
- *Caution:* The buoyant force is a consequence of the hydrostatic pressure distribution acting on the body and is *not* a "special force" to be included in addition to hydrostatic pressure effects.

EXAMPLE: A solid wooden cube (0.5 m by 0.5 m by 0.5 m) is floating in water and 0.2 m of the cube extends above the water surface. Determine the specific weight of the wood.

SOLUTION:

For equilibrium $\mathcal{W} = F_B$, where

$$\mathcal{W} = \gamma_{wood} \forall_{wood} = \gamma_{wood} (0.5 m)^3 = (0.125 m^3)\gamma_{wood}$$

and

$$F_B = \gamma_{H_2O} \forall_{displaced} = \gamma_{H_2O} (0.5m)(0.5m)(0.5m - 0.2m)$$

$$= (0.0750 m^3)\gamma_{H_2O}$$

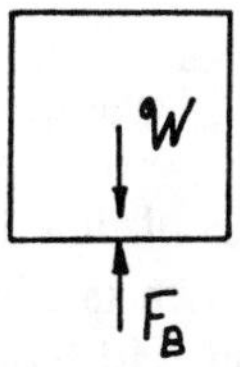

Thus, $\gamma_{wood} = [(0.075 m^3)/(0.125 m^3)]\gamma_{H_2O} = [(0.0750)/(0.125)](9.80 kN/m^3)$

$$= \underline{\underline{5.88 kN/m^3}}$$

2.12 Pressure Variation in a Fluid with Rigid-Body Motion

$$-\frac{\partial p}{\partial x} = \rho a_x \qquad -\frac{\partial p}{\partial y} = \rho a_y \qquad -\frac{\partial p}{\partial z} = \gamma + \rho a_z \qquad \textbf{(2.24)}$$

- If a mass of fluid moves as a rigid body there will be no shearing stresses present.
- Equation 2.24 is the general equation of motion for fluids in which there are no shearing stresses present.
- As shown in Eq. 2.24, for an accelerating fluid mass the pressure distribution will not be a hydrostatic distribution.

Linear Motion

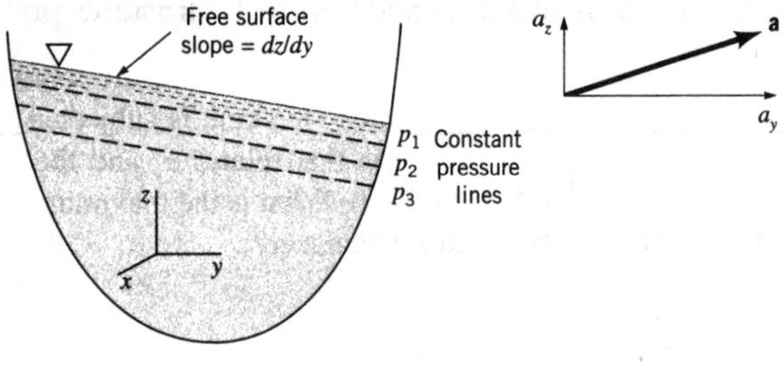

■ FIGURE 2.29 **Linear acceleration of a liquid with a free surface.**

$$\frac{dz}{dy} = -\frac{a_y}{g + a_z} \qquad \textbf{(2.28)}$$

- A mass of fluid that is translating along a straight path is considered to have *linear motion.*
- As shown in Fig. 2.29 and by Eq. 2.28, for an open container of a liquid that has linear motion with constant acceleration, the slope of the liquid surface is a function of the horizontal and vertical components of acceleration, a_y and a_z, respectively.

EXAMPLE: A square (5 ft by 5 ft) open tank of water is accelerated to the right along a horizontal surface with a constant acceleration, $a_y = 3$ ft/s^2. When at rest the depth of water in the tank is 3 ft. Determine the slope of the water surface and the maximum depth of water during the accelerated motion.

SOLUTION:

From Eq. 2.28,

$dz/dy = -a_y/(g + a_z)$, where

$a_z = 0$ and $a_y = 3$ ft/s^2.

Thus, $dz/dy = -3\text{ ft/s}^2/(32.2\text{ ft/s}^2)$

$= \underline{\underline{-0.0932}}$

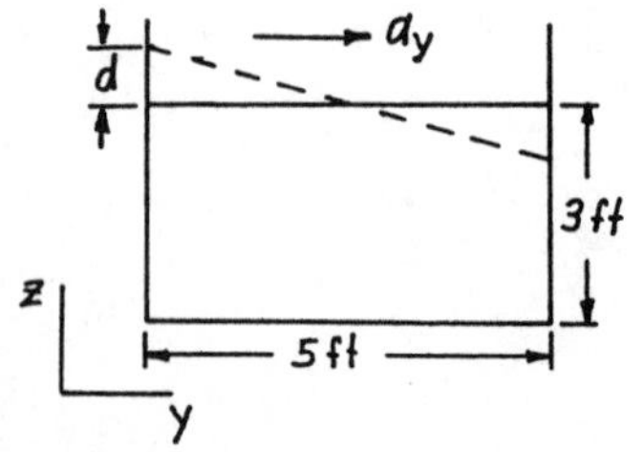

Hence,. $2d/5\text{ ft} = 0.0932$ or $d = 0.0932(5\text{ ft}) = 0.466$ ft

so that $z_{max} = 0.466\text{ ft} + 3\text{ ft} = \underline{\underline{3.466\text{ ft}}}$

Rigid-Body Rotation

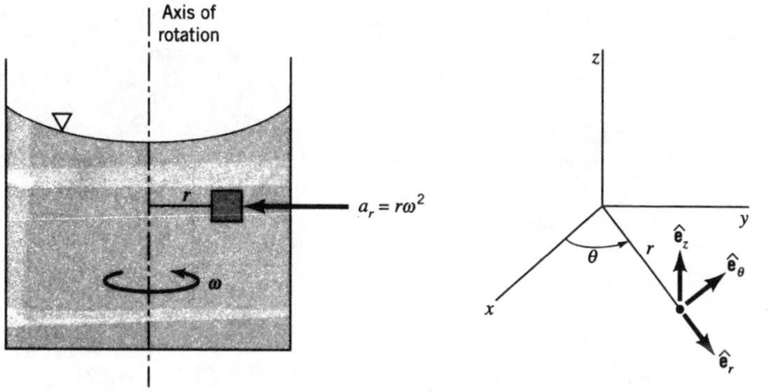

■ FIGURE 2.30 **Rigid-body rotation of a liquid in a tank.**

$$z = \frac{\omega^2 r^2}{2g} + \text{constant} \qquad \textbf{(2.32)}$$

- A fluid that is contained in a tank that is rotating with a constant angular velocity about its longitudinal axis will rotate as a rigid body.
- As shown in Fig. 2.30 and by Eq. 2.32, the free surface in a rotating tank is curved rather than flat.

EXAMPLE: A 2-ft diameter open tank containing a liquid rotates about its vertical axis with an angular velocity $\omega = 10$ rad/s. At this speed the depth of the liquid at the center of the tank is 5 ft. Determine the depth at the tank wall.

SOLUTION:

From Eq. 2.32, $z = \omega^2 r^2/(2g) + \text{constant}$.

Thus, with $z = 5$ ft at $r = 0$, the value of the constant is 5 ft.

At the wall $r = 1$ ft so that

$$z_{wall} = (10\,\text{rad/s})^2(1\,\text{ft})^2/[(2)(32.2\,\text{ft/s}^2)] + 5\,\text{ft} = \underline{\underline{6.55\,\text{ft}}}$$

Solutions

to Review Problems

Chapter 2

2.1R

2.1R (Pressure head) Compare the column heights of water, carbon tetrachloride, and mercury corresponding to a pressure of 50 kPa. Express your answer in meters.
(ANS: 5.10 m; 3.21 m; 0.376 m)

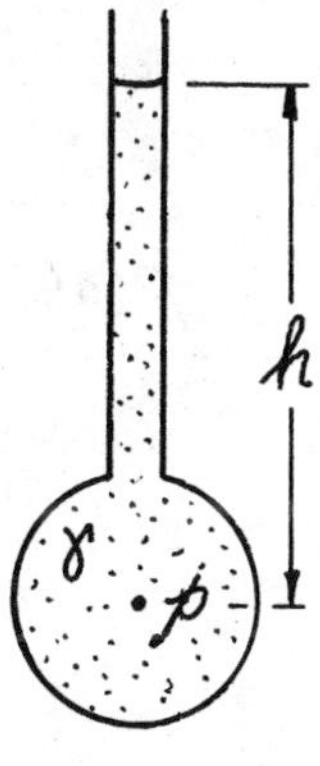

$p = \gamma h$

For water: $h = \dfrac{50 \times 10^3 \frac{N}{m^2}}{9.80 \times 10^3 \frac{N}{m^3}} = \underline{\underline{5.10\ m}}$

For carbon tetrachloride: $h = \dfrac{50 \times 10^3 \frac{N}{m^2}}{15.6 \times 10^3 \frac{N}{m^3}} = \underline{\underline{3.21\ m}}$

For mercury: $h = \dfrac{50 \times 10^3 \frac{N}{m^2}}{133 \times 10^3 \frac{N}{m^3}} = \underline{\underline{0.376\ m}}$

2.2R

2.2R (Pressure-depth relationship) A closed tank is partially filled with glycerin. If the air pressure in the tank is 6 lb/in.^{2} and the depth of glycerin is 10 ft, what is the pressure in lb/ft^2 at the bottom of the tank?
(ANS: 1650 lb/ft^2)

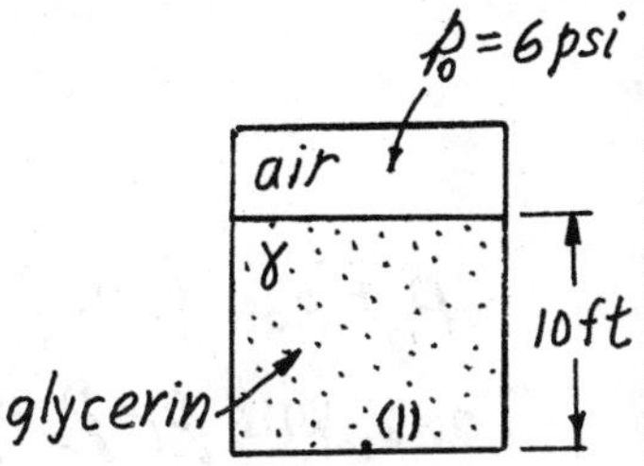

$p_1 = \gamma h + p_0 = \left(78.6\ \frac{lb}{ft^3}\right)(10\ ft) + \left(6\ \frac{lb}{in.^2}\right)\left(\frac{144\ in.^2}{ft^2}\right)$

$= \underline{\underline{1650\ \frac{lb}{ft^2}}}$

2.3R

2.3R (Gage-absolute pressure) On the inlet side of a pump a Bourdon pressure gage reads 600 lb/ft^2 vacuum. What is the corresponding absolute pressure if the local atmospheric pressure is 14.7 psia?

(ANS: 10.5 psia)

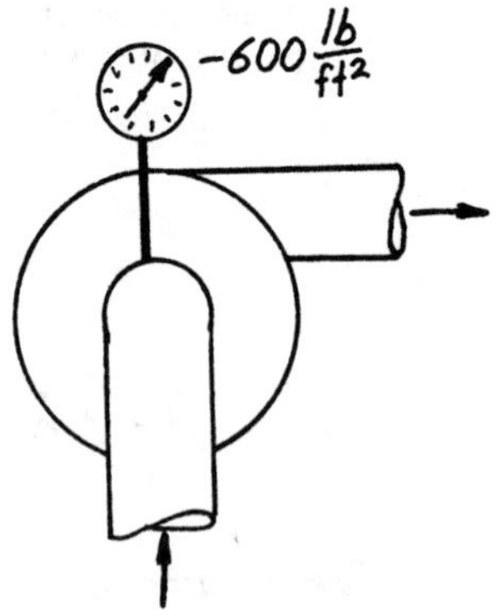

$$p(abs) = p(gage) + p(atm)$$

$$= -600 \frac{lb}{ft^2}\left(\frac{1\ ft^2}{144\ in.^2}\right) + 14.7\ psia$$

$$= \underline{\underline{10.5\ psia}}$$

2.4R

2.4R (Manometer) A tank is constructed of a series of cylinders having diameters of 0.30, 0.25, and 0.15 m as shown in Fig. P2.4R. The tank contains oil, water, and glycerin and a mercury manometer is attached to the bottom as illustrated. Calculate the manometer reading, h.

(ANS: 0.0327 m)

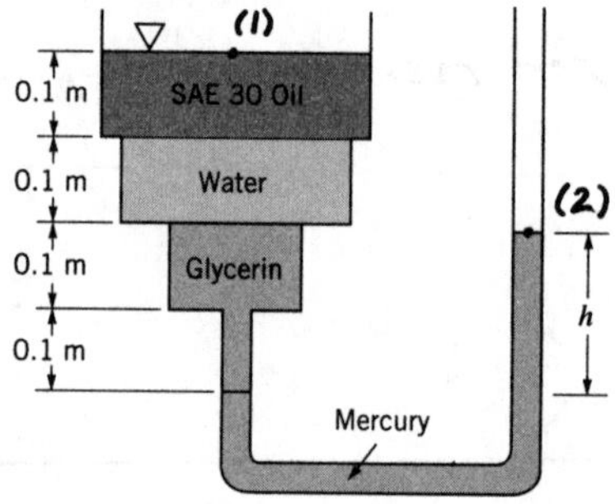

■ FIGURE P2.4R

$$p_1 + \gamma_{oil}(0.1m) + \gamma_{H_2O}(0.1m) + \gamma_{gly}(0.2m) - \gamma_{Hg}\, h = p_2$$

Thus, with $p_1 = p_2 = 0$,

$$h = \frac{\left(8.95 \frac{kN}{m^3}\right)(0.1m) + \left(9.80 \frac{kN}{m^3}\right)(0.1m) + \left(12.4 \frac{kN}{m^3}\right)(0.2m)}{133 \frac{kN}{m^3}}$$

$$= \underline{\underline{0.0327\ m}}$$

2.5R

2.5R (Manometer) A mercury manometer is used to measure the pressure difference in the two pipelines of Fig. P2.5R. Fuel oil (specific weight = 53.0 lb/ft^3) is flowing in A and SAE 30 lube oil (specific weight = 57.0 lb/ft^3) is flowing in B. An air pocket has become entrapped in the lube oil as indicated. Determine the pressure in pipe B if the pressure in A is 15.3 psi.

(ANS: 18.2 psi)

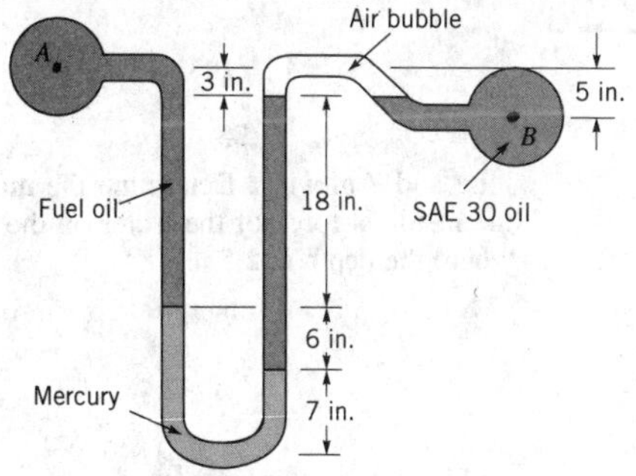

■ FIGURE P2.5R

$$p_A + \gamma_{\text{fuel oil}}\left(\frac{3+18}{12}\text{ ft}\right) + \gamma_{Hg}\left(\frac{6}{12}\text{ ft}\right) - \gamma_{SAE\,30}\left(\frac{6+18}{12}\text{ ft}\right) + \gamma_{SAE\,30}\left(\frac{2}{12}\text{ ft}\right) = p_B$$

Thus,

$$p_B = \left(15.3\ \frac{\text{lb}}{\text{in.}^2}\right)\left(144\ \frac{\text{in}^2}{\text{ft}^2}\right) + \left(53.0\ \frac{\text{lb}}{\text{ft}^3}\right)\left(\frac{21}{12}\text{ ft}\right) + \left(847\ \frac{\text{lb}}{\text{ft}^3}\right)\left(\frac{6}{12}\text{ ft}\right) - \left(57.0\ \frac{\text{lb}}{\text{ft}^3}\right)\left(\frac{22}{12}\text{ ft}\right)$$

$$= 2615\ \frac{\text{lb}}{\text{ft}^2} = \left(2615\ \frac{\text{lb}}{\text{ft}^2}\right)\left(\frac{1\text{ ft}^2}{144\text{ in.}^2}\right) = \underline{18.2\text{ psi}}$$

2.6R

2.6R (Manometer) Determine the angle θ of the inclined tube shown in Fig. P2.6R if the pressure at A is 1 psi greater than that at B.

(ANS: 19.3 deg)

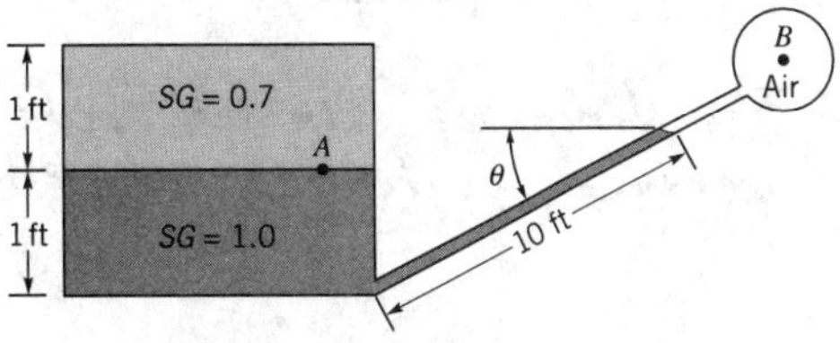

■ FIGURE P2.6R

$$p_A + (1.0)\left(62.4\ \frac{\text{lb}}{\text{ft}^3}\right) - (1.0)\left(62.4\ \frac{\text{lb}}{\text{ft}^3}\right)(10\text{ ft})\sin\theta = p_B$$

Thus,

$$p_A - p_B = (1.0)\left(62.4\ \frac{\text{lb}}{\text{ft}^3}\right)\left[(10\text{ ft})\sin\theta - 1\right]$$

Since $p_A - p_B = 1\text{ psi}$

$$(10\text{ ft})\sin\theta - 1 = \frac{\left(1\ \frac{\text{lb}}{\text{in.}^2}\right)\left(144\ \frac{\text{in.}^2}{\text{ft}^2}\right)}{(1.0)\left(62.4\ \frac{\text{lb}}{\text{ft}^3}\right)}$$

So that

$$\sin\theta = 0.331 \quad \text{or} \quad \underline{\theta = 19.3^\circ}$$

2.7R

2.7R (Force on plane surface) A swimming pool is 18 m long and 7 m wide. Determine the magnitude and location of the resultant force of the water on the vertical end of the pool where the depth is 2.5 m.

(ANS: 214 kN on centerline, 1.67 m below surface)

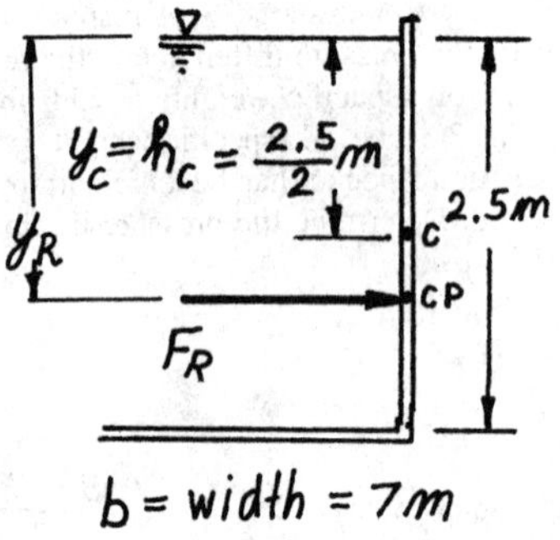

$$F_R = \gamma h_c A = \left(9.80 \frac{kN}{m^3}\right)\left(\frac{2.5m}{2}\right)(7m \times 2.5m) = \underline{\underline{214\ kN}}$$

$$y_R = \frac{I_{xc}}{y_c A} + y_c \qquad \text{where} \quad I_{xc} = \frac{1}{12}(7m)(2.5m)^3$$

Thus,

$$y_R = \frac{\frac{1}{12}(7m)(2.5m)^3}{\left(\frac{2.5m}{2}\right)(7m \times 2.5m)} + \frac{2.5m}{2} = \underline{\underline{1.67\ m}}$$

The force of 214 kN acts 1.67 m below surface along vertical centerline of end.

2.8R

2.8R (Force on plane surface) The vertical cross section of a 7-m-long closed storage tank is shown in Fig. P2.8R. The tank contains ethyl alcohol and the air pressure is 40 kPa. Determine the magnitude of the resultant fluid force acting on one end of the tank.

(ANS: 847 kN)

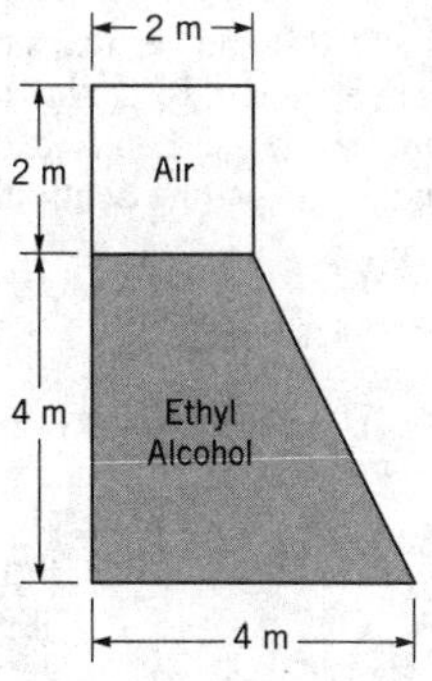

■ FIGURE P2.8R

Break area into three parts as shown in figure.

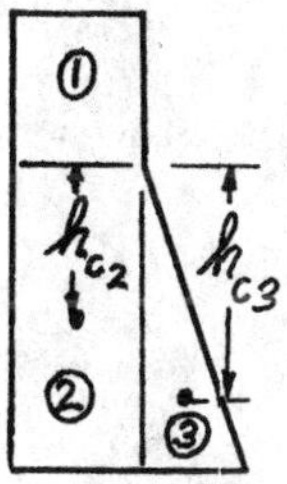

For area 1:

$$F_{R1} = p_{air} A_1 = \left(40 \frac{kN}{m^2}\right)(2m \times 2m) = 160 \; kN$$

For area 2: (From Table 1.6 $\gamma_{\text{ethyl alcohol}} = 7.74 \frac{kN}{m^3}$)

$$F_{R2} = p_{air} A_2 + \gamma h_{c2} A_2$$

$$= \left(40 \frac{kN}{m^2}\right)(2m \times 4m) + \left(7.74 \frac{kN}{m^3}\right)\left(\frac{4m}{2}\right)(2m \times 4m)$$

$$= 444 \; kN$$

For area 3:

$$F_{R3} = p_{air} A_3 + \gamma h_{c3} A_3$$

$$= \left(40 \frac{kN}{m^2}\right)\left(\frac{1}{2}\right)(2m \times 4m) + \left(7.74 \frac{kN}{m^3}\right)\left(\frac{2}{3}\right)(4m)\left(\frac{1}{2}\right)(2m \times 4m)$$

$$= 243 \; kN$$

Thus,

$$F_R = F_{R1} + F_{R2} + F_{R3}$$

$$= 160 \; kN + 444 \; kN + 243 \; kN = \underline{\underline{847 \; kN}}$$

2.9R

2.9R (Center of pressure) A 3-ft-diameter circular plate is located in the vertical side of an open tank containing gasoline. The resultant force that the gasoline exerts on the plate acts 3.1 in. below the centroid of the plate. What is the depth of the liquid above the centroid?

(ANS: 2.18 ft)

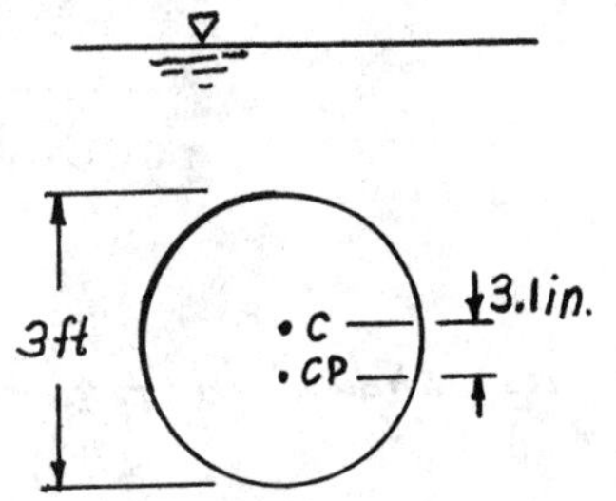

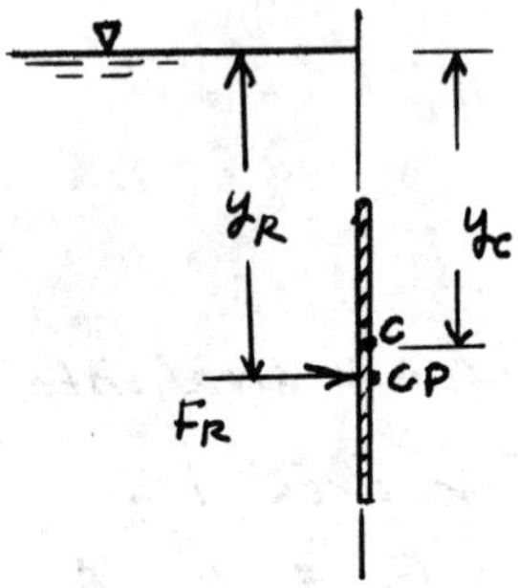

$$y_R = \frac{I_{xc}}{y_c A} + y_c \quad \text{where } I_{xc} = \frac{\pi}{4}\left(\frac{3}{2}\,ft\right)^4$$

Thus,

$$y_R - y_c = \frac{3.1}{12}\,ft = \frac{\frac{\pi}{4}\left(\frac{3}{2}\,ft\right)^4}{y_c\,\frac{\pi}{4}(3\,ft)^2}$$

so that

$$y_c = 2.18\,ft$$

2.10R

2.10R (Force on plane surface) A gate having the triangular shape shown in Fig. P2.10R is located in the vertical side of an open tank. The gate is hinged about the horizontal axis *AB*. The force of the water on the gate creates a moment with respect to the axis *AB*. Determine the magnitude of this moment.

(ANS: 3890 kN·m)

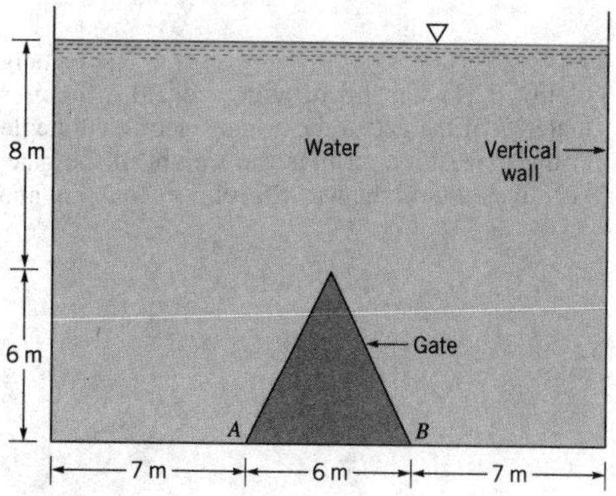

■ FIGURE P2.10R

$F_R = \gamma h_c A$ where $h_c = 8m + \frac{2}{3}(6m) = 12m$

Thus,

$$F_R = \left(9800\frac{N}{m^3}\right)(12m)\left(\frac{1}{2}\right)(6m \times 6m) = 2120\ kN$$

To locate F_R,

$$y_R = \frac{I_{xc}}{y_c A} + y_c \quad \text{with } y_c = h_c \text{ so that}$$

$$y_R = \frac{\frac{1}{36}(6m)(6m)^3}{(12m)\left(\frac{1}{2}\right)(6m)^2} + 12m = 12.167m$$

Thus, to determine the moment about AB

$$M_{AB} = (2120 \times 10^3 N)(14m - 12.167m)$$

$$= \underline{\underline{3890\ kN\cdot m}}$$

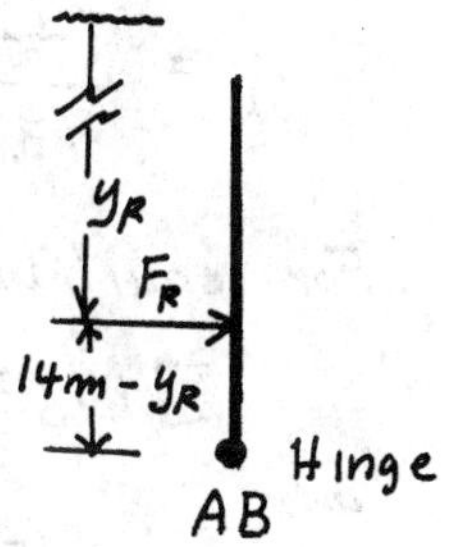

2.11R

2.11R (Force on plane surface) The rectangular gate *CD* of Fig P2.11R is 1.8 m wide and 2.0 m long. Assuming the material of the gate to be homogeneous and neglecting friction at the hinge *C*, determine the weight of the gate necessary to keep it shut until the water level rises to 2.0 m above the hinge.

(ANS: 180 kN)

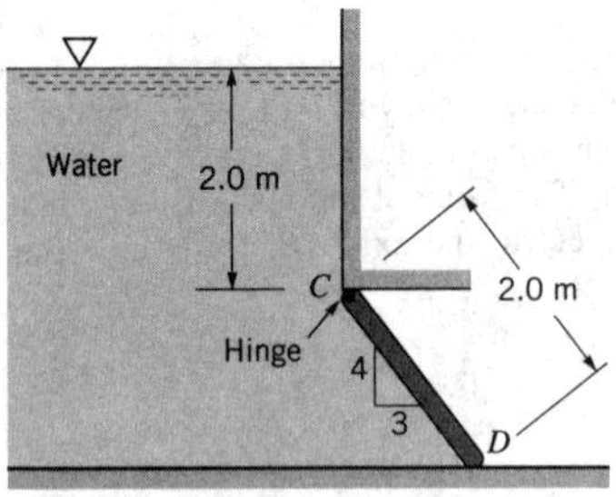

■ FIGURE P2.11R

$F_R = \gamma h_c A$

where $h_c = 2m + \frac{1}{2}\left[\left(\frac{4}{5}\right)(2m)\right] = 2.8m$

Thus,

$F_R = \left(9.80 \frac{kN}{m^3}\right)(2.8\,m)(1.8\,m \times 2\,m)$

$= 98.8\,kN$

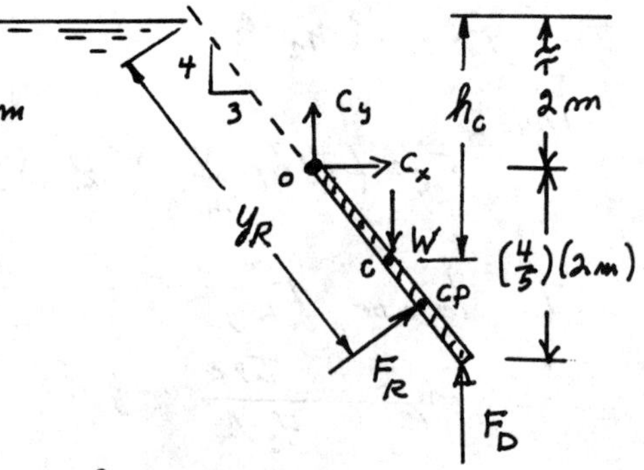

Also,

$y_R = \frac{I_{xc}}{y_c A} + y_c$ where $y_c = \frac{2\,m}{\left(\frac{4}{5}\right)} + 1\,m = 3.5m$

so that

$y_R = \frac{\left(\frac{1}{12}\right)(1.8m)(2m)^3}{(3.5m)(1.8m \times 2m)} + 3.5m = 3.595m$

For equilibrium,

$\sum M_o = 0$ (Note: Set $F_D = 0$ to obtain minimum weight)

and

$W\left(\frac{1}{2}\right)\left[\left(\frac{3}{5}\right)(2m)\right] - F_R\left(y_R - \frac{2m}{\left(\frac{4}{5}\right)}\right) = 0$

or

$W = \frac{(98.8\,kN)(3.595m - 2.5m)}{\left(\frac{1}{2}\right)\left[\left(\frac{3}{5}\right)(2m)\right]} = \underline{\underline{180\,kN}}$

2.12R

2.12R (Force on curved surface) A gate in the form of a partial cylindrical surface (called a *Tainter gate*) holds back water on top of a dam as shown in Fig. P2.12R. The radius of the surface is 22 ft, and its length is 36 ft. The gate can pivot about point *A*, and the pivot point is 10 ft above the seat, *C*. Determine the magnitude of the resultant water force on the gate. Will the resultant pass through the pivot? Explain.

(ANS: 118,000 lb)

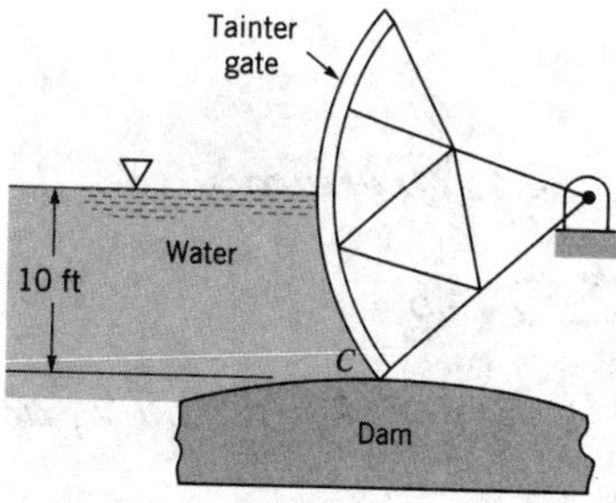

■ FIGURE P2.12R

Let F_G be force of gate on fluid

and

$$F_1 = \gamma h_c A$$
$$= \left(62.4 \frac{lb}{ft^3}\right)\left(\frac{10\ ft}{2}\right)(10\,ft \times 36\,ft)$$
$$= 112{,}000\ lb$$

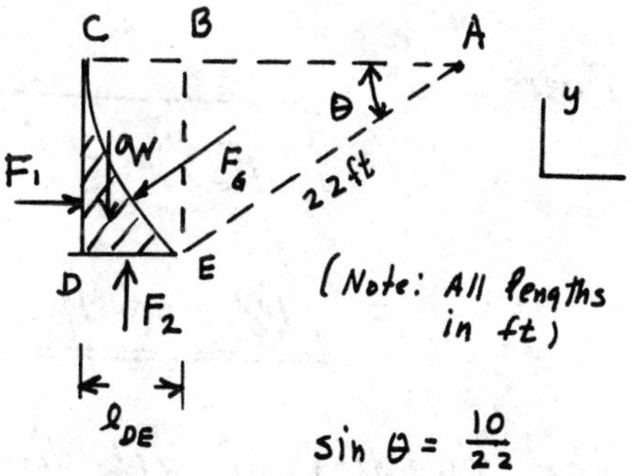

(Note: All lengths in ft)

$$\sin\theta = \frac{10}{22}$$
$$\therefore\ \theta = 27.0°$$
$$\ell_{DE} = \ell_{BC} = 22 - \ell_{AB}$$
$$= 22 - 22\cos\theta$$
$$= 2.40\ ft$$

Also,

$$F_2 = \left(62.4 \frac{lb}{ft^3}\right)(10\,ft)(\ell_{DE} \times 36\,ft)$$
$$= \left(62.4 \frac{lb}{ft^3}\right)(10\,ft)(2.40\,ft \times 36\,ft)$$
$$= 53{,}900\ lb$$

and

$$\mathcal{W} = \gamma \forall_{CDE} = \gamma(A_{CDE} \times 36\,ft)$$

where:

$$A_{CDE} = A_{BCDE} - A_{BCE} = (10 \times \ell_{DE}) - A_{BCE}$$

$$A_{BCE} = A_{ACE} - A_{ABE} = \pi(22\,ft)^2\left(\frac{27.0°}{360°}\right) - \frac{1}{2}\left(10\,ft \times [22\cos 27.0°]\,ft\right)$$
$$= 16.0\ ft^2$$

Thus,

$$A_{CDE} = (10\,ft \times 2.40\,ft) - 16.0\,ft^2 = 8.00\,ft^2$$

and

$$W = \left(62.4 \frac{lb}{ft^3}\right)(8.00\,ft^2)(36\,ft) = 18{,}000\ lb$$

continued

2.12R continued

For equilibrium,

$$\Sigma F_x = 0$$

or

$$F_{Gx} = F_1 = 112{,}000\ lb$$

Also,

$$\Sigma F_y = 0$$

or

$$F_{Gy} = F_2 - W = 53{,}900\ lb - 18{,}000\ lb = 35{,}900\ lb$$

Thus,

$$F_G = \sqrt{(F_{Gx})^2 + (F_{Gy})^2} = \sqrt{(112{,}000\ lb)^2 + (35{,}900\ lb)^2}$$

$$= \underline{\underline{118{,}000\ lb}}$$

The direction of all differential forces acting on the gate is perpendicular to the gate surface, and therefore, the resultant must pass through the intersection of all these forces which is at point A. Yes.

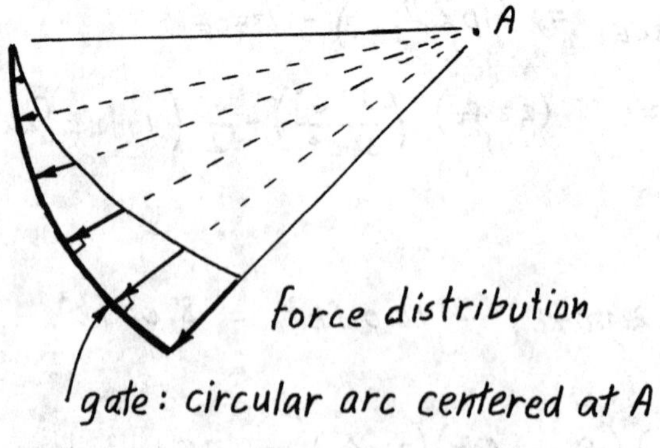

2.13 R

2.13R (Force on curved surface) A conical plug is located in the side of a tank as shown in Fig. 2.13R. **(a)** Show that the horizontal component of the force of the water on the plug does not depend on *h*. **(b)** For the depth indicated, what is the magnitude of this component?

(ANS: 735 lb)

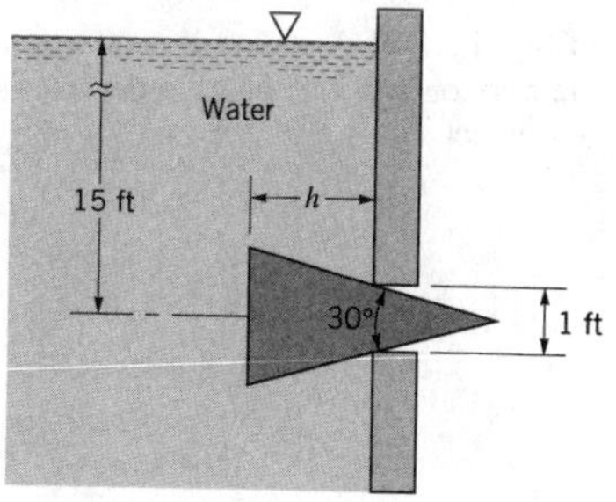

■ FIGURE P2.13R

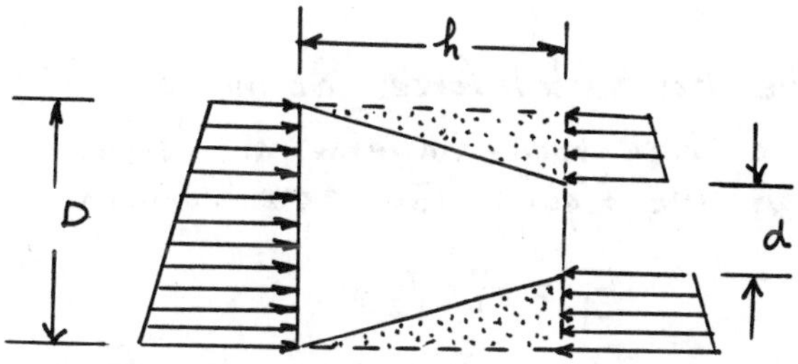

(a) Consider a cylinder of fluid of diameter, D, and length, h, with the plug removed (see figure). The pressure distributions over the right and left surfaces are shown. We note that the pressures cancel except for the center area of diameter, d. The pressure distribution over this center area will yield a resultant which is independent of h and depends only on the fluid specific weight, the fluid depth, and the hole diameter, d.

(b) For a circular area of diameter, $d = 1\,ft$,

$$F_R = \gamma h_c A$$

$$= \left(62.4\,\frac{lb}{ft^3}\right)(15\,ft)\left(\frac{\pi}{4}\right)(1\,ft)^2 = \underline{\underline{735\,lb}}$$

2.14R

2.14R (Force on curved surface) The 9-ft-long cylinder of Fig. P2.14R floats in oil and rests against a wall. Determine the horizontal force the cylinder exerts on the wall at the point of contact, *A*.

(ANS: 2300 lb)

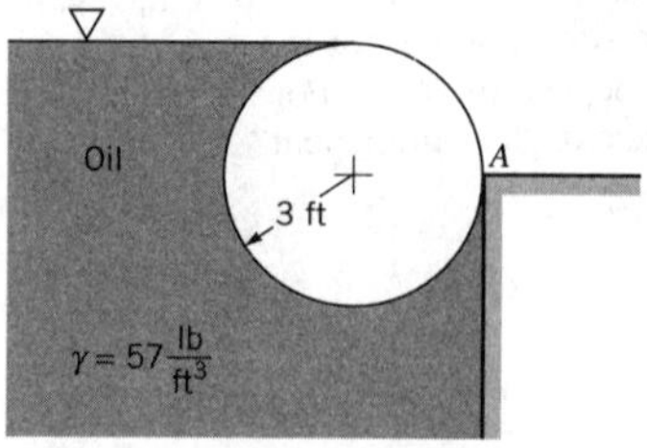

■ FIGURE P2.14R

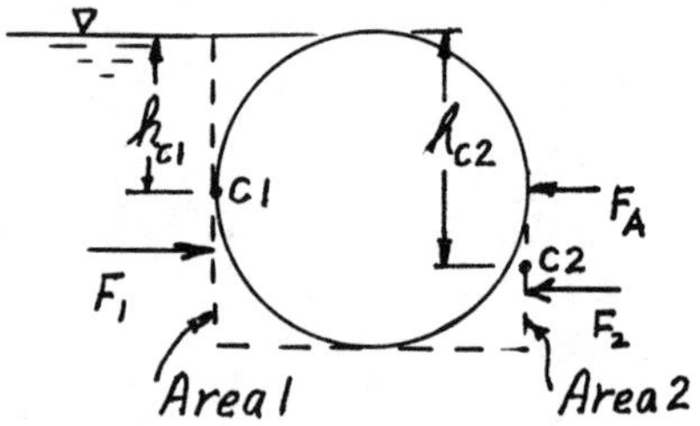

The horizontal forces acting on the free-body-diagram are shown on the figure. For equilibrium,

$$F_A = F_1 - F_2$$

where F_A is the horizontal force the wall exerts on the cylinder.

Since,

$$F_1 = \gamma h_{c_1} A_1$$

$$= \left(57.0 \frac{lb}{ft^3}\right)\left(\frac{6 ft}{2}\right)(6 ft \times 9 ft)$$

$$= 9230 \text{ lb}$$

and

$$F_2 = \gamma h_{c2} A_2$$

$$= \left(57.0 \frac{lb}{ft^3}\right)\left(3 ft + \frac{3}{2} ft\right)(3 ft \times 9 ft)$$

$$= 6930 \text{ lb}$$

then

$$F_A = 9230 \text{ lb} - 6930 \text{ lb} = \underline{\underline{2300 \text{ lb} \longrightarrow \text{on the wall}}}$$

2.15R

2.15R (Buoyancy) A hot-air balloon weighs 500 lb, including the weight of the balloon, the basket, and one person. The air outside the balloon has a temperature of 80 °F, and the heated air inside the balloon has a temperature of 150 °F. Assume the inside and outside air to be at standard atmospheric pressure of 14.7 psia. Determine the required volume of the balloon to support the weight. If the balloon had a spherical shape, what would be the required diameter?

(ANS: 59,200 ft^3; 48.3 ft)

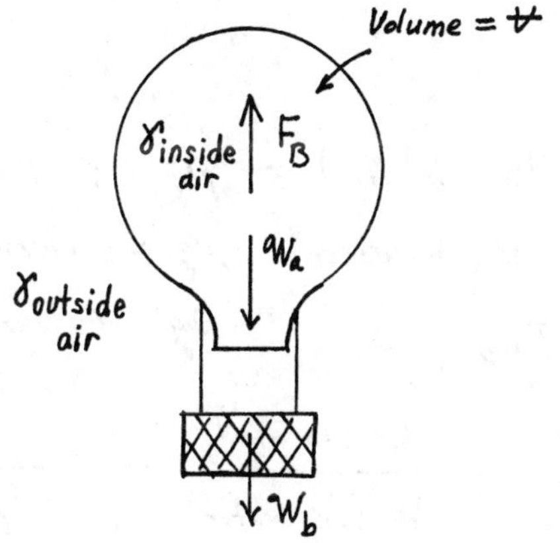

For equilibrium,

$$\Sigma F_{vertical} = 0$$

so that

$$F_B = \mathcal{W}_a + \mathcal{W}_b$$

where:

F_B = buoyant force

$\mathcal{W}_a$ = weight of air inside balloon

$\mathcal{W}_b$ = weight of basket and load

Thus,

$$(\gamma_{\substack{outside \\ air}})\,\forall = (\gamma_{\substack{inside \\ air}})\,\forall + \mathcal{W}_b \qquad (1)$$

From the ideal gas law $p = \rho R T = \frac{\gamma}{g} R T$ or

$$\gamma = \frac{g\,p}{RT}$$

For outside air with $T = 80°F + 460 = 540°R$,

$$\gamma_{\substack{outside \\ air}} = \frac{\left(32.2\ \frac{ft}{s^2}\right)\left(14.7\ \frac{lb}{in.^2}\right)\left(144\ \frac{in.^2}{ft^2}\right)}{\left(1716\ \frac{ft\cdot lb}{slug\cdot °R}\right)(540\ °R)} = 0.07356\ \frac{lb}{ft^3}$$

Similarly for inside air with $T = 150°F + 460 = 610°R$,

$$\gamma_{\substack{inside \\ air}} = \left(\frac{540\ °R}{610\ °R}\right)\left(0.07356\ \frac{lb}{ft^3}\right) = 0.06512\ \frac{lb}{ft^3}$$

Thus, from Eq.(1)

$$\forall = \frac{\mathcal{W}_b}{\gamma_{\substack{outside \\ air}} - \gamma_{\substack{inside \\ air}}} = \frac{500\ lb}{0.07356\ \frac{lb}{ft^3} - 0.06512\ \frac{lb}{ft^3}} = \underline{\underline{59{,}200\ ft^3}}$$

For spherical shape (with d = diameter),

$$\frac{\pi}{6} d^3 = 59{,}200\ ft^3 \quad \text{so that} \quad \underline{\underline{d = 48.3\ ft}}$$

2.16R

2.16R (Buoyancy) An irregularly shaped piece of a solid material weighs 8.05 lb in air and 5.26 lb when completely submerged in water. Determine the density of the material.

(ANS: 5.60 slugs/ft^3)

$W(\text{in air}) = \rho g \times (\text{volume})$ where $\rho \sim$ density of material

$W(\text{in water}) = \rho g \times (\text{volume}) -$ buoyant force

$= \rho g \times (\text{volume}) - \rho_{H_2O}\, g \times (\text{volume})$

Thus,

$$\frac{W(\text{in air})}{W(\text{in water})} = \frac{\rho}{\rho - \rho_{H_2O}} = \frac{1}{1 - \frac{\rho_{H_2O}}{\rho}}$$

or

$$\rho = \frac{\rho_{H_2O}}{1 - \frac{W(\text{in water})}{W(\text{in air})}} = \frac{1.94\ \frac{\text{slugs}}{\text{ft}^3}}{1 - \frac{5.26\ \text{lb}}{8.05\ \text{lb}}} = \underline{\underline{5.60\ \frac{\text{slugs}}{\text{ft}^3}}}$$

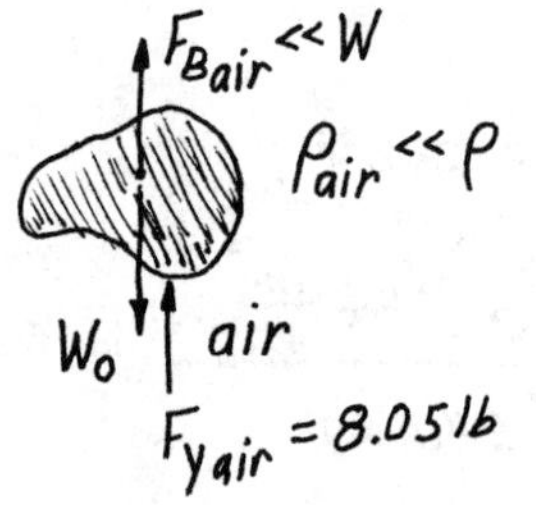

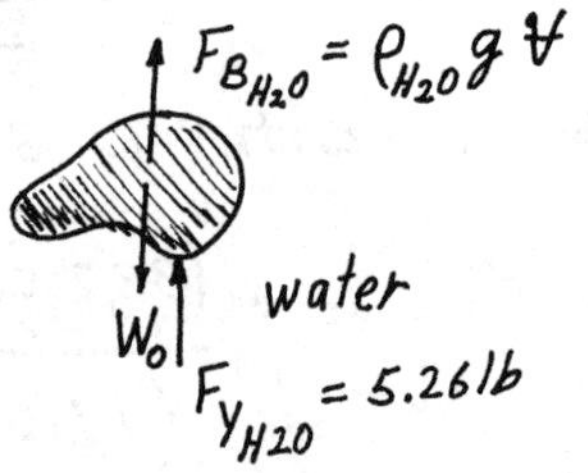

2.17R

2.17R (Buoyancy, force on plane surface) A cube, 4 ft on a side, weighs 3000 lb and floats half-submerged in an open tank as shown in Fig. P2.17R. For a liquid depth of 10 ft, determine the force of the liquid on the inclined section *AB* of the tank wall. The width of the wall is 8 ft. Show the magnitude, direction, and location of the force on a sketch.

(ANS: 75,000 lb on centerline, 13.33 ft along wall from free surface)

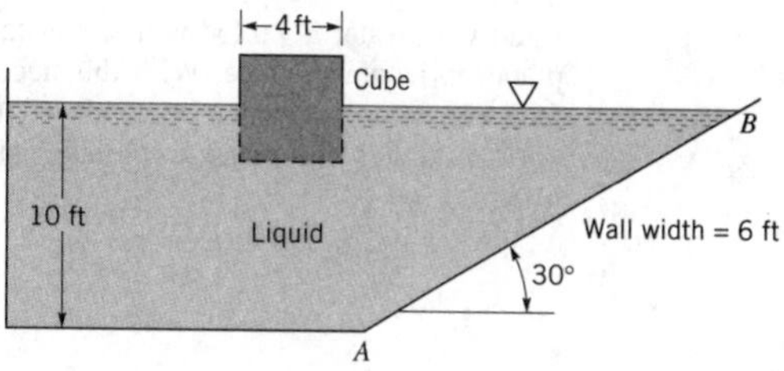

■ FIGURE P2.17R

Since the cube is floating,

$$\Sigma F_{vertical} = 0$$

or

$$\mathcal{W} = F_{buoyant} = \gamma \times \forall ol$$

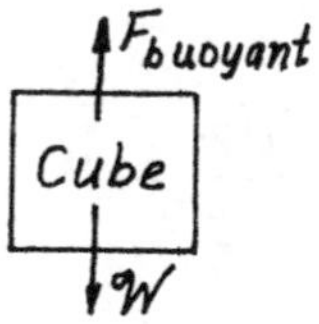

so that

$$\gamma = \frac{3000\ lb}{\frac{1}{2}(4 ft \times 4 ft \times 4 ft)} = 93.8\ \frac{lb}{ft^3}$$

For the tank wall AB,

$$F_R = \gamma h_c A$$

Where $h_c = 5\ ft$

so that

$$F_R = \left(93.8\ \frac{lb}{ft^3}\right)(5\ ft)\left(\frac{10\ ft}{\sin 30°} \times 8\ ft\right)$$

$$= \underline{\underline{75{,}000\ lb}}$$

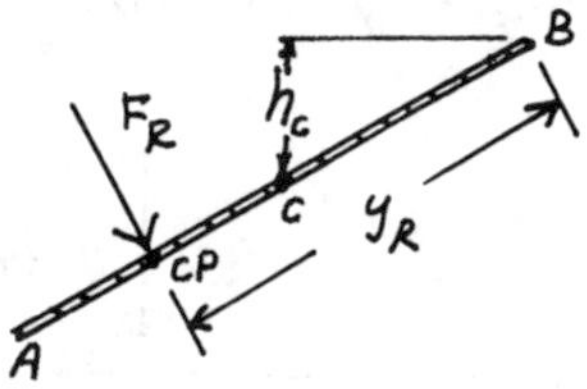

Also,

$$y_R = \frac{I_{xc}}{y_c A} + y_c \qquad \text{where } y_c = \frac{h_c}{\sin 30°} = \frac{5\ ft}{\sin 30°} = 10\ ft$$

and

$$y_R = \frac{\frac{1}{12}(8\ ft)(20\ ft)^3}{(10\ ft)(20\ ft \times 8\ ft)} + 10\ ft = \underline{\underline{13.33\ ft}}$$

Thus,

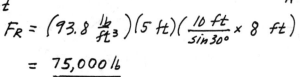

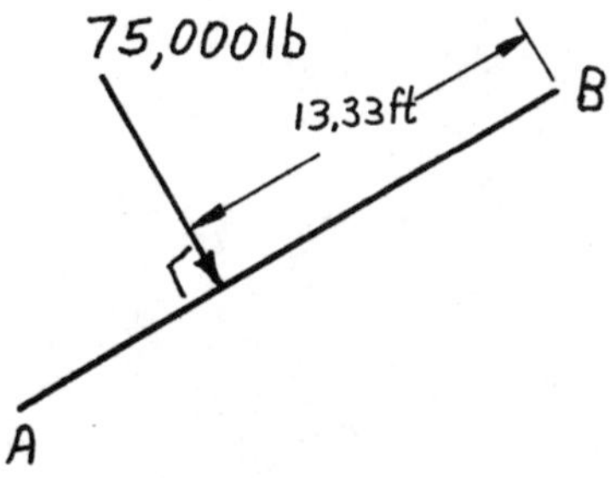

2.18R

2.18R (Rigid body motion) A container that is partially filled with water is pulled with a constant acceleration along a plane horizontal surface. With this acceleration the water surface slopes downward at an angle of 40° with respect to the horizontal. Determine the acceleration. Express your answer in m/s^2.

(ANS: $8.23\ m/s^2$)

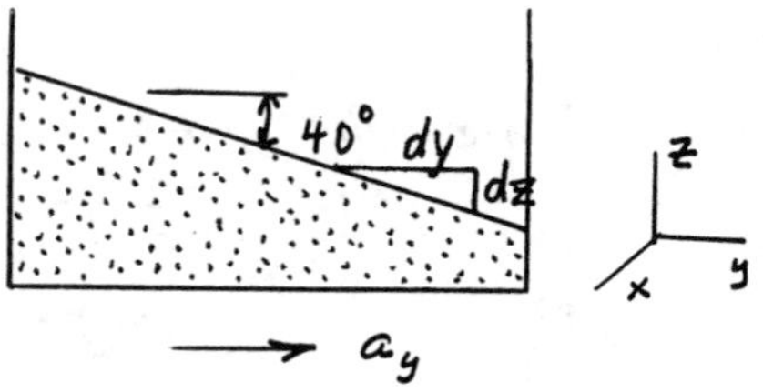

$$\frac{dz}{dy} = -\frac{a_y}{g + a_z} \qquad \text{(Eq. 2.28)}$$

Since

$$\frac{dz}{dy} = -\tan 40°$$

and $a_z = 0$. Thus,

$$a_y = g \tan 40° = \left(9.81\ \frac{m}{s^2}\right) \tan 40° = \underline{\underline{8.23\ \frac{m}{s^2}}}$$

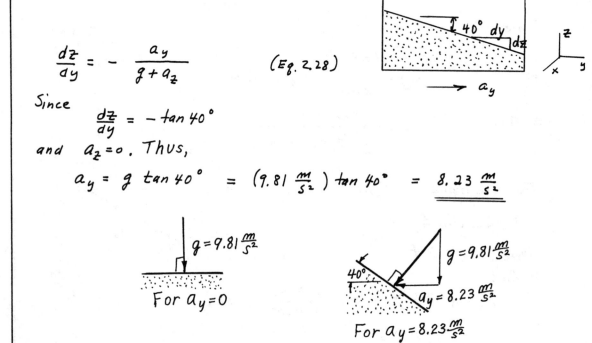

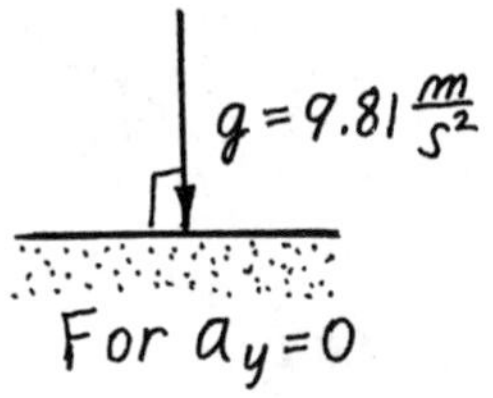

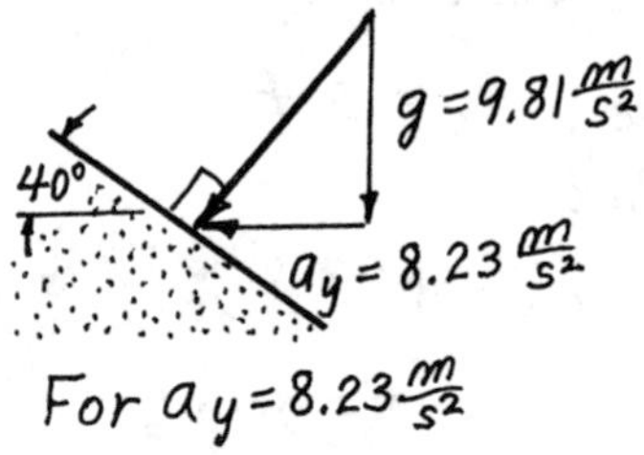

2.19R

2.19R (Rigid-body motion) An open, 2-ft-diameter tank contains water to a depth of 3 ft when at rest. If the tank is rotated about its vertical axis with an angular velocity of 160 rev/min, what is the minimum height of the tank walls to prevent water from spilling over the sides?

(ANS: 5.18 ft)

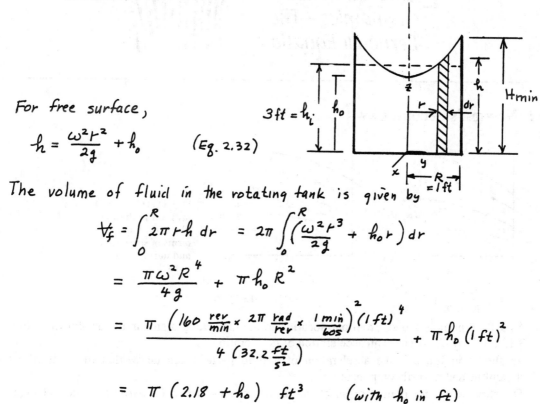

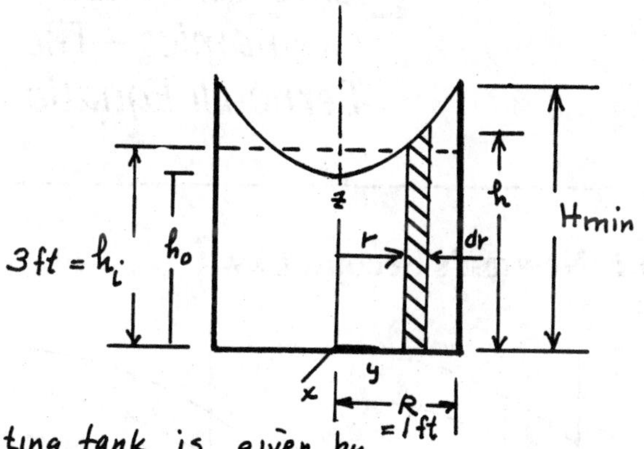

For free surface,

$$h = \frac{\omega^2 r^2}{2g} + h_0 \qquad \text{(Eq. 2.32)}$$

The volume of fluid in the rotating tank is given by

$$\forall_f = \int_0^R 2\pi r h \, dr = 2\pi \int_0^R \left(\frac{\omega^2 r^3}{2g} + h_0 r \right) dr$$

$$= \frac{\pi \omega^2 R^4}{4g} + \pi h_0 R^2$$

$$= \frac{\pi \left(160 \frac{\text{rev}}{\text{min}} \times 2\pi \frac{\text{rad}}{\text{rev}} \times \frac{1\,\text{min}}{60\,\text{s}}\right)^2 (1\,\text{ft})^4}{4\left(32.2 \frac{\text{ft}}{\text{s}^2}\right)} + \pi h_0 (1\,\text{ft})^2$$

$$= \pi\,(2.18 + h_0)\ \text{ft}^3 \qquad (\text{with } h_0 \text{ in ft})$$

Since the initial volume,

$$\forall_i = \pi R^2 h_i = \pi (1\,\text{ft})^2 (3\,\text{ft}) = 3\pi\ \text{ft}^3$$

and the final volume must be equal,

$$\forall_f = \forall_i$$

or

$$\pi\,(2.18 + h_0)\ \text{ft}^3 = 3\pi\ \text{ft}^3$$

and

$$h_0 = 0.820\ \text{ft}$$

Thus, from the first equation (Eq. 2.32)

$$h = \frac{\omega^2 r^2}{2g} + 0.820\ \text{ft}$$

and

$$H_{min} = \frac{\left(160 \frac{\text{rev}}{\text{min}} \times 2\pi \frac{\text{rad}}{\text{rev}} \times \frac{1\,\text{min}}{60\,\text{s}}\right)^2 (1\,\text{ft})^2}{2\left(32.2 \frac{\text{ft}}{\text{s}^2}\right)} + 0.820\ \text{ft} = \underline{\underline{5.18\ \text{ft}}}$$

3

Elementary Fluid Dynamics—The Bernoulli Equation

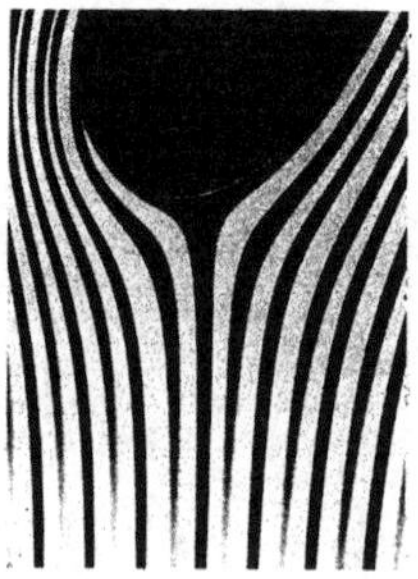

3.1 Newton's Second Law

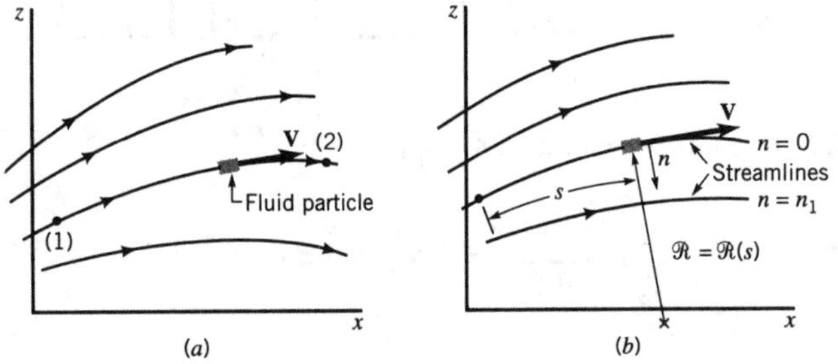

■ **FIGURE 3.1** **(*a*) Flow in the *x–z* plane. (*b*) Flow in terms of streamline and normal coordinates.**

$$a_s = V\frac{\partial V}{\partial s}, \qquad a_n = \frac{V^2}{\mathcal{R}} \qquad \textbf{(3.1)}$$

- As a fluid particle moves from one location to another along a streamline as shown in Fig. 3.1, it usually experiences an acceleration or deceleration.
- As shown in Eq. 3.1, the acceleration of a fluid particle can be written in terms of its streamline and normal components.
- The streamwise component of acceleration, a_s, is non-zero if the particle's speed changes along the streamline.
- The normal component of acceleration, a_n, is non-zero if the particle travels along a curved streamline.
- *Caution:* Even steady flows can have non-zero acceleration.

EXAMPLE: A fluid particle travels on a streamline that has a radius of curvature of 4 ft. The particle's speed increases with distance along the streamline so that $V = 10\,s$ ft/s, where s is in feet. Determine the streamwise and normal components of acceleration when the particle is at $s = 2$ ft.

SOLUTION:

From Eq. 3.1, $a_s = V\frac{dV}{ds} = (10s \text{ ft/s})(10 \text{ 1/s}) = 100s \text{ ft/s}^2$ since $\frac{dV}{ds} = 10$

and $a_n = \frac{V^2}{R} = \frac{(10s \text{ ft/s})^2}{4 \text{ ft}} = 25\,s^2 \text{ ft/s}^2$

Thus, with $s = 2$ ft $a_s = 100(2) = \underline{\underline{200 \text{ ft/s}^2}}$, $a_n = 25(2)^2 = \underline{\underline{100 \text{ ft/s}^2}}$

3.2 F = *ma* Along a Streamline

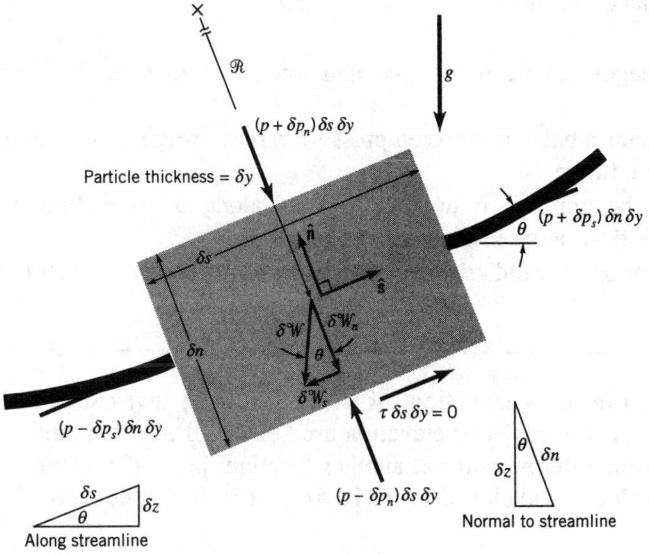

■ **FIGURE 3.3** **Free-body diagram of a fluid particle for which the important forces are those due to pressure and gravity.**

$$\sum \delta F_s = \delta m\, a_s = \delta m\, V \frac{\partial V}{\partial s} = \rho\, \delta\forall\, V \frac{\partial V}{\partial s} \qquad (3.2)$$

$$-\gamma \sin\theta - \frac{\partial p}{\partial s} = \rho V \frac{\partial V}{\partial s} = \rho a_s \qquad (3.4)$$

- For steady flow, the component of Newton's second law in the streamline direction can be written as shown in Eq. 3.2.
- As shown by Fig. 3.3 and Eq. 3.4, for steady, inviscid, incompressible flow, a change in fluid speed is accomplished by a combination of pressure gradient and/or weight *along* the streamline.
- *Caution:* Equation 3.4 involves pressure changes *along* a streamline, not *across* streamlines.

EXAMPLE: Water accelerates at 3 m/s^2 as it flows down a pipe that is tilted 30 degrees below the horizontal. Determine the pressure gradient along the streamline.

SOLUTION:

From Eq. 3.4, $\frac{dp}{ds} = -\rho a_s - \gamma \sin\theta$, where $\theta = -30^\circ$ and $a_s = 3\ m/s^2$

Thus, $\frac{dp}{ds} = -999\ kg/m^3(3\ m/s^2) - 9.80\times10^3\ N/m^3(\sin(-30^\circ)) = \underline{1900\ N/m^3}$

Recall that $1\ kg\cdot m/s^2 = 1\ N$

The Bernoulli Equation

$$p + \frac{1}{2}\rho V^2 + \gamma z = \text{constant along streamline} \qquad \textbf{(3.7)}$$

- Equation 3.4 (see previous page) can be rearranged and integrated to give the Bernoulli equation, Eq. 3.7.
- The Bernoulli equation represents a balance between pressure forces, weight, and velocity for steady, inviscid, incompressible flows.
- The Bernoulli equation can be applied to any two points along a streamline without knowledge of the details of the flow between the two points.
- *Caution:* The Bernoulli equation is valid *along* a streamline; it may not be valid *across* streamlines.

EXAMPLE: At point (1) in the steady flow of water (assumed inviscid and incompressible) the pressure, velocity, and elevation are 1000 lb/ft^2, 20 ft/s, and 40 ft, respectively. Determine the pressure at another location, point (2), along the streamline where the velocity and elevation are 10 ft/s and 50 ft, respectively.
SOLUTION:

From Eq. 3.7, $p_2 = p_1 + \gamma(z_1 - z_2) + \frac{1}{2}\rho(V_1^2 - V_2^2)$

Thus, $p_2 = 1000\ \text{lb/ft}^2 + 62.4\ \text{lb/ft}^3(40-50)\ \text{ft} + \frac{1}{2}(1.94\ \text{slug/ft}^3)(20^2 - 10^2)\ \text{ft}^2/\text{s}^2$

or $p_2 = 667\ \text{lb/ft}^2$

Recall that $1\ \text{slug}\cdot\text{ft/s}^2 = 1\ \text{lb}$

3.3 F = *ma* Normal to a Streamline

$$-\gamma \frac{dz}{dn} - \frac{\partial p}{\partial n} = \frac{\rho V^2}{\mathcal{R}} \quad \textbf{(3.10)}$$

$$p + \rho \int \frac{V^2}{\mathcal{R}} \, dn + \gamma z = \text{constant across the streamline} \quad \textbf{(3.12)}$$

- As shown by Eq. 3.10, for steady, inviscid, incompressible flow, a change in direction of flow (i.e., flow along a curved path) is accomplished by a combination of the pressure gradient and/or weight *normal* to the streamline.
- As shown by Eq. 3.12, the velocity distribution across the streamlines must be known in order to determine the relationship among pressure, elevation, and velocity across a streamline.
- *Caution:* The positive *n*-direction points from the fluid particle *toward* the center of curvature of the streamline.

EXAMPLE: Water flows through the curved pipe shown in the figure with a velocity of 2 m/s. If the radius of curvature of the pipe bend is 0.5 m, determine the pressure gradient normal to the streamline at point (1).

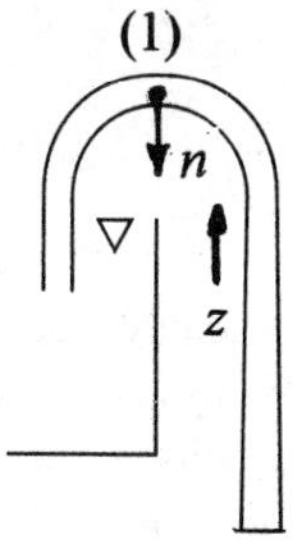

SOLUTION:

From Eq. 3.10, $\frac{dp}{dn} = -\gamma \frac{dz}{dn} - \frac{\rho V^2}{\mathcal{R}}$, where $\mathcal{R} = 0.5\,m$, $V = 2\,m/s$, and $\frac{dz}{dn} = -1$

Thus, $\frac{dp}{dn} = -9.80 \times 10^3 (-1)\,N/m^3 - \frac{999\,kg/m^3\,(2\,m/s)^2}{0.5\,m}$

$= \underline{\underline{1810\,N/m^3}}$

3.5 Static, Stagnation, Dynamic, and Total Pressure

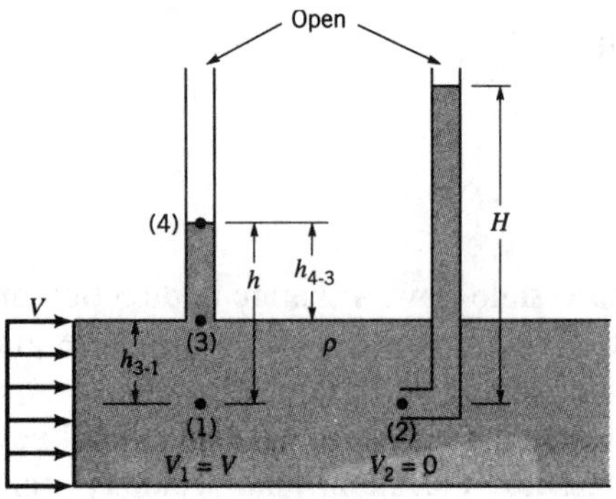

■ **FIGURE 3.4 Measurement of static and stagnation pressures.**

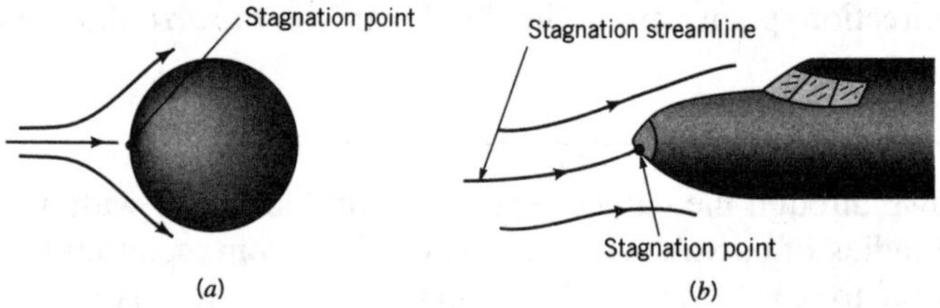

■ **FIGURE 3.5 Stagnation points on bodies in flowing fluids.**

$$p_2 = p_1 + \tfrac{1}{2}\rho V_1^2$$

- The stagnation and dynamic pressures arise from the conversion of kinetic energy of a flowing fluid into a "pressure rise".
- The velocity at a stagnation point on an object (see Fig. 3.5) is zero relative to the object.
- As shown by the above equation, the pressure at the stagnation point is greater than the static pressure by an amount equal to the dynamic pressure, $\rho V^2/2$.
- *Caution:* Application of the Bernoulli equation between points (1) and (4) in Fig. 3.4 is incorrect because these two points do not lie on the same streamline.

EXAMPLE: Determine the maximum pressure on you when you stand in a hurricane force wind (i.e., a 70 mph = 103 ft/s wind).
SOLUTION:

From the Bernoulli equation, $p_1 + \frac{1}{2}\rho V_1^2 + \gamma z_1 = p_2 + \frac{1}{2}\rho V_2^2 + \gamma z_2$
With $p_1 = 0$ and $V_1 = 103$ ft/s (free stream), $V_2 = 0$ (stagnation point), and $z_1 = z_2$ this gives

$$p_2 = \frac{1}{2}\rho V_1^2 = \frac{1}{2}(0.00238\ \text{slugs/ft}^3)(103\ \text{ft/s})^2 = \underline{\underline{12.6\ \text{lb/ft}^2}}$$

3.6 Examples of Use of the Bernoulli Equation

Free Jets

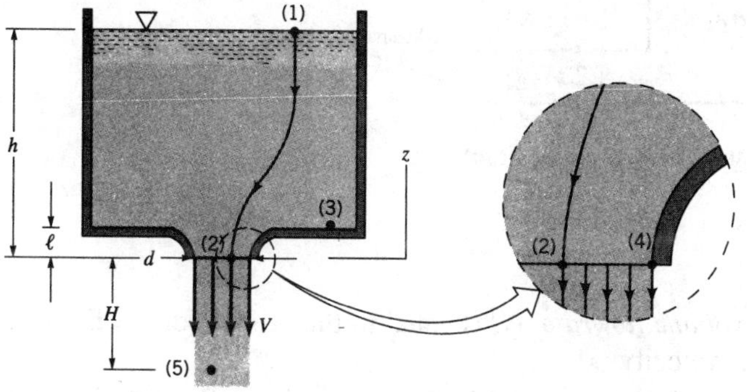

■ FIGURE 3.11
Vertical flow from a tank.

$$V = \sqrt{2\frac{\gamma h}{\rho}} = \sqrt{2gh} \qquad \textbf{(3.18)}$$

- The pressure in the exit plane of an incompressible fluid flowing from a pipe or tank as shown in Fig. 3.11 is equal to the surrounding pressure. This type of flow is termed a *free jet*.
- As shown by the above equation, the velocity of a liquid draining from a large open tank is proportional to the square root of the difference in elevation between the free surface and the exit plane.
- *Caution:* If the tank shown in Fig. 3.11 were pressurized, the above equation would not be valid since the flow would not then be merely an interchange between potential and kinetic energies of the fluid.

EXAMPLE: Water flows from the tank shown in Fig. 3.11. **(a)** Determine the velocity at the exit, point (2), 10 ft below the free surface. **(b)** Determine the velocity in the stream at point (5), 4 ft below the exit.
SOLUTION:

(a) From the Bernoulli equation with $p_1 = p_2 = 0$, $V_1 = 0$, and $z_2 = 0$,

$$\gamma h = \frac{1}{2}\rho V_2^2$$

Thus, $V_2 = \sqrt{2\frac{\gamma}{\rho}h} = \sqrt{2gh} = \left[2\,(32.2\,\text{ft/s}^2)(10\,\text{ft})\right]^{1/2} = \underline{\underline{25.4\,\text{ft/s}}}$

(b) Similarly, for point (5) $p_5 = 0$ so that with $h = (10+4)\,\text{ft} = 14\,\text{ft}$

$$V_5 = \sqrt{2gh} = \left[2\,(32.2\,\text{ft/s}^2)(14\,\text{ft})\right]^{1/2} = \underline{\underline{30.0\,\text{ft/s}}}$$

Confined Flows

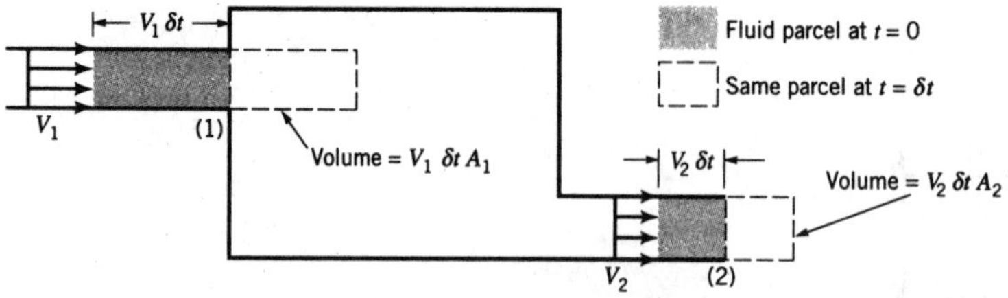

■ FIGURE 3.15 **Steady flow into and out of a tank.**

$$A_1V_1 = A_2V_2, \text{ or } Q_1 = Q_2 \qquad \textbf{(3.19)}$$

- As indicated in Fig. 3.15, the *volume flowrate*, Q, is equal to the product of the fluid velocity and the flow area normal to the velocity.
- As indicated by Eq 3.19, for steady incompressible flow the volume flowrate into a "box", A_1V_1, is equal to that out of the "box", A_2V_2. This is a consequence of the conservation of mass.
- *Caution:* Equation 3.19 is restricted to incompressible flows for which the inlet and outlet velocities are uniform and normal to the inlet and outlet areas.

EXAMPLE: Water flows with a velocity of 10 ft/s in a 6-in.-diameter pipe. Determine the velocity at a location where the pipe diameter is reduced to 4 in.
SOLUTION:

From Eq. 3.19, $V_2 = \frac{A_1}{A_2} V_1 = \frac{\pi D_1^2/4}{\pi D_2^2/4} V_1 = \left(\frac{D_1}{D_2}\right)^2 V_1$

Thus, $V_2 = \left(\frac{6\text{ in.}}{4\text{ in.}}\right)^2 10\text{ ft/s} = \underline{\underline{22.5\text{ ft/s}}}$

Flowrate Measurement

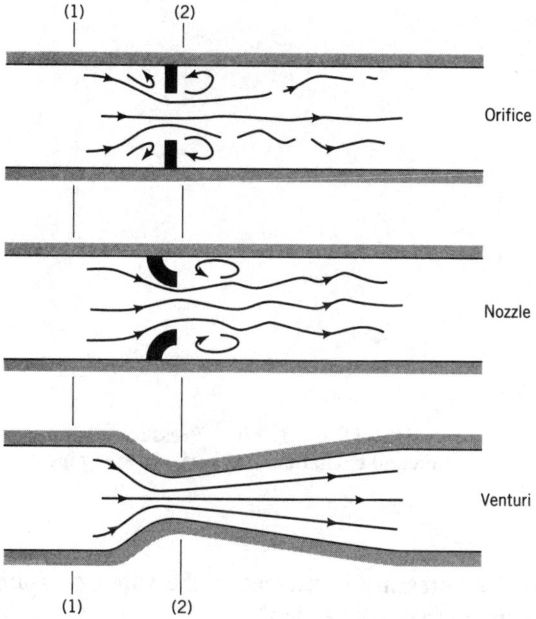

■ FIGURE 3.18 Typical devices for measuring flowrate in pipes.

$$Q = A_2 \sqrt{\frac{2(p_1 - p_2)}{\rho[1 - (A_2/A_1)^2]}} \quad (3.20)$$

- An effective way to measure the flowrate through a pipe is to place some type of restriction within the pipe as shown in Fig. 3.18.
- Orifice, nozzle, and Venturi flow meters are all based on the same principle--low speed means high pressure; high speed means low pressure.
- As shown by Eq. 3.20, for orifice, nozzle, and Venturi flow meters, the volume flowrate is proportional to the square root of the pressure difference across them.

EXAMPLE: Water flows through the Venturi meter shown in Fig. 3.18. Determine the pressure drop, $p_1 - p_2$, if the flowrate is 0.01 m^3/s and the diameters at sections (1) and (2) are 0.04 m and 0.03 m, respectively.

SOLUTION:

From Eq. 3.20 with $\left(\frac{A_2}{A_1}\right)^2 = \left(\frac{\pi D_2^2/4}{\pi D_1^2/4}\right)^2 = \left(\frac{D_2}{D_1}\right)^4 = \left(\frac{0.03\,m}{0.04\,m}\right)^4 = 0.316$

it follows that

$$0.01\,m^3/s = \frac{\pi(0.03\,m)^2}{4}\sqrt{\frac{2(p_1-p_2)}{(999\,kg/m^3)[1-0.316]}}$$

Hence, $p_1 - p_2 = \underline{\underline{6.84 \times 10^4\,N/m^2}}$

Recall that $1\,kg/s^2m = 1\,N/m^2$

Cavitation

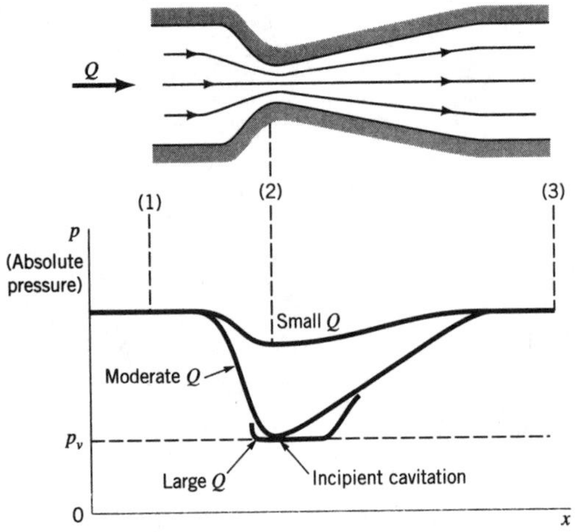

■ FIGURE 3.16 **Pressure variation and cavitation in a variable area pipe.**

- *Cavitation* occurs in a flowing liquid when the pressure is reduced to the vapor pressure, p_v, causing the liquid to boil, even though the liquid may not be "hot".
- According to the Bernoulli equation, if the liquid velocity is increased (for example, by a reduction in flow area as shown in Fig. 3.16) the pressure will decrease and cavitation may occur.
- *Caution:* The vapor pressure is usually given in terms of absolute pressure, not gage pressure.

EXAMPLE: At one location on a horizontal streamline the pressure and velocity of the water are 10 psia and 20 ft/s, respectively. Determine the velocity at another point on the streamline where cavitation begins if the vapor pressure of water is 0.25 psia.

SOLUTION:

From the Bernoulli equation, $p_1 + \frac{1}{2}\rho V_1^2 + \gamma z_1 = p_2 + \frac{1}{2}\rho V_2^2 + \gamma z_2$

where $z_1 = z_2$, $p_1 = 10\,\text{psia}$, $V_1 = 20\,\text{ft/s}$, and $p_2 = 0.25\,\text{psia}$.

Thus, $10\,\text{lb/in.}^2(144\,\text{in.}^2/\text{ft}^2) + \frac{1}{2}(1.94\,\text{slugs/ft}^3)(20\,\text{ft/s})^2$

$$= 0.25\,\text{lb/in.}^2(144\,\text{in.}^2/\text{ft}^2) + \frac{1}{2}(1.94\,\text{slugs/ft}^3)V_2^2$$

so that $V_2 = \underline{\underline{43.0\,\text{ft/s}}}$

3.7 The Energy Line and the Hydraulic Grade Line

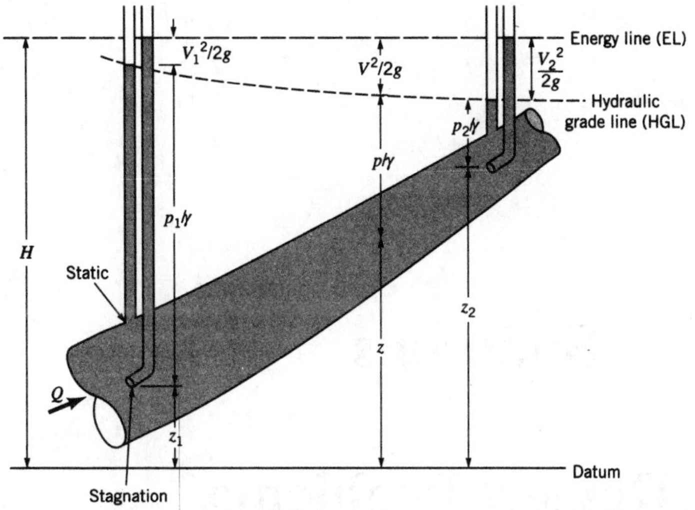

■ **FIGURE 3.21 Representation of the energy line and the hydraulic grade line.**

- The *hydraulic grade line* (HGL) and the *energy line* (EL) concepts shown in Fig. 3.21 provide a useful physical interpretation of the Bernoulli equation..
- In the absence of viscous effects, energy is conserved and the energy line is horizontal.
- *Caution:* The hydraulic grade line is not horizontal if the fluid speed is not constant.

EXAMPLE: Water flows steadily, with negligible viscous effects, from a lake as shown in the figure. The constant-area pipe has a diffuser (an enlargement) attached at the end. Sketch the energy line and hydraulic grade line for this flow.

SOLUTION:

At the exit, $p_3 = 0$, $z_3 = 0$ (arbitrary datum), and $V_3^2/2g = 20\,m$. In the pipe, $A_2 < A_3$ so that $V_2 > V_3$. Hence, $p_2 < p_3$ (i.e. $p_2 < 0$). At the free surface $p_1 = V_1 = 0$. The EL and HGL are as shown.

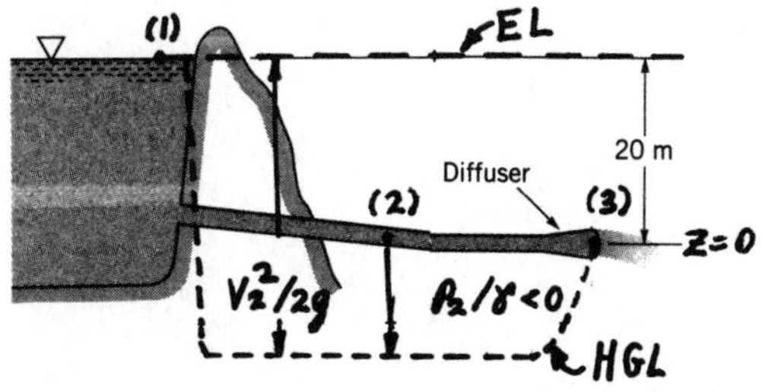

Solutions

to Review Problems

Chapter 3

3.1R

3.1R (F = ma along streamline) What pressure gradient along the streamline, dp/ds, is required to accelerate air at standard temperature and pressure in a horizontal pipe at a rate of 300 ft/s^2?

(ANS: -0.714 lb/ft^3)

$$\frac{\partial p}{\partial s} = -\gamma \sin\theta - \rho V \frac{\partial V}{\partial s}$$ where $\theta = 0$ and $V\frac{\partial V}{\partial s} = a_s = 300 \frac{ft}{s^2}$

Thus,

$$\frac{\partial p}{\partial s} = -\rho a_s = -2.38 \times 10^{-3} \frac{slug}{ft^3}\left(300 \frac{ft}{s^2}\right) = \underline{\underline{-0.714 \left(\frac{lb}{ft^2}\right)/ft}}$$

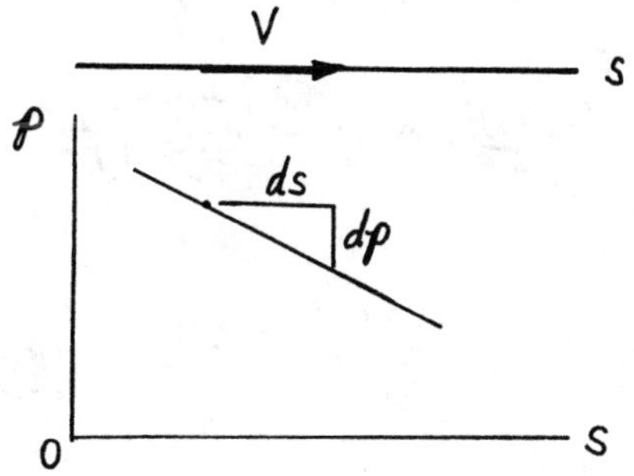

3.2R

3.2R (F = ma normal to streamline) An incompressible, inviscid fluid flows steadily with circular streamlines around a horizontal bend as shown in Fig. P3.2R. The radial variation of the velocity profile is given by $rV = r_0V_0$, where V_0 is the velocity at the inside of the bend which has radius $r = r_0$. Determine the pressure variation across the bend in terms of V_0, r_0, ρ, r, and p_0, where p_0 is the pressure at $r = r_0$. Plot the pressure distribution, $p = p(r)$, if $r_0 = 1.2$ m, $r_1 = 1.8$ m, $V_0 = 12$ m/s, $p_0 = 20$ kN/m^2, and the fluid is water. Neglect gravity.

(ANS: $p_0 + 0.5\rho V_0^2[1 - (r_0/r)^2]$)

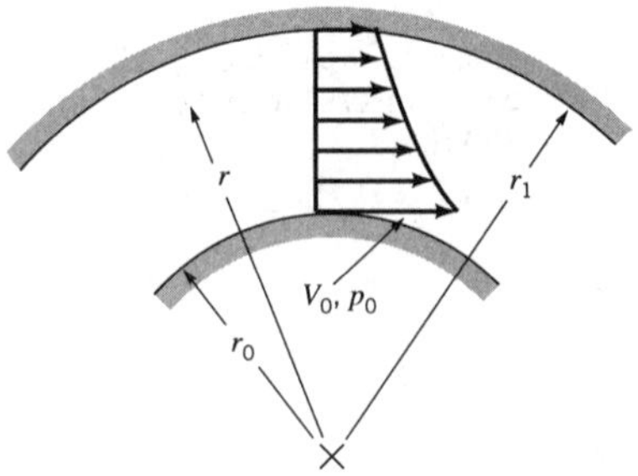

■ FIGURE P3.2R

$$-\gamma\frac{dz}{dn} - \frac{\partial p}{\partial n} = \frac{\rho V^2}{\mathcal{R}} \quad \text{with } \frac{dz}{dn} = 0 \text{ and } r = \mathcal{R} = r_1 - n, \text{ or } \frac{\partial}{\partial n} = -\frac{\partial}{\partial r}$$

Thus,

$$\frac{dp}{dr} = +\frac{\rho V^2}{r} \quad \text{or} \quad \int_{p_0}^{p} dp = +\rho\int_{r_0}^{r}\frac{V^2}{r}dr$$

But

$$\int_{r_0}^{r}\frac{V^2}{r}dr = \int_{r_0}^{r}\frac{(r_0V_0)^2}{r^3}dr = -\frac{r_0^2V_0^2}{2}\left[\frac{1}{r^2} - \frac{1}{r_0^2}\right]$$

Hence,

$$p = p_0 - \tfrac{1}{2}\rho r_0^2V_0^2\left[\frac{1}{r^2} - \frac{1}{r_0^2}\right] = \underline{\underline{p_0 + \tfrac{1}{2}\rho V_0^2\left[1 - \left(\frac{r_0}{r}\right)^2\right]}}$$

For the given data:

$$p = 20\,\frac{kN}{m^2} + \tfrac{1}{2}\left(999\,\frac{kg}{m^3}\right)\left(12\,\frac{m}{s}\right)^2\left[1 - \left(\frac{1.2}{r}\right)^2\right]$$

$$= 20 + 71.9\left[1 - \left(\frac{1.2}{r}\right)^2\right]\,\frac{kN}{m^2}, \text{ where } r \sim m$$

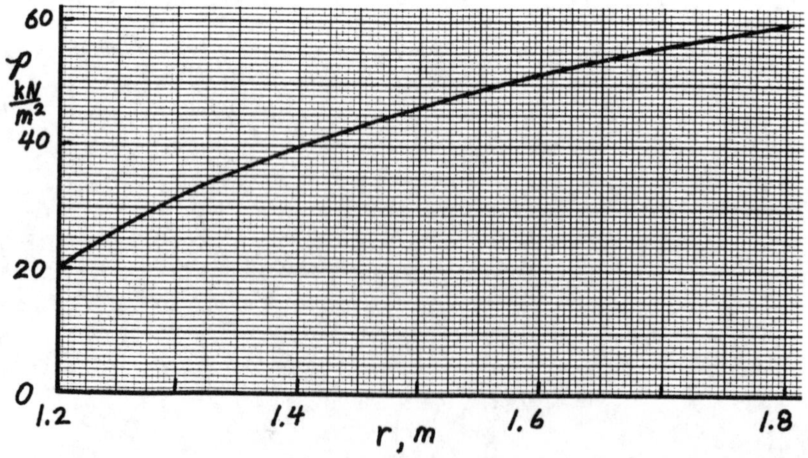

3.3R

3.3R (Stagnation pressure) A hang glider soars through standard sea level air with an airspeed of 10 m/s. What is the gage pressure at a stagnation point on the structure?

(ANS: 61.5 Pa)

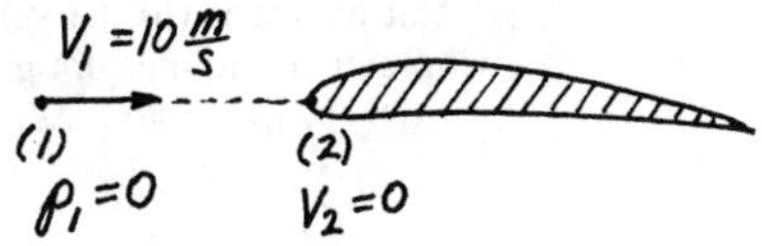

$$\frac{p_1}{\gamma} + \frac{V_1^2}{2g} + z_1 = \frac{p_2}{\gamma} + \frac{V_2^2}{2g} + z_2$$ with $z_1 = z_2$, $p_1 = 0$, $V_1 = 10\,\frac{m}{s}$ and $V_2 = 0$

Thus, $p_2 = \frac{1}{2}\rho V_1^2 = \frac{1}{2}(1.23\,\frac{kg}{m^3})(10\,\frac{m}{s})^2 = 61.5\,\frac{kg\,m}{s^2}/m^2$

$= \underline{\underline{61.5\ Pa}}$

3.4R

3.4R (Bernoulli equation) The pressure in domestic water pipes is typically 60 psi above atmospheric. If viscous effects are neglected, determine the height reached by a jet of water through a small hole in the top of the pipe.

(ANS: 138 ft)

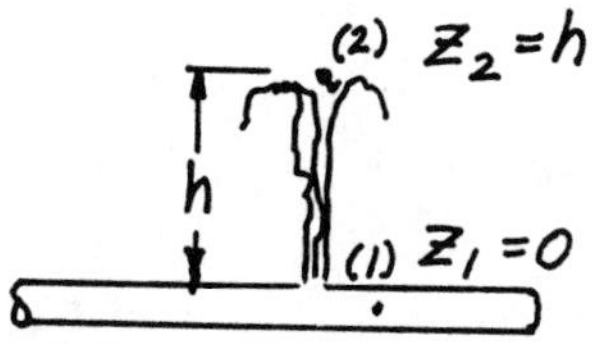

$$\frac{p_1}{\gamma} + \frac{V_1^2}{2g} + z_1 = \frac{p_2}{\gamma} + \frac{V_2^2}{2g} + z_2$$

but $p_1 = 60\ psi$
$p_2 = 0$
$V_2 = 0$
$V_1 \approx 0$ if the hole diameter is much smaller than the pipe diameter

Thus,

$\frac{p_1}{\gamma} = h$

or

$$h = \frac{60\,\frac{lb}{in.^2}(144\,\frac{in.^2}{ft^2})}{62.4\,\frac{lb}{ft^3}}$$

$= \underline{\underline{138\ ft}}$

Note: Because of viscous effects between the water and the pipe and the water and the air, the actual value of would be less than 138 ft.

3.5R

3.5R (Heads) A 4-in.-diameter pipe carries 300 gal/min of water at a pressure of 30 psi. Determine **(a)** the pressure head in feet of water, **(b)** the velocity head, and **(c)** the total head with reference to a datum plane 20 ft below the pipe.

(ANS: 69.2 ft; 0.909 ft; 90.1 ft)

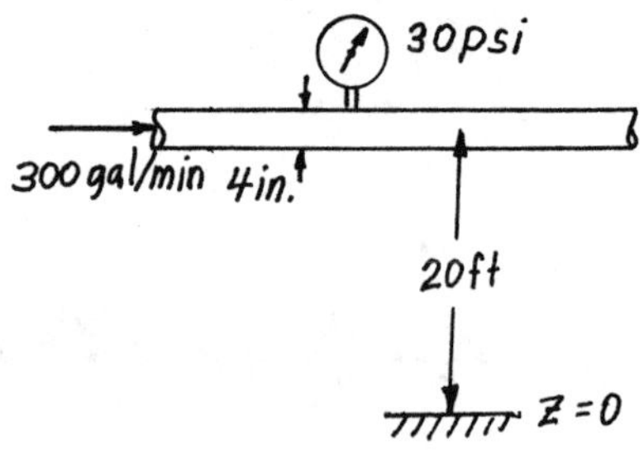

(a) $\frac{p}{\gamma} = \frac{30 \frac{lb}{in.^2}\left(144 \frac{in^2}{ft^2}\right)}{62.4 \frac{lb}{ft^3}} = \underline{\underline{69.2 ft}}$

(b) $Q = 300 \frac{gal}{min}\left(231 \frac{in^3}{gal}\right)\left(\frac{1 min}{60 s}\right)\left(\frac{1 ft^3}{1728 in.^3}\right) = 0.668 \frac{ft^3}{s}$

so that

$V = \frac{Q}{A} = \frac{0.668 \frac{ft^3}{s}}{\frac{\pi}{4}\left(\frac{4}{12} ft\right)^2} = 7.65 \frac{ft}{s}$

or

$\frac{V^2}{2g} = \frac{\left(7.65 \frac{ft}{s}\right)^2}{2\left(32.2 \frac{ft}{s^2}\right)} = \underline{\underline{0.909 ft}}$

(c) $\frac{p}{\gamma} + \frac{V^2}{2g} + z = 69.2 + 0.909 + 20 = \underline{\underline{90.1 ft}}$

3.6R

3.6R (Free jet) Water flows from a nozzle of triangular cross section as shown in Fig. P3.6R. After it has fallen a distance of 2.7 ft, its cross section is circular (because of surface tension effects) with a diameter $D = 0.11$ ft. Determine the flowrate, Q.

(ANS: 0.158 ft^3/s)

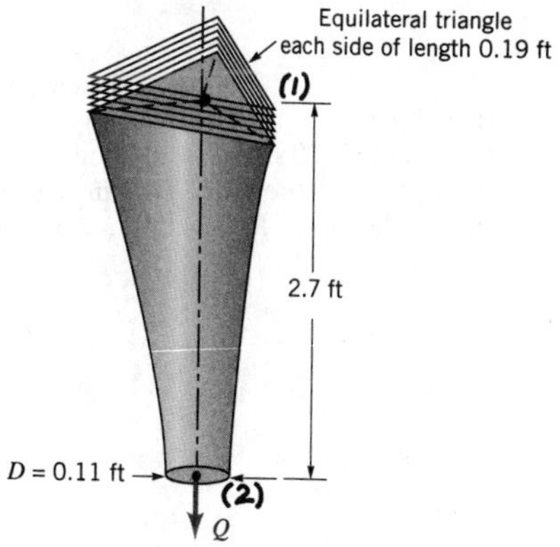

■ FIGURE P3.6R

$$\frac{p_1}{\gamma} + \frac{V_1^2}{2g} + z_1 = \frac{p_2}{\gamma} + \frac{V_2^2}{2g} + z_2$$

where $p_1 = p_2 = 0$, $z_2 = 0$, $z_1 = 2.7\ ft$

and

$$V_1 = \frac{Q}{A_1}, \quad V_2 = \frac{Q}{A_2}$$

Thus,

$$\left(\frac{Q}{A_1}\right)^2 + 2gz_1 = \left(\frac{Q}{A_2}\right)^2 \quad \text{or} \quad Q = \left[\frac{2gz_1}{\left(\frac{1}{A_2^2} - \frac{1}{A_1^2}\right)}\right]^{1/2} = \frac{A_2\sqrt{2gz_1}}{\sqrt{1-(A_2/A_1)^2}}$$

but

$$A_2 = \frac{\pi}{4}(0.11\ ft)^2 = 0.00950\ ft^2$$

and

$$A_1 = \frac{1}{2}(0.19\ ft)(0.1645\ ft) = 0.0156\ ft^2$$

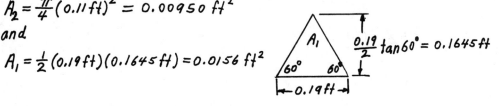

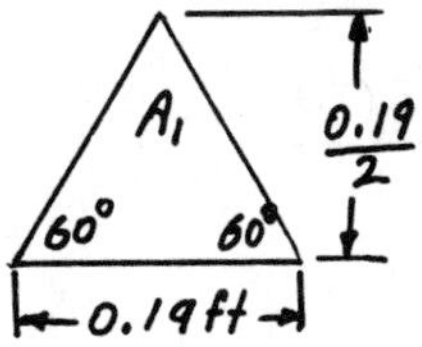

$$\frac{0.19}{2}\tan 60° = 0.1645\ ft$$

Thus,

$$Q = \frac{(0.00950\ ft^2)\sqrt{2(32.2\frac{ft}{s^2})(2.7\ ft)}}{\left[1 - \frac{(0.00950\ ft^2)^2}{(0.0156\ ft^2)^2}\right]^{1/2}} = \underline{\underline{0.158\ \frac{ft^3}{s}}}$$

3.7R

3.7R (Bernoulli/continuity) Water flows into a large tank at a rate of 0.011 m^3/s as shown in Fig. P3.7R. The water leaves the tank through 20 holes in the bottom of the tank, each of which produces a stream of 10-mm diameter. Determine the equilibrium height, h, for steady state operation.

(ANS: 2.50 m)

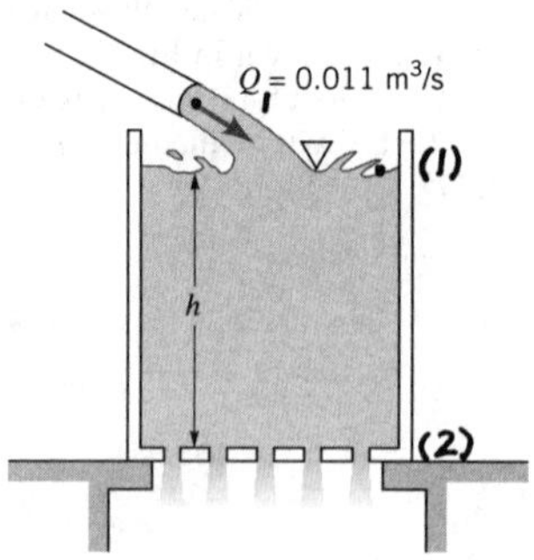

■ FIGURE P3.7R

$Q_1 = Q_2$ where $Q_1 = 0.011\ \frac{m^3}{s}$

and

$Q_2 = 20 A_2 V_2 = 20 \frac{\pi}{4} D_2^2 V_2$

but

$\frac{p_1}{\gamma} + \frac{V_1^2}{2g} + z_1 = \frac{p_2}{\gamma} + \frac{V_2^2}{2g} + z_2$ where $p_1 = p_2 = 0$, $V_1 = 0$, and $z_1 - z_2 = h$

Thus,

$V_2 = \sqrt{2gh}$

so that

$0.011\ \frac{m^3}{s} = 20 \frac{\pi}{4}(0.01\,m)^2 \sqrt{2(9.81\ \frac{m}{s^2})h}$

or

$h = \underline{\underline{2.50\ m}}$

3.8R

3.8R (Bernoulli/continuity) Gasoline flows from a 0.3-m-diameter pipe in which the pressure is 300 kPa into a 0.15-m-diameter pipe in which the pressure is 120 kPa. If the pipes are horizontal and viscous effects are negligible, determine the flowrate.

(ANS: 0.420 m^3/s)

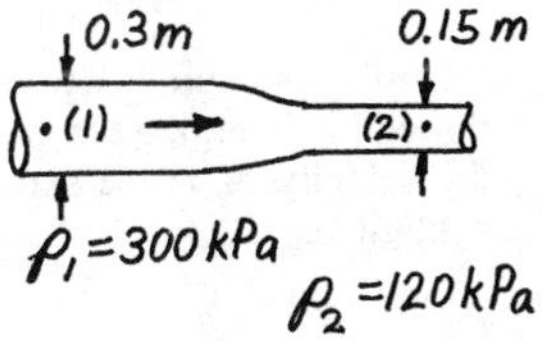

$$\frac{p_1}{\gamma}+\frac{V_1^2}{2g}+z_1=\frac{p_2}{\gamma}+\frac{V_2^2}{2g}+z_2 \quad \text{with } z_1=z_2 \text{ and } \gamma=6.67\,\frac{kN}{m^3}$$

Also, $A_1V_1=A_2V_2$ or $V_2=\left(\frac{D_1}{D_2}\right)^2V_1=\left(\frac{0.3\,m}{0.15\,m}\right)^2V_1=4V_1$

Thus,

$$\frac{p_1}{\gamma}+\frac{V_1^2}{2g}=\frac{p_2}{\gamma}+\frac{16V_1^2}{2g} \quad \text{or} \quad 15V_1^2=2g\,\frac{p_1-p_2}{\gamma}$$

so that

$$V_1=\left[\frac{2(9.81\,\frac{m}{s^2})(300\,kPa-120\,kPa)}{15\,(6.67\,\frac{kN}{m^3})}\right]^{1/2}=5.94\,\frac{m}{s}$$

Thus,

$$Q=A_1V_1=\frac{\pi}{4}(0.3m)^2(5.94\,\frac{m}{s})=\underline{\underline{0.420\,\frac{m^3}{s}}}$$

3.9R

3.9R (Bernoulli/continuity) Water flows steadily through the pipe shown in Fig. P3.9R such that the pressures at sections (1) and (2) are 300 kPa and 100 kPa, respectively. Determine the diameter of the pipe at section (2), D_2, if the velocity at section 1 is 20 m/s and viscous effects are negligible.

(ANS: 0.0688 m)

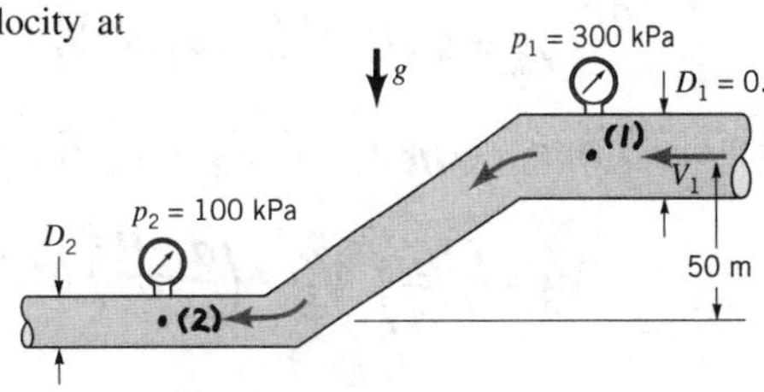

■ FIGURE P3.9R

$$\frac{p_1}{\gamma}+\frac{V_1^2}{2g}+z_1=\frac{p_2}{\gamma}+\frac{V_2^2}{2g}+z_2$$

or

$$\frac{300\times10^3\,N/m^2}{9.80\times10^3\,N/m^3}+\frac{(20\,m/s)^2}{2(9.81\,m/s^2)}+50\,m=\frac{100\times10^3\,N/m^2}{9.80\times10^3\,N/m^3}+\frac{V_2^2}{2(9.81\,m/s^2)}$$

Thus, $V_2=42.2\,m/s$

so that since $V_1A_1=V_2A_2$ or $V_1\frac{\pi}{4}D_1^2=V_2\frac{\pi}{4}D_2^2$, then

$$D_2=\left[\frac{V_1}{V_2}\right]^{1/2}D_1=\left[\frac{20\,m/s}{42.2\,m/s}\right]^{1/2}(0.1\,m)=\underline{\underline{0.0688\,m}}$$

3.10R

3.10R (Bernoulli/continuity) Water flows steadily through a diverging tube as shown in Fig. P3.10R. Determine the velocity, V, at the exit of the tube if frictional effects are negligible.
(ANS: 1.04 ft/s)

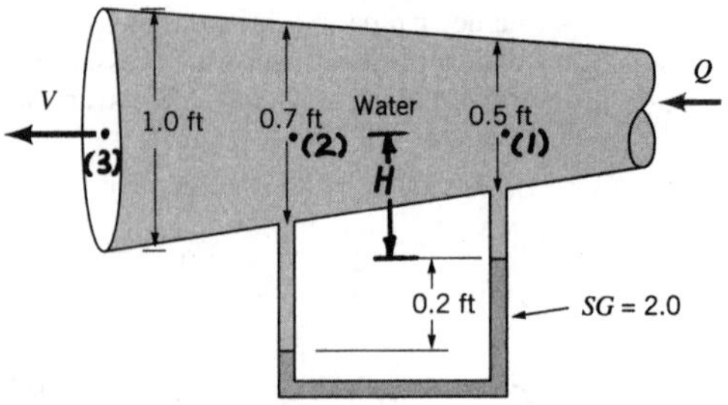

■ FIGURE P3.10R

$$\frac{p_1}{\gamma}+\frac{V_1^2}{2g}+z_1=\frac{p_2}{\gamma}+\frac{V_2^2}{2g}+z_2 \qquad (z_1=z_2) \qquad (1)$$

where $A_1V_1=A_2V_2$, or

$$V_1=\left(\frac{D_2}{D_1}\right)^2V_2=\left(\frac{0.7\,m}{0.5\,m}\right)^2V_2=1.96\,V_2 \qquad (2)$$

Also, $p_2+\gamma(H+0.2\,ft)-SG\gamma(0.2\,ft)-\gamma H=p_1$

or $p_2=p_1+2\gamma(0.2)-\gamma(0.2)=p_1+62.4\frac{lb}{ft^3}[2-1](0.2\,ft)$

$$=p_1+12.48\frac{lb}{ft^2} \qquad (3)$$

By combining (1), (2), and (3):

$$\frac{p_1}{\gamma}+\frac{(1.96\,V_2)^2\,ft^2/s^2}{2(32.2\,ft/s^2)}=\frac{p_1}{\gamma}+\frac{12.48\,lb/ft^2}{62.4\,lb/ft^3}+\frac{V_2^2\,ft^2/s^2}{2(32.2\,ft/s^2)}$$

or $V_2=2.13\,ft/s$ and $V_1=1.96(2.13\,ft/s)=4.17\,ft/s$

Thus, since $V_3A_3=V_2A_2$, then

$$V_3=\left(\frac{D_2}{D_3}\right)^2V_2=\left(\frac{0.7\,ft}{1\,ft}\right)^2(2.13\,ft/s)=\underline{\underline{1.04\,ft/s}}$$

3.11R

3.11R (Bernoulli/continuity/Pitot tube) Two Pitot tubes and two static pressure taps are placed in the pipe contraction shown in Fig. P3.11R. The flowing fluid is water, and viscous effects are negligible. Determine the two manometer readings, h and H.

(ANS: 0; 0.252 ft)

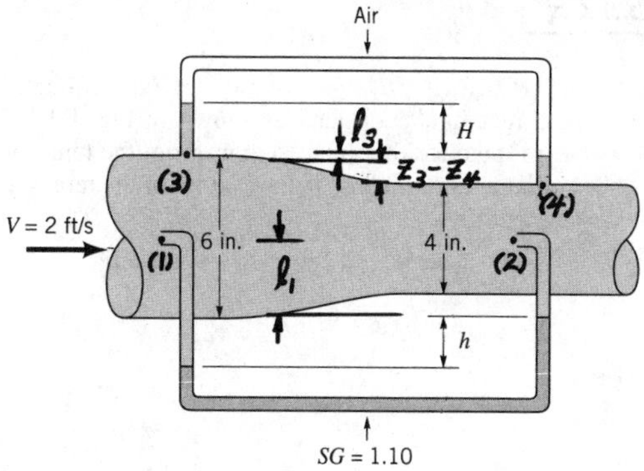

■ FIGURE P3.11R

$$\frac{p_1}{\gamma} + \frac{V_1^2}{2g} + z_1 = \frac{p_2}{\gamma} + \frac{V_2^2}{2g} + z_2 \quad \text{with } z_1 = z_2 \text{ and } V_1 = V_2 = 0$$

Thus, $p_1 = p_2$ so that manometer considerations give

$p_1 + \gamma(\ell_1 + h) = p_2 + \gamma \ell_1 + 1.10\gamma h$, or with $p_1 = p_2$ this gives $\underline{\underline{h = 0}}$

Also,

$$\frac{p_3}{\gamma} + \frac{V_3^2}{2g} + z_3 = \frac{p_4}{\gamma} + \frac{V_4^2}{2g} + z_4 \quad \text{where } z_3 = \tfrac{3}{12}\text{ ft},\ z_4 = \tfrac{2}{12}\text{ ft},\ V_3 = 2\,\tfrac{\text{ft}}{\text{s}},$$

and $A_3 V_3 = A_4 V_4$ or

$$V_4 = \frac{A_3}{A_4} V_3 = \frac{\frac{\pi}{4} D_3^2}{\frac{\pi}{4} D_4^2} V_3 = \left(\frac{6\text{ in.}}{4\text{ in.}}\right)^2 \left(2\,\frac{\text{ft}}{\text{s}}\right) = 4.50\,\frac{\text{ft}}{\text{s}}$$

Thus, $$\frac{p_3 - p_4}{\gamma} = \left(\frac{2}{12} - \frac{3}{12}\right)\text{ft} + \frac{(4.50\,\frac{\text{ft}}{\text{s}})^2 - (2\,\frac{\text{ft}}{\text{s}})^2}{2(32.2\,\frac{\text{ft}}{\text{s}^2})} = 0.169\text{ ft} \qquad (1)$$

But $p_3 - \gamma \ell_3 - \gamma H = p_4 - \gamma(z_3 - z_4 + \ell_3)$

or $$\frac{p_3 - p_4}{\gamma} = H - (z_3 - z_4) \qquad (2)$$

From Eqs.(1) and (2) we obtain

$0.169\text{ ft} = H - \left(\frac{3-2}{12}\text{ ft}\right)$ or $\underline{\underline{H = 0.252\text{ ft}}}$

3.12R

3.12R (Bernoulli/continuity) Water collects in the bottom of a rectangular oil tank as shown in Fig. P3.12R. How long will it take for the water to drain from the tank through a 0.02-m-diameter drain hole in the bottom of the tank? Assume quasi-steady flow.

(ANS: 2.45 hr)

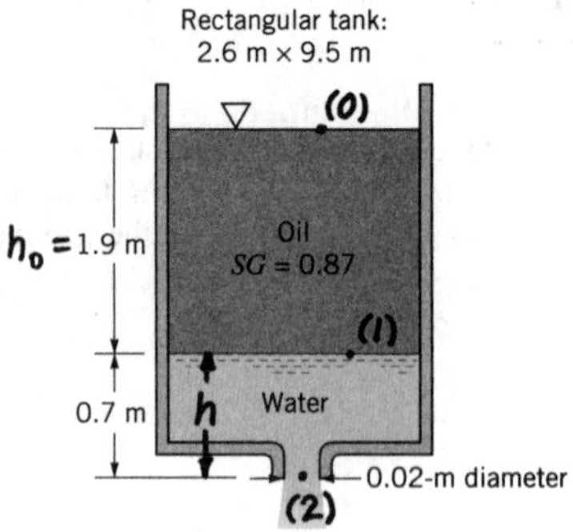

■ FIGURE P3.12R

$$\frac{p_1}{\gamma} + \frac{V_1^2}{2g} + z_1 = \frac{p_2}{\gamma} + \frac{V_2^2}{2g} + z_2$$

where $p_1 = p_0 + \gamma_0 h_0$, $p_0 = 0$, $p_2 = 0$, $z_1 = h$, $z_2 = 0$, and $V_1 = 0$

Thus, since $\frac{\gamma_0}{\gamma} = SG = 0.87$,

$$\frac{\gamma_0 h_0}{\gamma} + h = \frac{V_2^2}{2g} \quad \text{or} \quad V_2 = \sqrt{2g(h + SG\,h_0)}$$

or

$$V_2 = \sqrt{2\left(9.81\frac{m}{s^2}\right)(h + 0.87(1.9\,m))} = 4.43\sqrt{h + 1.653}\ \frac{m}{s},$$

where $h \sim m$

Also,

$$Q = A_2 V_2 = \frac{\pi}{4} D_2^2 V_2 = \frac{\pi}{4}(0.02\,m)^2\left[4.43\sqrt{h + 1.653}\ \frac{m}{s}\right] \qquad (1)$$

$$= 1.39 \times 10^{-3}\sqrt{h + 1.653}\ \frac{m^3}{s}$$

and

$$Q = A_1\left(-\frac{dh}{dt}\right),$$

where

$A_1 = 2.6\,m\,(9.5\,m) = 24.7\,m^2$ Hence,

$$Q = -24.7\frac{dh}{dt} \qquad (2)$$

Combine Eqs. (1) and (2) to give

$$\frac{dh}{dt} = -5.63 \times 10^{-5}\sqrt{h + 1.653}$$

or

$$\int_{h=0.7}^{h=0} \frac{dh}{\sqrt{h + 1.653}} = -5.63 \times 10^{-5}\int_{t=0}^{t=t_f} dt$$

where t_f = time to drain the water

or

$$2\sqrt{h + 1.653}\,\Big|_{0.7}^{0} = -5.63 \times 10^{-5}\, t_f$$

Thus,

$$t_f = 8.83 \times 10^3\ s = \underline{\underline{2.45\ hr}}$$

3.13R

3.13R (Cavitation) Water flows past the hydrofoil shown in Fig. P3.13R with an upstream velocity of V_0. A more advanced analysis indicates that the maximum velocity of the water in the entire flow field occurs at point *B* and is equal to $1.1V_0$. Calculate the velocity, V_0, at which cavitation will begin if the atmospheric pressure is 101 kPa (abs) and the vapor pressure of the water is 3.2 kPa (abs).

(ANS: 31.4 m/s)

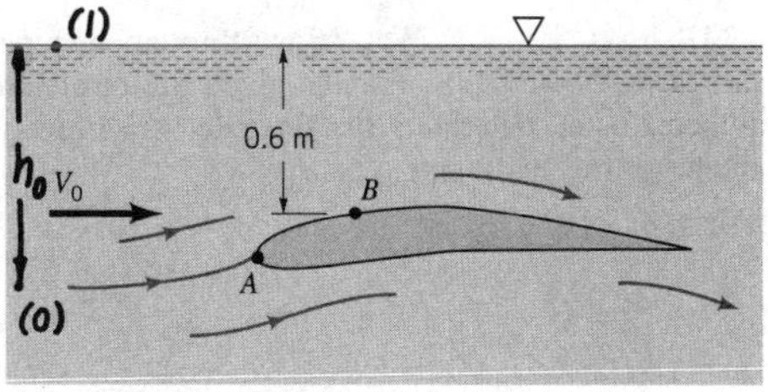

■ FIGURE P3.13R

$$\frac{p_0}{\gamma} + \frac{V_0^2}{2g} + z_0 = \frac{p_B}{\gamma} + \frac{V_B^2}{2g} + z_B \quad \text{where } p_B = p_v = 3.2 \text{ kPa (abs)}$$

$$p_0 = p_1 + \gamma h_0, \; z_0 = -h_0,$$

Thus, and $z_B = -0.6\text{m}$

$$\frac{p_1 + \gamma h_0}{\gamma} + \frac{V_0^2}{2g} - h_0 = \frac{p_v}{\gamma} + \frac{V_B^2}{2g} + z_B \qquad (1)$$

But $V_B = 1.1 V_0$ and $p_1 = 101 \text{ kPa (abs)}$ so that Eq. (1) gives

$$\frac{101 \frac{kN}{m^2}}{9.80 \frac{kN}{m^3}} + \frac{V_0^2}{2(9.81 \frac{m}{s^2})} = \frac{3.2 \frac{kN}{m^2}}{9.80 \frac{kN}{m^3}} + \frac{(1.1 V_0)^2}{2(9.81 \frac{m}{s^2})} - 0.6\text{m}$$

or

$$V_0 = \underline{\underline{31.4 \frac{m}{s}}}$$

3.14R

3.14R (Flowrate) Water flows through the pipe contraction shown in Fig. P3.14R. For the given 0.2-m difference in manometer level, determine the flowrate as a function of the diameter of the small pipe, D.

(ANS: 0.0156 m^3/s)

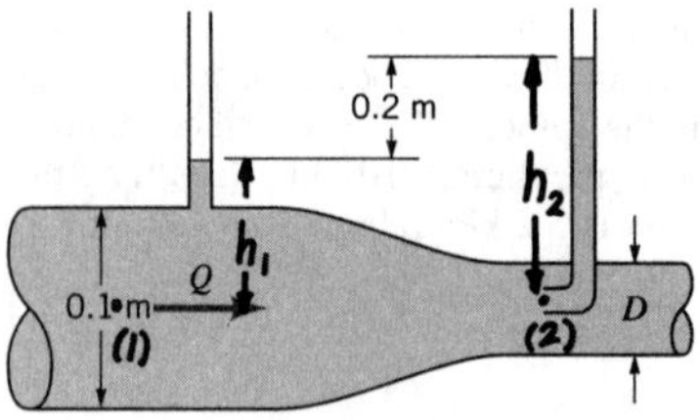

■ FIGURE P3.14R

$$\frac{p_1}{\gamma} + \frac{V_1^2}{2g} + z_1 = \frac{p_2}{\gamma} + \frac{V_2^2}{2g} + z_2 \text{ , where } z_1 = z_2 \text{ and } V_2 = 0$$

Thus,

$$V_1 = \sqrt{2g\frac{(p_2 - p_1)}{\gamma}}$$

But

$p_1 = \gamma h_1$ and $p_2 = \gamma h_2$, so that $p_2 - p_1 = \gamma(h_2 - h_1) = 0.2\gamma$

Thus,

$$V_1 = \sqrt{2g\frac{0.2\gamma}{\gamma}} = \sqrt{2g(0.2)}$$

or

$$Q = A_1 V_1 = \frac{\pi}{4}(0.1\,m)^2\sqrt{2(9.81\tfrac{m}{s^2})(0.2\,m)} = \underline{\underline{0.0156\,\tfrac{m^3}{s}}}$$

3.15R

3.15R (Channel flow) Water flows down the ramp shown in the channel of Fig. P3.15R. The channel width decreases from 15 ft at section (1) to 9 ft at section (2). For the conditions shown, determine the flowrate.

(ANS: 509 ft^3/s)

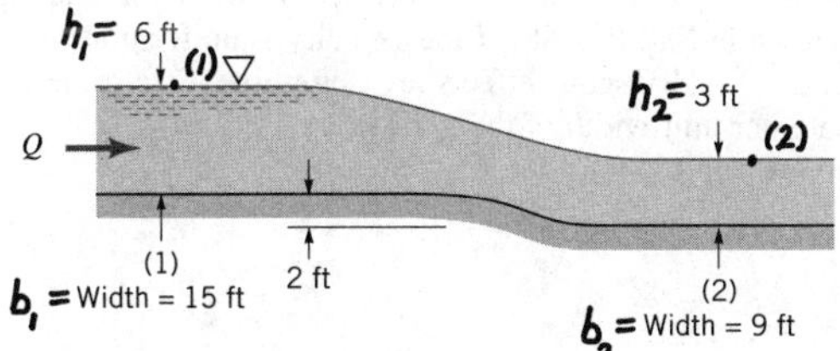

FIGURE P3.15R

$$\frac{p_1}{\gamma} + \frac{V_1^2}{2g} + z_1 = \frac{p_2}{\gamma} + \frac{V_2^2}{2g} + z_2 \qquad (1)$$

where $p_1 = 0$, $p_2 = 0$, $z_2 = 3\,ft$, and $z_1 = (6+2)\,ft = 8\,ft$

Also, $A_1 V_1 = A_2 V_2$

or

$$V_2 = \frac{h_1 b_1}{h_2 b_2} V_1 = \frac{(6\,ft)(15\,ft)}{(3\,ft)(9\,ft)} V_1 = 3.33\,V_1$$

Thus, Eq. (1) becomes

$$\left[3.33^2 - 1\right] V_1^2 = 2(32.2\,\tfrac{ft}{s^2})(8-3)\,ft, \text{ or } V_1 = 5.65\,\tfrac{ft}{s}$$

Hence,

$$Q = A_1 V_1 = (6\,ft)(15\,ft)(5.65\,\tfrac{ft}{s}) = \underline{\underline{509\,\tfrac{ft^3}{s}}}$$

3.16R

3.16R (Channel flow) Water flows over the spillway shown in Fig. P3.16R. If the velocity is uniform at sections (1) and (2) and viscous effects are negligible, determine the flowrate per unit width of the spillway.

(ANS: 7.44 m^2/s)

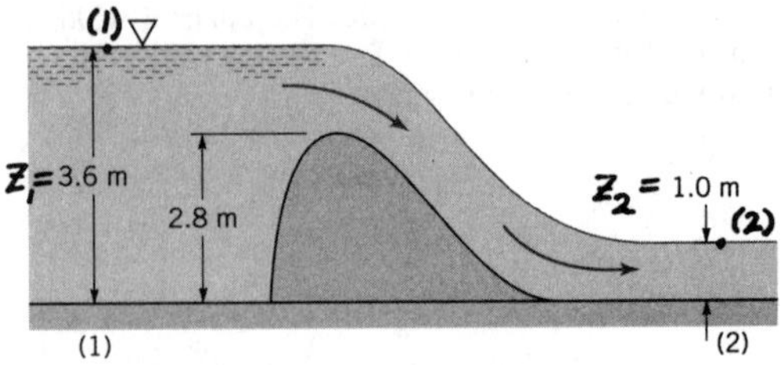

■ FIGURE P3.16R

$$\frac{p_1}{\gamma} + \frac{V_1^2}{2g} + z_1 = \frac{p_2}{\gamma} + \frac{V_2^2}{2g} + z_2$$

where, if points (1) and (2) are located on the free surface, $p_1 = 0$, $p_2 = 0$, $z_1 = 3.6\,m$, and $z_2 = 1.0\,m$.

Also, $A_1 V_1 = A_2 V_2$

or

$$V_1 = \frac{z_2}{z_1} V_2 = \frac{1.0\,m}{3.6\,m} V_2 = 0.278\,V_2$$

Thus, Eq. (1) becomes

$$\frac{V_2^2}{2(9.81\,\frac{m}{s^2})}\left[1 - (0.278)^2\right] = 3.6\,m - 1.0\,m, \quad \text{or} \quad V_2 = 7.44\,\frac{m}{s}$$

Hence,

$$q = V_2 z_2 = (7.44\,\tfrac{m}{s})(1.0\,m) = \underline{\underline{7.44\,\tfrac{m^2}{s}}}$$

3.17R

3.17R (Energy line/hydraulic grade line) Draw the energy line and hydraulic grade line for the flow shown in Problem 3.43.

3.43 A smooth plastic, 10-m-long garden hose with an inside diameter of 20 mm is used to drain a wading pool as is shown in Fig. P3.43. If viscous effects are neglected, what is the flowrate from the pool?

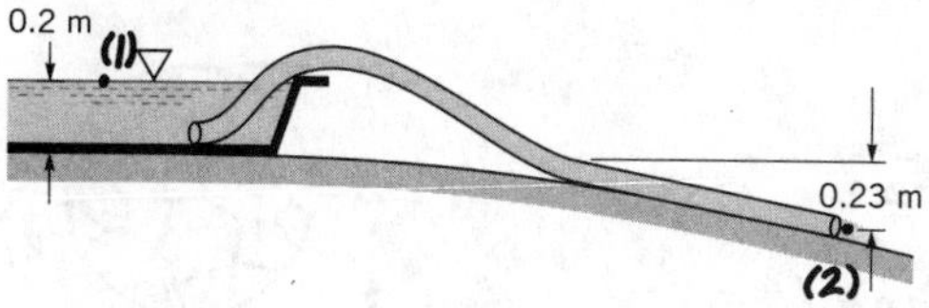

■ FIGURE P3.43

Since $\frac{p_1}{\gamma} + \frac{V_1^2}{2g} + z_1 = \frac{p_2}{\gamma} + \frac{V_2^2}{2g} + z_2$, with $p_1 = p_2 = 0$, $V_1 = 0$, $z_2 = 0$ and $z_1 = 0.2m + 0.23m = 0.43m$, it follows that

$$\frac{V_2^2}{2g} = 0.43\,m$$

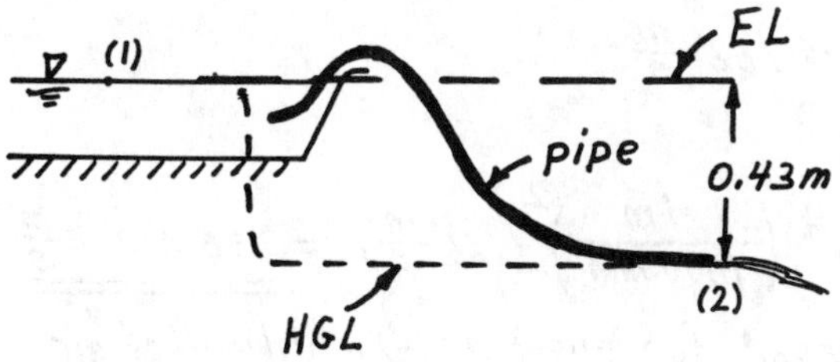

For inviscid flow with no pumps or turbines, the energy line (EL) is horizontal, at an elevation of the free surface. The hydraulic grade line (HGL) is one velocity head lower—even with the pipe outlet. Since the fluid velocity is constant throughout the pipe with $\frac{V^2}{2g} = 0.43m$, the above diagram is obtained.

3.18R

3.18R (Restrictions on Bernoulli equation) A 0.3-m-diameter soccer ball, pressurized to 20 kPa, develops a small leak with an area equivalent to 0.006 mm². If viscous effects are neglected and the air is assumed to be incompressible, determine the flowrate through the hole. Would the ball become noticeably softer during a 1-hr soccer game? Explain. Is it reasonable to assume incompressible flow for this situation? Explain.

(ANS: 9.96×10^{-7} m³/s; yes; no, Ma > 0.3)

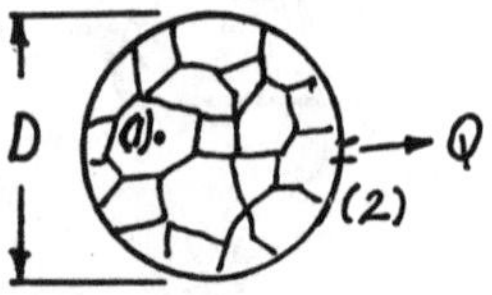

$$\frac{p_1}{\gamma} + \frac{V_1^2}{2g} + z_1 = \frac{p_2}{\gamma} + \frac{V_2^2}{2g} + z_2$$

where $z_1 = z_2$, $p_2 = 0$ and $V_1 = 0$

Thus,

$$\frac{V_2^2}{2g} = \frac{p_1}{\gamma} \quad \text{or} \quad V_2 = \sqrt{\frac{2p_1}{\rho}}$$

where $\rho = \frac{p_1}{RT_1}$

Assume $T_1 = 15\,^\circ\text{C}$ so that

$$\rho = \frac{(101.+20)\times 10^3 \frac{N}{m^2}}{\left(286.9 \frac{N\cdot m}{kg\cdot K}\right)(273+15)K} = 1.46 \frac{kg}{m^3}$$

Thus,

$$V_2 = \sqrt{\frac{2\left(20\times 10^3 \frac{N}{m^2}\right)}{1.46 \frac{kg}{m^3}}} = 166 \frac{m}{s}$$

so that

$$Q = A_2 V_2 = (0.006\ mm^2)\left(\frac{1\,m}{1000\,mm}\right)^2\left(166 \frac{m}{s}\right) = \underline{\underline{9.96 \times 10^{-7} \frac{m^3}{s}}}$$

If this flowrate continued for one hour, the volume of air leaving the ball would be

$$\forall = Qt = \left(9.96\times 10^{-7} \frac{m^3}{s}\right)(3{,}600\ s) = 3.59\times 10^{-3}\ m^3$$

Since the volume of the ball is

$$\forall_{ball} = \frac{\pi}{6} D^3 = \frac{\pi}{6}(0.3\,m)^3 = 14.1\times 10^{-3}\ m^3,$$

approximately

$$\frac{\forall}{\forall_{ball}} = \frac{3.59\times 10^{-3}\ m^3}{14.1\times 10^{-3}\ m^3} = 0.255 = 25.5\% \text{ of the air would escape.}$$

<u>The ball would become noticably softer</u>

Note that $Ma_2 = \frac{V_2}{c_2}$ where $c_2 = \sqrt{kRT_2}$ or if $T_2 \approx 15\,^\circ\text{C}$

$$c_2 = \left[1.40\left(286.9 \frac{N\cdot m}{kg\cdot K}\right)(273+15)K\right]^{1/2} = 340 \frac{m}{s}$$

so that the Mach number is

$$Ma_2 = \frac{166 \frac{m}{s}}{340 \frac{m}{s}} = 0.488 > 0.3$$

Thus, should assume <u>compressible flow.</u>

3.19R

3.19R (Restrictions on Bernoulli equation) Niagara Falls is approximately 167 ft high. If the water flows over the crest of the falls with a velocity of 8 ft/s and viscous effects are neglected, with what velocity does the water strike the rocks at the bottom of the falls? What is the maximum pressure of the water on the rocks? Repeat the calculations for the 1430-ft-high Upper Yosemite Falls in Yosemite National Park. Is it reasonable to neglect viscous effects for these falls? Explain.
(ANS: 104 ft/s, 72.8 psi; 304 ft/s, 620 psi; no)

$$\frac{p_1}{\gamma} + \frac{V_1^2}{2g} + z_1 = \frac{p_2}{\gamma} + \frac{V_2^2}{2g} + z_2$$

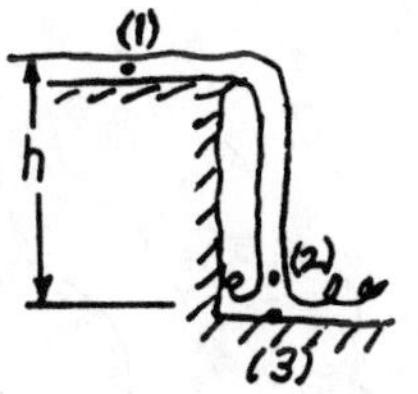

and

$$\frac{p_1}{\gamma} + \frac{V_1^2}{2g} + z_1 = \frac{p_3}{\gamma} + \frac{V_3^2}{2g} + z_3$$

with $p_1 = p_2 = 0$, $V_1 = 8\ \frac{ft}{s}$, $V_3 = 0$, $z_2 \approx z_3 = 0$, and $z_1 = h$

Thus,

$$V_2 = \sqrt{2g\left(h + \frac{V_1^2}{2g}\right)} \quad \text{and} \quad p_3 = \tfrac{1}{2}\rho V_1^2 + \gamma h$$

a) With $h = 167\ ft$,

$$V_2 = \left[2\left(32.2\ \frac{ft}{s^2}\right)\left(167\ ft + \frac{\left(8\ \frac{ft}{s}\right)^2}{2\left(32.2\ \frac{ft}{s^2}\right)}\right)\right]^{1/2} = \underline{\underline{104\ \frac{ft}{s}}}$$

and

$$p_3 = \tfrac{1}{2}\left(1.94\ \frac{slugs}{ft^3}\right)\left(8\ \frac{ft}{s}\right)^2 + 62.4\ \frac{lb}{ft^3}\,(167\ ft) = 10{,}500\ \frac{lb}{ft^2} = \underline{\underline{72.8\ psi}}$$

b) With $h = 1430\ ft$,

$$V_2 = \left[2\left(32.2\ \frac{ft}{s^2}\right)\left(1430\ ft + \frac{\left(8\ \frac{ft}{s}\right)^2}{2\left(32.2\ \frac{ft}{s^2}\right)}\right)\right]^{1/2} = \underline{\underline{304\ \frac{ft}{s}}}$$

and

$$p_3 = \tfrac{1}{2}\left(1.94\ \frac{slugs}{ft^3}\right)\left(8\ \frac{ft}{s}\right)^2 + 62.4\ \frac{lb}{ft^3}\,(1430\ ft) = 89{,}300\ \frac{lb}{ft^2} = \underline{\underline{620\ psi}}$$

Aerodynamic drag on the water would reduce the values of V_2 and p_3 (especially for the $h = 1{,}430$ ft case).

4
Fluid Kinematics

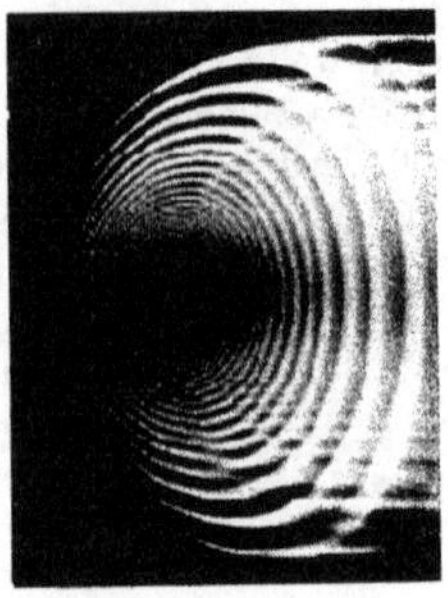

4.1 The Velocity Field

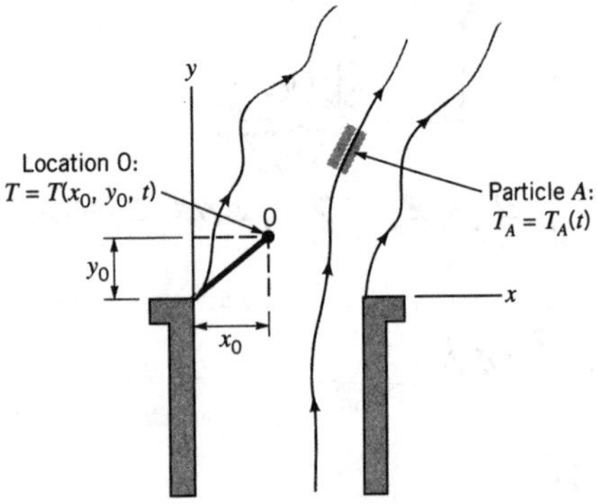

■ FIGURE 4.2 **Eulerian and Lagrangian descriptions of temperature of a flowing fluid.**

- A flow can be described by its *velocity field*, $\mathbf{V} = \mathbf{V}(x, y, z, t)$.
- For *steady flow*, the velocity at a given point in space does not vary with time.
- The *Eulerian method* of describing a flow involves flow properties specified as a function of space and time. For example, the measurement of temperature, T, at location $x = x_0$ and $y = y_0$, denoted $T = T(x_0, y_0, t)$, is indicated in Fig. 4.2. The use of numerous temperature measuring devices fixed at various locations throughout the flow would provide the temperature field, $T = T(x, y, z, t)$.
- The *Lagrangian method* involves prescribing flow properties of individual particles as they move about. The measurement of temperature for particle A as it moves about, denoted as $T_A = T_A(t)$, is indicated in Fig. 4.2. The use of numerous temperature measuring devices attached to numerous particles would provide the temperature of the particles as they flow. The temperature would not be known as a function of position unless the location of the particles was known as a function of time.
- It is usually easier to use the Eulerian method (rather than Lagrangian method) to describe a flow.

Streamlines, Streaklines, and Pathlines

$$\frac{dy}{dx} = \frac{v}{u} \qquad \textbf{(4.1)}$$

- A *streamline* is a line everywhere tangent to the velocity field. Its slope, dy/dx, is given by the ratio of the y- and x-components of velocity as indicated in Eq. 4.1.
- A *streakline* consists of all particles in a flow that previously passed through a common point.
- A *pathline* is the line traced out by a given particle as it flows from one point to another.
- *Caution:* Pathlines, streaklines, and streamlines are identical if the flow is steady. However, if the flow is unsteady none of these three lines need be the same.

EXAMPLE: The unsteady wind velocity is given by $\mathbf{V} = 0$ for $t < 0$; $\mathbf{V} = 10\,\hat{\mathbf{i}}$ mph for $0 < t < 1$ hr; and $\mathbf{V} = 5\,\hat{\mathbf{j}}$ mph for $t > 1$ hr. For time $t = 2$ hr plot the pathline, streakline, and streamline that pass through the origin.

SOLUTION:

(a) Pathline: From $0<t<1\,hr$ the particle travels from $(x,y)=(0,0)$ to $(x,y)=(10mi,0)$; from $1\,hr<t<2\,hr$ from $(10mi,0)$ to $(10mi,5mi)$.

(b) Streakline: Particles from the origin for $0<t<1\,hr$ stretch out along the x-axis from $0<x<10mi$. From $1\,hr<t<2\,hr$ these particles move upward to $y=5mi$ while others lie along $x=0$ from $0<y<5mi$.

(c) Streamline: At $t=2\,hr$, $\vec{V}=5\hat{j}$, or $u=0$, $v=5$.
Thus, $\frac{dy}{dx}=\frac{v}{u}=\frac{5}{0}=\infty$. Streamlines are vertical lines.

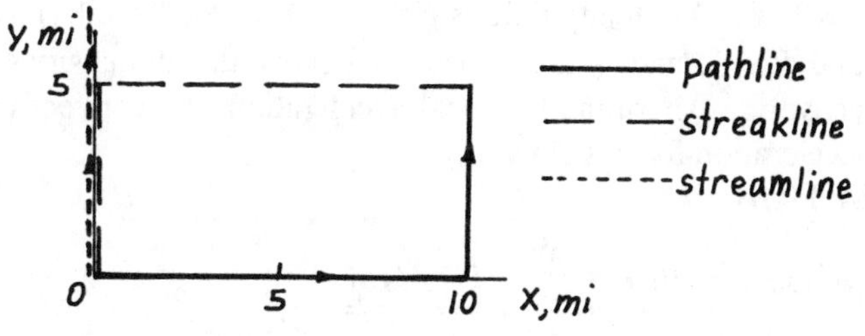

4.2 The Acceleration Field

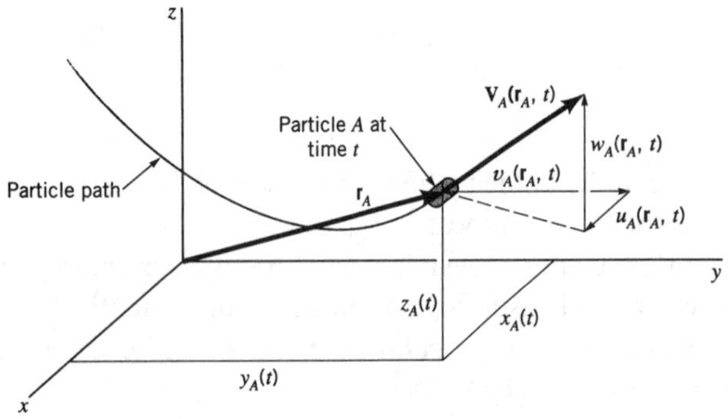

■ FIGURE 4.4 **Velocity and position of particle *A* at time *t*.**

$$\mathbf{a} = \frac{\partial \mathbf{V}}{\partial t} + u\frac{\partial \mathbf{V}}{\partial x} + v\frac{\partial \mathbf{V}}{\partial y} + w\frac{\partial \mathbf{V}}{\partial z} \tag{4.3}$$

$$\frac{D(\;)}{Dt} \equiv \frac{\partial(\;)}{\partial t} + u\frac{\partial(\;)}{\partial x} + v\frac{\partial(\;)}{\partial y} + w\frac{\partial(\;)}{\partial z} \tag{4.5}$$

- The acceleration, $\mathbf{a} = \mathbf{a}(x, y, z, t)$, is the time rate of change of a particle's velocity.
- The *material derivative* provides a means of determining the time rate of change of any fluid property for a given particle. As indicated by Eq. 4.5, the material derivative is the sum of the unsteady and the convective effects.
- As shown by Eq. 4.3, the acceleration is the sum of the *local acceleration*, $\partial\mathbf{V}/\partial t$, and the *convective acceleration*, $u\partial\mathbf{V}/\partial x + v\partial\mathbf{V}/\partial y + w\partial\mathbf{V}/\partial z$. The local acceleration is a result of the fact that the fluid velocity at a fixed point is a function of time. The convective acceleration results from the fact that a particle's velocity is a function of position (see Fig. 4.4).
- *Caution:* Although for steady flow, the local acceleration is zero, the convective acceleration is usually not zero because the velocity of the particles changes as they move from one location to another.

EXAMPLE: A velocity field is given by $\mathbf{V} = (V_0/\ell)x\,\hat{\mathbf{i}} + (V_0/t_0)t\,\hat{\mathbf{j}}$, where $\mathbf{V}$ is in ft/s and V_0, ℓ, and t_0 are constants having the dimensions of ft/s, ft, and s, respectively. Determine the local acceleration, the convective acceleration, and the acceleration for this flow.

SOLUTION:

Local acceleration $= \frac{\partial \vec{V}}{\partial t} = (V_0/t_0)\hat{j}$

Convective acceleration $= u\frac{\partial \vec{V}}{\partial x} + v\frac{\partial \vec{V}}{\partial y} + w\frac{\partial \vec{V}}{\partial z} = u\left(\frac{\partial u}{\partial x}\hat{i} + \frac{\partial v}{\partial x}\hat{j} + \frac{\partial w}{\partial x}\hat{k}\right)$ (with $v\,\partial\vec{V}/\partial y = 0$, $w\,\partial\vec{V}/\partial z = 0$, $\partial v/\partial x = 0$, $\partial w/\partial x = 0$)

$= (V_0/\ell)x(V_0/\ell)\hat{i} = (V_0/\ell)^2 x\,\hat{i}$

Thus, $\vec{a} = (V_0/\ell)^2 x\,\hat{i} + (V_0/t_0)\hat{j}$

Streamline Coordinates

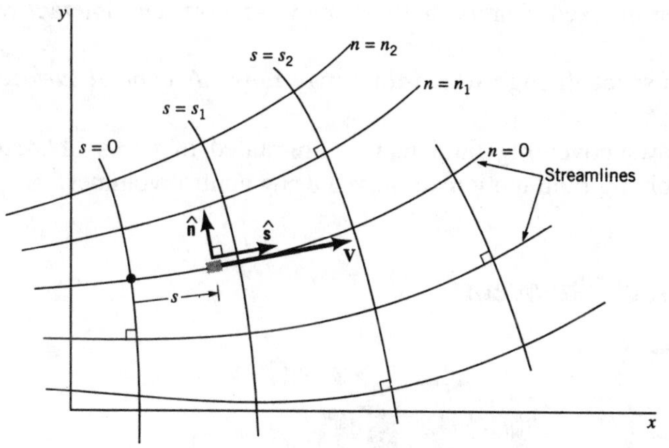

■ **FIGURE 4.8 Streamline coordinate system for two-dimensional flow.**

$$\mathbf{a} = V\frac{\partial V}{\partial s}\hat{\mathbf{s}} + \frac{V^2}{\mathcal{R}}\hat{\mathbf{n}} \quad \text{or} \quad a_s = V\frac{\partial V}{\partial s}, \quad a_n = \frac{V^2}{\mathcal{R}} \qquad \textbf{(4.7)}$$

- It is often convenient to use a coordinate system defined in terms of the streamlines of the flow, rather than the standard *x*, *y* Cartesian coordinate system, for example.
- As shown in Fig. 4.8, a two-dimensional flow can be given in terms of one coordinate along the streamlines, denoted *s*, and the second coordinate normal to the streamlines, denoted *n*.
- The flow net consists of coordinate lines s = constant and n = constant. These lines are generally curved and always orthogonal.
- As shown by Eq. 4.7, for steady, two-dimensional flow the acceleration, **a**, can be determined in terms of its streamwise and normal components, a_s and a_n, respectively.
- *Caution:* The acceleration given by Eq. 4.7 is valid only for steady flows.

EXAMPLE: A fluid flows with a constant speed of 10 ft/s. Determine the radius of curvature of the streamline at a point where the magnitude of the fluid acceleration is 20 ft/s^2.

SOLUTION:

The magnitude of the acceleration is $|\vec{a}| = \sqrt{a_s^2 + a_n^2}$, where from Eq. 4.7, $a_s = V\,\partial V/\partial s = 0$ since V = constant, and $a_n = V^2/\mathcal{R}$.

Thus, $|\vec{a}| = a_n$, or $20\text{ ft/s}^2 = V^2/\mathcal{R} = (10\text{ ft/s})^2/\mathcal{R}$.

Hence, $\mathcal{R} = (10\text{ ft/s})^2/(20\text{ ft/s}^2) = \underline{\underline{5.00\text{ ft}}}$

4.3 Control Volume and System Representations

- A *system* is a collection of matter of fixed identity that may move, distort, and interact with its surroundings.
- A *control volume* is a volume in space through which fluid may flow. A *control surface* is the surface of the control volume.
- *Caution:* The basic physical laws governing fluid motion are stated in terms of system behavior, but most problems involving fluid motion are solved using control volumes.

4.4 The Reynolds Transport Theorem

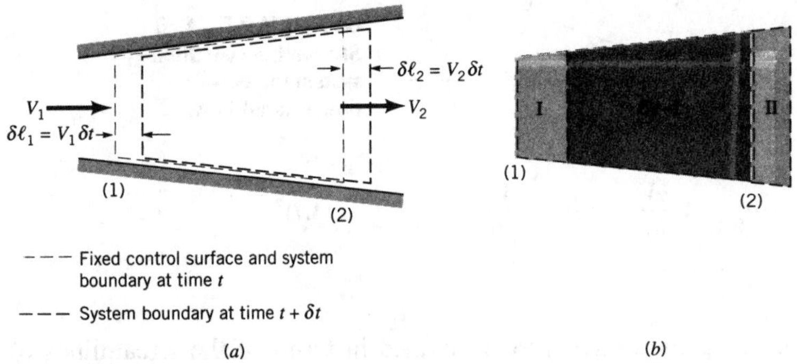

■ FIGURE 4.11 Control volume and system for flow through a variable area pipe.

$$\frac{DB_{\text{sys}}}{Dt} = \frac{\partial B_{\text{cv}}}{\partial t} + \rho_2 A_2 V_2 b_2 - \rho_1 A_1 V_1 b_1 \qquad \textbf{(4.15)}$$

- Fluid flow analysis involves various extensive properties, B (i.e., mass, momentum, energy), and the corresponding intensive properties, $b = B/m$, which are the amount of B per unit mass.
- The *Reynolds transport theorem* provides an analytical tool to shift from the system representation to the control volume representation.
- For a simple control volume containing one inlet and one outlet with uniform velocity normal to the inlet and outlet areas as shown in Fig. 4.11, the Reynolds transport theorem is given by Eq. 4.15.
- *Caution:* The time rate of change of a property for a system is not equal to the time rate of change of that property for the corresponding control volume if the rate of inflow of that property into the control volume is not equal to the rate of outflow.

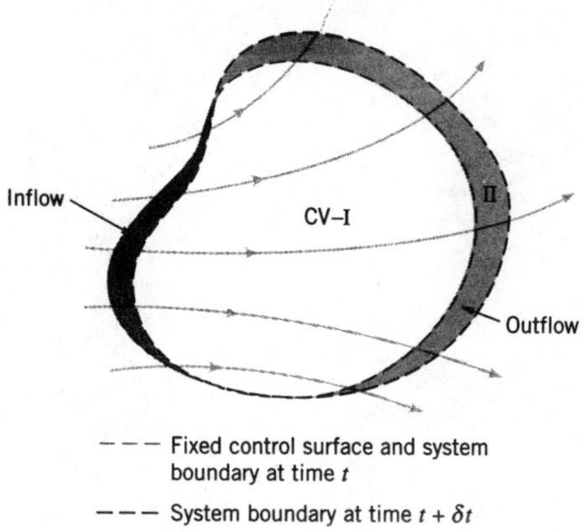

■ **FIGURE 4.12 Control volume and system for flow through an arbitrary, fixed control volume.**

$$\frac{DB_{sys}}{Dt} = \frac{\partial}{\partial t}\int_{cv} \rho b \, d\forall + \int_{cs} \rho b \, \mathbf{V} \cdot \hat{\mathbf{n}} \, dA \qquad \textbf{(4.19)}$$

- The Reynolds transport theorem for a general flow as shown in Fig. 4.12 is given by Eq. 4.19.
- The Reynolds transport theorem for a moving, nondeforming control volume is the same as for a fixed control volume if the *absolute velocity*, **V**, is replaced by the *relative velocity*, **W**.
- *Caution:* The unit normal vector $\hat{\mathbf{n}}$ always points out from the control volume, regardless of the flow direction. Hence, $\mathbf{V}\cdot\hat{\mathbf{n}} > 0$ for outflows, and $\mathbf{V}\cdot\hat{\mathbf{n}} < 0$ for inflows.

EXAMPLE: Water flows over a flat plate with a velocity profile given by $\mathbf{V} = u(y)\,\hat{\mathbf{i}}$ where, as shown in the figure, $u = 2y$ ft/s for $0 < y < 0.5$ ft and $u = 1$ ft/s for $y > 0.5$ ft. The control volume $ABCD$ coincides with the system at time $t = 0$. Make a sketch to indicate **(a)** the boundary of the system (denoted by $A'B'C'D'$) at $t = 0.5$ s, **(b)** the fluid that moved out of the control volume in the interval $0 < t < 0.5$ s, and **(c)** the fluid that moved into the control volume during that time interval.

SOLUTION:

(a) Since $u = 0$ on AD, that portion of of the system boundary remains fixed. From $t = 0$ to $t = 0.5$s all of the fluid for $y > 0.5$ ft moves to the right a distant $\ell = ut = 1$ ft/s $(0.5\text{s}) = 0.5$ft. Similarly, for $0 \le y 0.5$ ft, $\ell = ut = 2y$ ft/s $(0.5\text{s}) = y$ ft. Hence, at $t = 0.5$s points A, B, C, and D have moved to A', B', C', and D' as shown in the figure.

(b) and (c) The fluid that has moved out of or into the control volume is shown in the figure.

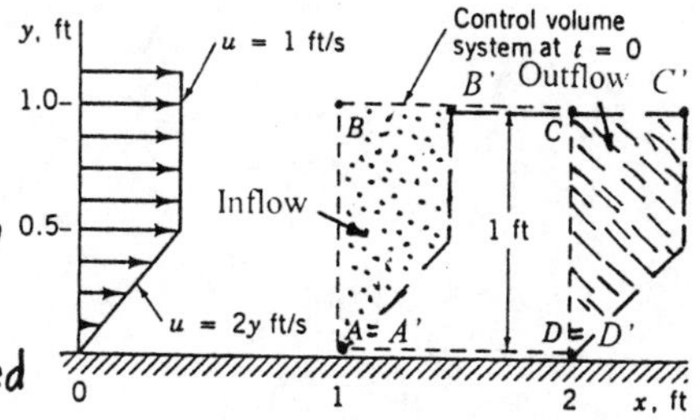

Solutions

to Review Problems

Chapter 4

4.1R (Streamlines) The velocity field in a flow is given by $\mathbf{V} = x^2y\hat{\mathbf{i}} + x^2t\hat{\mathbf{j}}$. **(a)** Plot the streamline through the origin at times $t = 0$, $t = 1$, and $t = 2$. **(b)** Do the streamlines plotted in part (a) coincide with the path of particles through the origin? Explain.

(ANS: $y^2/2 = tx + C$; no)

a) $u = x^2y$ and $v = x^2t$ where the streamlines are obtained from

$\frac{dy}{dx} = \frac{v}{u} = \frac{x^2t}{x^2y}$ or $y\,dy = t\,dx$ which, for a given time t, can be integrated to give

$\frac{1}{2}y^2 = tx + C$, where C is a constant.

For streamlines through the origin $(x=0, y=0)$, $C=0$.

Thus: $y^2 = 2tx$ or

$y = 0$ for $t=0$

$y = \sqrt{2x}$ for $t=1$, and

$y = 2\sqrt{x}$ for $t=2$. These streamlines are plotted below.

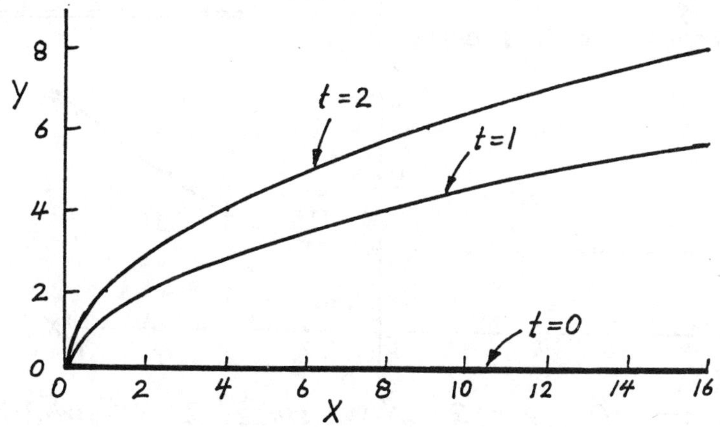

b) Since this is an unsteady flow (i.e. $\frac{\partial v}{\partial t} = x^2 \neq 0$) and since $\frac{dy}{dx} = \frac{t}{y}$ (i.e. at a given location the slope is a function of time), streamlines and pathlines do not coincide.

4.2R (Streamlines) A velocity field is given by $u = y - 1$ and $v = y - 2$, where u and v are in m/s and x and y are in meters. Plot the streamline that passes through the point $(x, y) = (4, 3)$. Compare this streamline with the streakline through the point $(x, y) = (4, 3)$.

(ANS: $x = y + \ln(y - 2) + 1$)

$u = y-1$, $v = y-2$ where the streamlines are obtained from

$$\frac{dy}{dx} = \frac{v}{u} = \frac{y-2}{y-1}$$

or

$$\int \frac{(y-1)}{(y-2)}\,dy = \int dx \quad \text{or} \quad \int \frac{y\,dy}{(y-2)} - \int \frac{dy}{(y-2)} = x + \tilde{C}, \text{ where } \tilde{C} \text{ is a constant.}$$

From integral tables:

$$\int \frac{y\,dy}{(y-2)} = y-2+2\ln(y-2) \quad \text{and} \quad \int \frac{dy}{(y-2)} = \ln(y-2)$$

Thus, the streamlines are given by

$$y-2+2\ln(y-2)-\ln(y-2) = x+\tilde{C}$$

or

$$y+\ln(y-2) = x+C \text{, where } C \text{ is a constant} \qquad (1)$$

For the streamline that passes through $x=4$ and $y=3$, the value of C is found from Eq.(1) as:

$3+\ln(3-2) = 4+C$ or $C=-1$ Thus, $\underline{\underline{x = y+\ln(y-2)+1}}$

This streamline is plotted below:

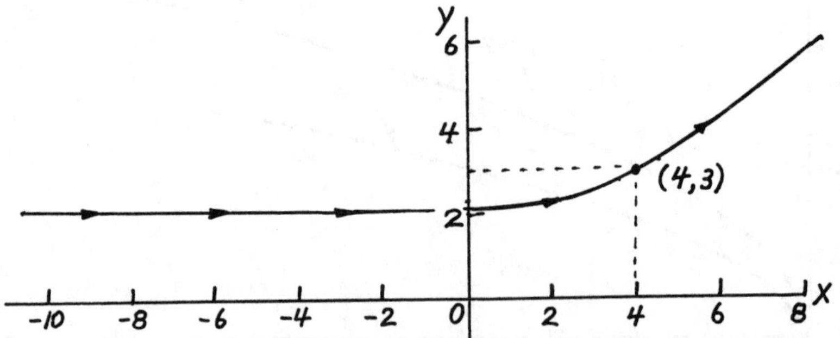

Note: As $x \to -\infty$, $y \to 2$. Also, since $u = y-1$, with $y \geq 2$ anywhere on this streamline, it follows that $u > 0$. The flow is from left to right.

Since the flow is steady, <u>streamlines are the same as streaklines.</u>

4.3R

4.3R (Material derivative) The pressure in the pipe near the discharge of a reciprocating pump fluctuates according to $p = [200 + 40 \sin(8t)]$ kPa, where t is in seconds. If the fluid speed in the pipe is 5 m/s, determine the maximum rate of change of pressure experienced by a fluid particle.

(ANS: 320 kPa/s)

Since $u = 5\,\frac{m}{s}$, $v = 0$, $w = 0$ it follows that with $p = p(t)$

$$\frac{Dp}{Dt} = \frac{\partial p}{\partial t} + u\frac{\partial p}{\partial x} + v\frac{\partial p}{\partial y} + w\frac{\partial p}{\partial z} = \frac{\partial p}{\partial t}$$

or

$$\frac{Dp}{Dt} = 40(8)\cos(8t)\,\frac{kPa}{s} \quad \text{Thus,} \quad \left.\frac{Dp}{Dt}\right)_{max} = 40(8)\,\frac{kPa}{s} = \underline{\underline{320\,\frac{kPa}{s}}}$$

Note: Since $\frac{\partial p}{\partial x} = 0$ the value of u is not important.

4.4R

4.4R (Acceleration) A shock wave is a very thin layer (thickness $= \ell$) in a high-speed (supersonic) gas flow across which the flow properties (velocity, density, pressure, etc.) change from state (1) to state (2) as shown in Fig. P4.4R. If $V_1 = 1800$ fps, $V_2 = 700$ fps, and $\ell = 10^{-4}$ in., estimate the average deceleration of the gas as it flows across the shock wave. How many g's deceleration does this represent?

(ANS: -1.65×10^{11} ft/s^2; -5.12×10^9)

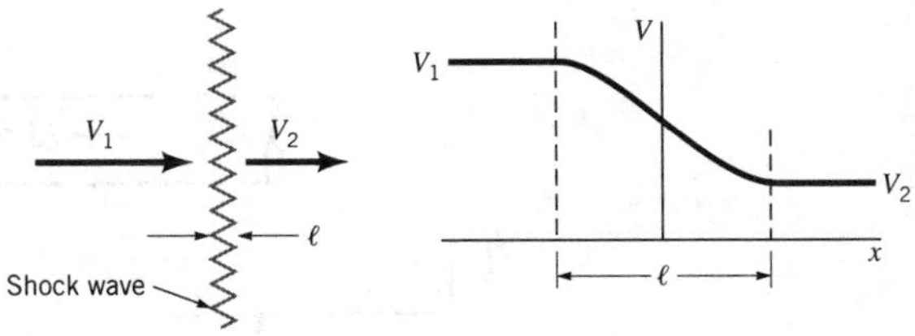

■ FIGURE P4.4R

$$\vec{a} = \frac{\partial \vec{V}}{\partial t} + \vec{V}\cdot\nabla\vec{V} \quad \text{so with} \quad \vec{V} = u(x)\hat{i}\,, \quad \vec{a} = a_x\hat{i} = u\frac{\partial u}{\partial x}\hat{i}$$

Without knowing the actual velocity distribution, $u = u(x)$, the acceleration can be approximated as

$$a_x = u\frac{\partial u}{\partial x} \approx \frac{(V_1 + V_2)}{2}\,\frac{(V_2 - V_1)}{\ell} = \frac{(1800 + 700)\,fps}{2}\,\frac{(700 - 1800)\,fps}{\left(\frac{10^{-4}}{12}\right)ft}$$

or

$$\underline{\underline{a_x = -1.65 \times 10^{11}\,\frac{ft}{s^2}}} \qquad \text{This is} \quad \frac{a_x}{g} = \frac{-1.65 \times 10^{11}\,\frac{ft}{s^2}}{32.2\,\frac{ft}{s^2}} = \underline{\underline{-5.12 \times 10^9}}$$

4.5R

4.5R (Acceleration) Air flows through a pipe with a uniform velocity of $\mathbf{V} = 5\,t^2\hat{\mathbf{i}}$ ft/s, where t is in seconds. Determine the acceleration at time $t = -1$, 0, and 1 s.
(ANS: $-10\,\hat{\mathbf{i}}$ ft/s^2; 0; $10\,\hat{\mathbf{i}}$ ft/s^2)

$\vec{a} = \frac{\partial \vec{V}}{\partial t} + \vec{V}\cdot\nabla\vec{V}$ With $u = 5t^2 \frac{ft}{s}$, $v=0$, $w=0$
this becomes

$\vec{a} = \left(\frac{\partial u}{\partial t} + u\frac{\partial u}{\partial x}\right)\hat{i} = \frac{\partial u}{\partial t}\hat{i} = 10t\,\hat{i}\,\frac{ft}{s^2}$ since $\frac{\partial u}{\partial x} = 0.$

Thus, $\vec{a} = -10\hat{i}\,\frac{ft}{s^2}$ at $t=-1\,s$

$\vec{a} = 0$ at $t = 0$

and

$\vec{a} = 10\hat{i}\,\frac{ft}{s^2}$ at $t = 1\,s$

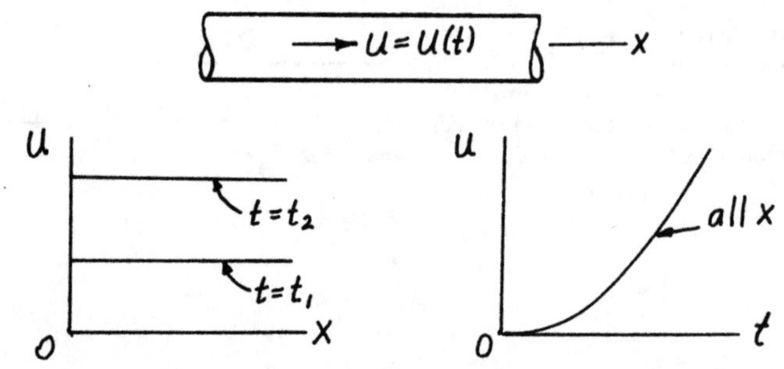

4.6R

4.6R (Acceleration) A fluid flows steadily along the streamline as shown in Fig. P4.6R. Determine the acceleration at point *A*. At point *A* what is the angle between the acceleration and the *x* axis? At point *A* what is the angle between the acceleration and the streamline?

(ANS: $10\,\hat{\mathbf{n}} + 30\,\hat{\mathbf{s}}$ ft/s²; 48.5 deg; 18.5 deg)

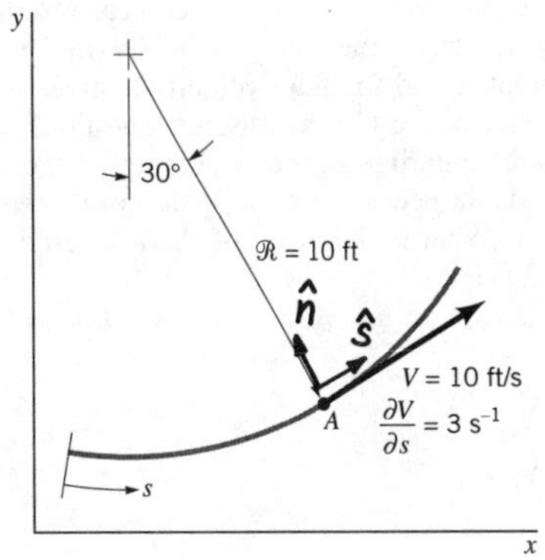

■ FIGURE P4.6R

$$\vec{a} = a_n\hat{n} + a_s\hat{s} = \frac{V^2}{\mathcal{R}}\hat{n} + V\frac{\partial V}{\partial s}\hat{s} = \frac{(10\,\frac{ft}{s})^2}{10\,ft}\hat{n} + (10\,\tfrac{ft}{s})(3\,\tfrac{1}{s})\hat{s}$$

or

$$\vec{a} = 10\hat{n} + 30\hat{s}\ \frac{ft}{s^2}$$

In terms of unit vectors $\hat{\imath}$ and $\hat{\jmath}$, $\hat{n} = -\sin 30°\,\hat{\imath} + \cos 30°\,\hat{\jmath}$

and $\hat{s} = \cos 30°\,\hat{\imath} + \sin 30°\,\hat{\jmath}$

Thus,

$$\vec{a} = 10(-0.5\hat{\imath} + 0.866\hat{\jmath}) + 30(0.866\hat{\imath} + 0.5\hat{\jmath}) = 21.0\hat{\imath} + 23.7\hat{\jmath}\ \frac{ft}{s^2}$$

Hence, $\theta = \tan^{-1}\frac{a_y}{a_x} = \tan^{-1}\frac{23.7}{21.0}$

or $\theta = \underline{48.5°}$

and

$\alpha = \theta - 30° = \underline{18.5°}$

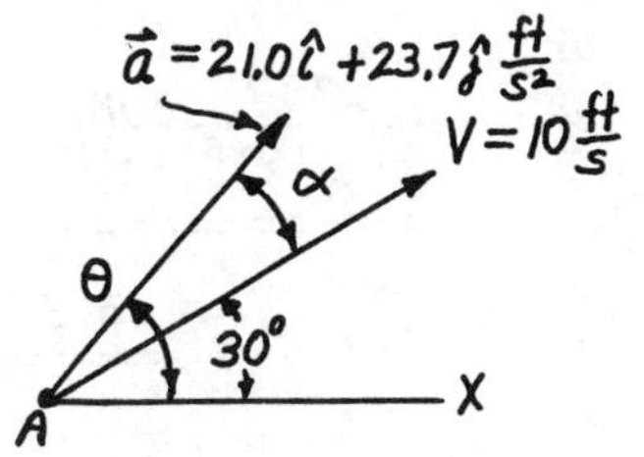

4.7R

4.7R (Acceleration) In the conical nozzle shown in Fig. P4.7R the streamlines are essentially radial lines emanating from point A and the fluid velocity is given approximately by $V = C/r^2$, where C is a constant. The fluid velocity is 2 m/s along the centerline at the beginning of the nozzle ($x = 0$). Determine the acceleration along the nozzle centerline as a function of x. What is the value of the acceleration at $x = 0$ and $x = 0.3$ m?

(ANS: $1.037/(0.6 - x)^5\ \hat{i}$ m/s²; $13.3\ \hat{i}$ m/s²; $427\ \hat{i}$ m/s²)

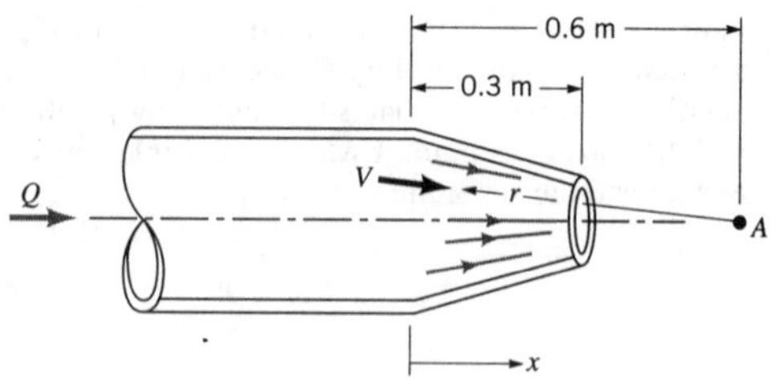

FIGURE P4.7R

Along the nozzle centerline, $\vec{a} = \frac{\partial \vec{V}}{\partial t} + \vec{V}\cdot\nabla\vec{V}$ becomes

$\vec{a} = u\frac{\partial u}{\partial x}\hat{i}$ where $u = \frac{C}{r^2}$ with $r + x = 0.6\,m$, or $r = 0.6 - x$.

Thus,

$$u = \frac{C}{(0.6-x)^2}$$

Since $u = 2\,\frac{m}{s}$ at $x = 0$ it follows that

$$2\,\frac{m}{s} = \frac{C}{(0.6\,m)^2} \quad \text{or} \quad C = 0.72\,\frac{m^3}{s}$$

Hence,

$$\vec{a} = \left[\frac{C}{(0.6-x)^2}\right]\frac{2C}{(0.6-x)^3}\hat{i} = \frac{2C^2}{(0.6-x)^5}\hat{i} = \frac{2(0.72)^2\,\frac{m^6}{s^2}}{(0.6-x)^5\,m^5}\hat{i}$$

or

$$\vec{a} = \frac{1.037}{(0.6-x)^5}\hat{i}\,\frac{m}{s^2} \quad \text{where } x \sim m$$

At $x = 0$, $\vec{a} = 13.3\,\hat{i}\,\frac{m}{s^2}$; at $x = 0.3\,m$, $\vec{a} = 427\,\hat{i}\,\frac{m}{s^2}$

4.8R

4.8R (Reynolds transport theorem) A sanding operation injects 10^5 particles/s into the air in a room as shown in Fig. P4.8R. The amount of dust in the room is maintained at a constant level by a ventilating fan that draws clean air into the room at section (1) and expels dusty air at section (2). Consider a control volume whose surface is the interior surface of the room (excluding the sander) and a system consisting of the material within the control volume at time $t = 0$. **(a)** If N is the number of particles, discuss the physical meaning of and evaluate the terms DN_{sys}/Dt and $\partial N_{cv}/\partial t$. **(b)** Use the Reynolds transport theorem to determine the concentration of particles (particles/m³) in the exhaust air for steady state conditions.

(ANS: 5×10^5 particles/m³)

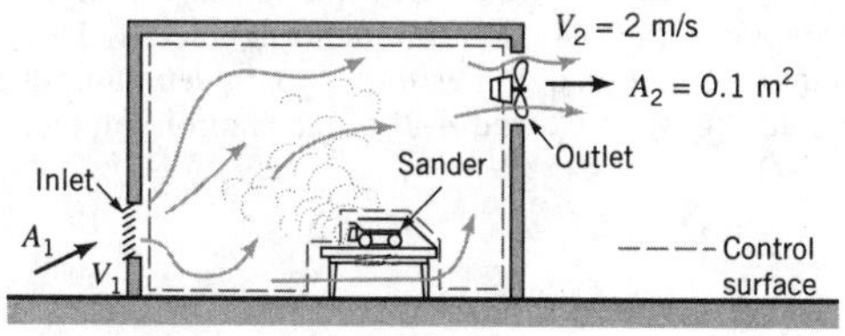

■ FIGURE P4.8R

a) $\frac{DN_{sys}}{Dt}$ = time rate of change of the number of particles in the system. At time $t=0$ the system consists of N particles. In fact, the system is these particles for all time $t>0$. Assuming that the particles do not get "glued" together, N remains constant.

Thus, $\frac{DN_{sys}}{Dt} = 0$

$\frac{\partial N_{cv}}{\partial t}$ = time rate of change of the number of particles in the control volume. Depending on the rate at which the sander creates particles and ejects them into the room compared to the rate at which the fan draws them from the room, we could have $\frac{\partial N_{cv}}{\partial t} \gtreqless 0$.

b) $\frac{DN_{sys}}{Dt} = \frac{\partial N_{cv}}{\partial t}$ + net rate of flow of particles out of control volume

or for steady state $\frac{\partial N_{cv}}{\partial t} = 0$ so that:

flow of particles into control volume (from sander, none enter A_1)
= flow of particles out of control volume (through fan exhaust, A_2)

Thus,

$10^5 \frac{\text{particles}}{s} = V_2 A_2 n_2$, where n_2 = particle concentration $\left(\frac{\text{particles}}{m^3}\right)$

Hence,

$$n_2 = \frac{10^5 \frac{\text{particles}}{s}}{(2\frac{m}{s})(0.1 m^2)} = 5 \times 10^5 \frac{\text{particles}}{m^3}$$

4.9R

4.9R (Flowrate) Water flows through the rectangular channel shown in Fig. P4.9R with a uniform velocity. Directly integrate Eqs. 4.16 and 4.17 with $b = 1$ to determine the mass flowrate (kg/s) across and A–B of the control volume. Repeat for C–D.

(ANS: 18,000 kg/s; 18,000 kg/s)

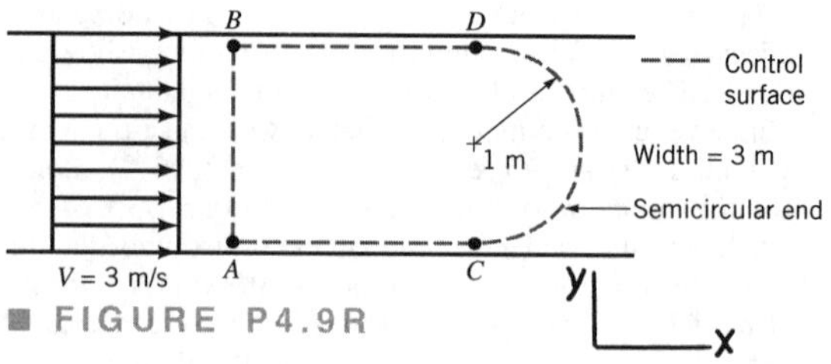

■ FIGURE P4.9R

Equation 4.17:

$$\dot{B}_{in} = -\int_{AB} \rho b \vec{V}\cdot\hat{n}\, dA \quad \text{or, with } \rho = 999\,\frac{kg}{m^3} \text{ and } b=1 \text{ this gives}$$

$$\dot{B}_{in} = -999\,\frac{kg}{m^3}\int (3\,\tfrac{m}{s})\hat{\imath}\cdot(\hat{\imath})\,dA = -(999\,\tfrac{kg}{m^3})(3\,\tfrac{m}{s})(2m)(3m)$$

$$= 18,000\,\frac{kg}{s}$$

and Equation 4.16:

$$\dot{B}_{out} = \int_{CD} \rho b \vec{V}\cdot\hat{n}\, dA = \rho\int_{CD} \vec{V}\cdot\hat{n}\, dA$$

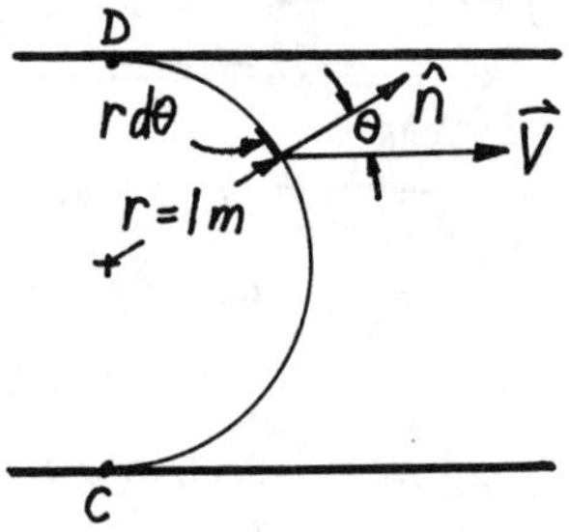

where $dA = 3\,r d\theta = 3\,d\theta$

and $\vec{V}\cdot\hat{n} = V\cos\theta = 3\cos\theta$

Thus,

$$\dot{B}_{out} = 999\,\frac{kg}{m^3}\int_{\theta=-90^\circ}^{\theta=90^\circ}(3\cos\theta\,\tfrac{m}{s})(3\,m)(1\,m)\,d\theta = 9(999)\sin\theta\Big|_{-90^\circ}^{90^\circ}$$

or

$$\dot{B}_{out} = 18,000\,\frac{kg}{s}$$

4.10R

4.10R (Flowrate) Air blows through two windows as indicated in Fig. P4.10R. Use Eq. 4.16 with $b = 1/\rho$ to determine the volume flowrate (ft^3/s) through each window. Explain the relationship between the two results you obtained.

(ANS: 80 ft^3/s; 160 ft^3/s)

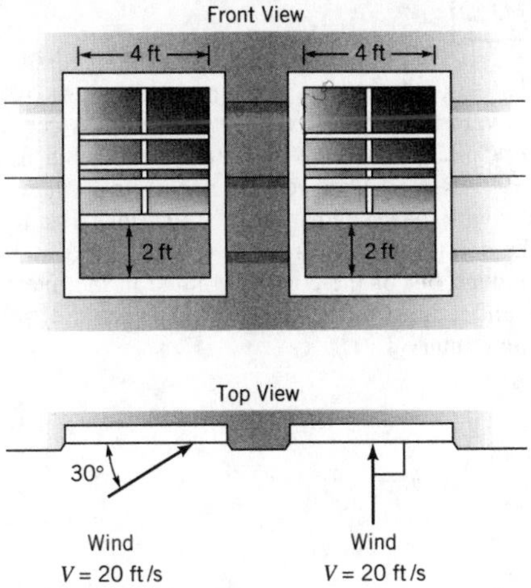

■ FIGURE P4.10R

$$\dot{B} = \int \rho b \vec{V} \cdot \hat{n} \, dA$$

For the left window $Q = \int \rho \left(\frac{1}{\rho}\right) \vec{V} \cdot \hat{n} \, dA = \int \vec{V} \cdot \hat{n} \, dA$

or $Q = \int 20 \sin 30^\circ \, dA = \left(20 \sin 30^\circ \frac{ft}{s}\right)(4 \, ft)(2 \, ft) = \underline{\underline{80 \, \frac{ft^3}{s}}}$

For the right window $Q = \int \vec{V} \cdot \hat{n} \, dA$

or $Q = \int 20 \, dA = \left(20 \frac{ft}{s}\right)(4 \, ft)(2 \, ft) = \underline{\underline{160 \, \frac{ft^3}{s}}}$

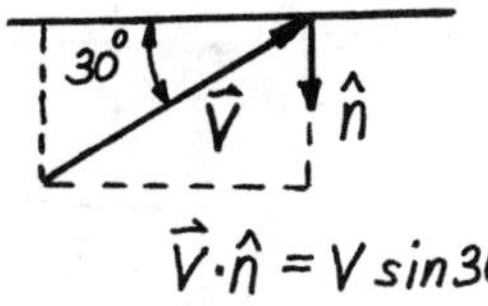

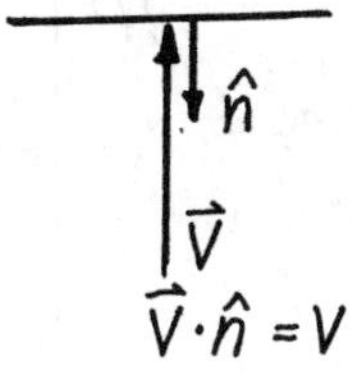

4.11R

4.11R (Control volume/system) Air flows over a flat plate with a velocity profile given by $\mathbf{V} = u(y)\hat{\mathbf{i}}$, where $u = 2y$ ft/s for $0 \leq y \leq 0.5$ ft and $u = 1$ ft/s for $y > 0.5$ ft as shown in Fig. P4.11R. The fixed rectangular control volume *ABCD* coincides with the system at time $t = 0$. Make a sketch to indicate **(a)** the boundary of the system at time $t = 0.1$ s, **(b)** the fluid that moved out of the control volume in the interval $0 \leq t \leq 0.1$ s, and **(c)** the fluid that moved into the control volume during that time interval.

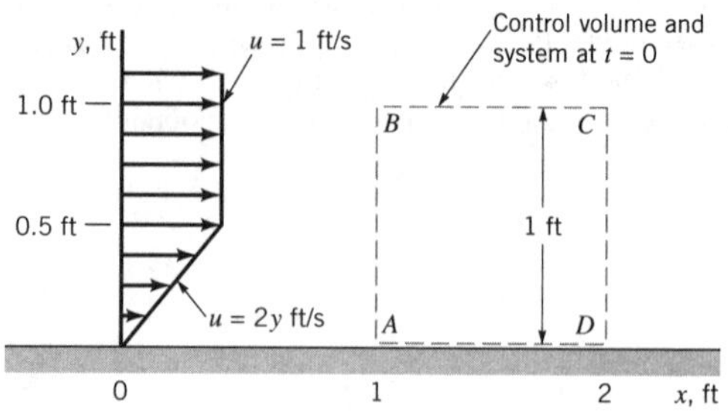

■ FIGURE P4.11R

Since $\vec{V} = u(y)\hat{i}$, each fluid particle travels only in the x-direction, with the distance of travel $\delta x = u\,\delta t$, where $\delta t = 0.1$ s.

Thus, $\delta x_A = \delta x_D = 0$ since $u = 0$ at $y = 0$.

Also $\delta x_B = \delta x_C = (1\,\frac{ft}{s})(0.1\,s) = 0.1$ ft.

The fluid originally along lines A-B and C-D move to positions A-E-B′ and D-F-C′ as shown it the figure below. The location of the system at $t = 0.1$ s and the fluid that moved into or out of the control volume is indicated.

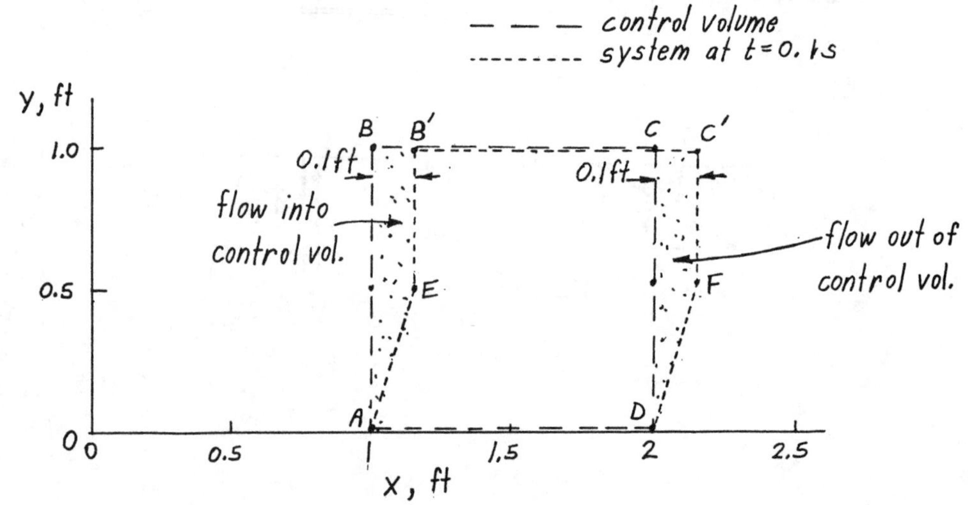

5
Finite Control Volume Analysis

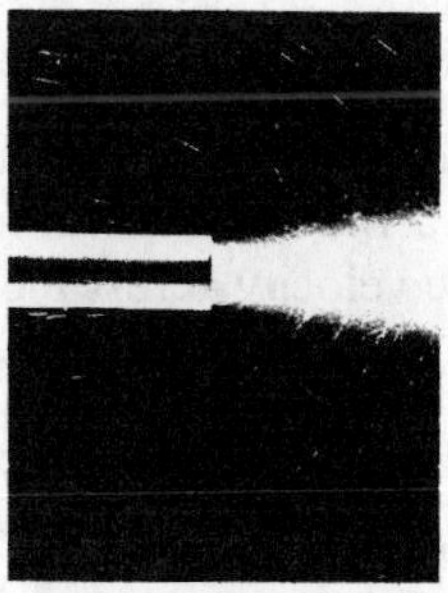

5.1 Conservation of Mass--The Continuity Equation

$$\frac{DM_{sys}}{Dt} = 0 \quad \textbf{(5.1)}$$

$$\frac{\partial}{\partial t}\int_{cv} \rho \, d\mathcal{V} + \int_{cs} \rho \mathbf{V} \cdot \hat{\mathbf{n}} \, dA = 0 \quad \textbf{(5.5)}$$

$$\dot{m} = \rho Q = \rho A V \quad \textbf{(5.6)}$$

$$\overline{V} = \frac{\int_A \rho \mathbf{V} \cdot \hat{\mathbf{n}} \, dA}{\rho A} \quad \textbf{(5.7)}$$

$$\dot{m} = \rho_1 A_1 \overline{V}_1 = \rho_2 A_2 \overline{V}_2 \quad \textbf{(5.12)}$$

$$Q = A_1 \overline{V}_1 = A_2 \overline{V}_2 \quad \textbf{(5.13)}$$

- Conservation of mass, Eq. 5.1, states that the time rate of change of the system mass is zero.
- The *continuity equation*, Eq. 5.5, relates the rate of change of mass within the control volume to the net flowrate of mass across the control surface.
- For steady flow the *mass flowrate* into a control volume equals the mass flowrate out of it.
- *Caution:* On outflow portions of the control volume $\mathbf{V} \cdot \hat{\mathbf{n}} > 0$; on inflow portions $\mathbf{V} \cdot \hat{\mathbf{n}} < 0$.
- As shown by Eq. 5.6, the mass flowrate, $\dot{m}$, equals the product of the density, ρ, area, A, and velocity, V, provided the density and velocity are uniform and the velocity is normal to the area.
- For nonuniform flows the average velocity can be determined by use of Eq. 5.7.
- For steady flow through a control volume with one inlet and one outlet having uniform properties across the inlet and outlet, the continuity equation is given by Eq. 5.12. If, in addition, the flow is incompressible, the continuity equation simplifies to Eq. 5.13.

EXAMPLE: (Uniform flow) Water flows steadily into a tank through a 0.1-m-diameter pipe with a velocity of 4 m/s. The water level in the tank remains constant. Determine the diameter of the outlet pipe if the outlet velocity is to be 2 m/s.

SOLUTION:

From Eq. 5.13, $A_1\overline{V}_1 = A_2\overline{V}_2$ or with $\overline{V}_1 = 4\,m/s$ and $\overline{V}_2 = 2\,m/s$,

$\frac{\pi}{4}(0.1m)^2(4m/s) = \frac{\pi}{4}D_2^2(2m/s)$ Thus, $D_2 = \underline{\underline{0.141\,m}}$

EXAMPLE: (Nonuniform flow) Air flows steadily into the 2-ft-by-2-ft inlet of an air scoop with the nonuniform velocity profile shown in the figure. Determine the average velocity across the inlet.

SOLUTION:

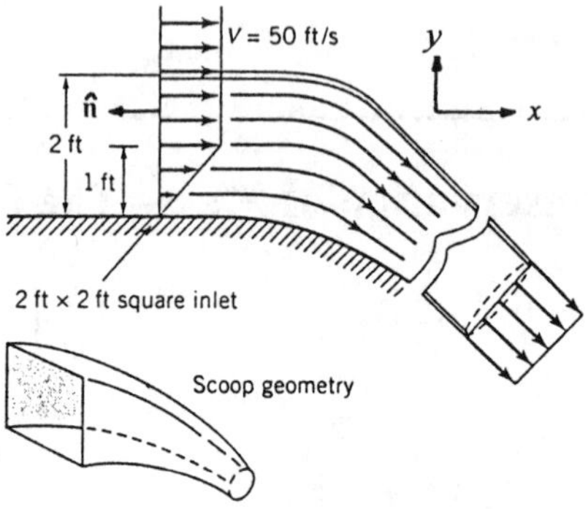

From Eq. 5.7, with uniform density

$\bar{V} = \int_A \vec{V}\cdot\hat{n}\, dA/A$, where $A = 4\,ft^2$

For $0 \le y \le 1\,ft$, $\vec{V} = 50y\hat{i}\ ft/s$;

for $y > 1\,ft$, $\vec{V} = 50\hat{i}\ ft/s$. Also,

$\hat{n} = -\hat{i}$ and $dA = 2\,dy$

Hence,

$$A\bar{V} = \int_{y=0}^{y=1}(-50y)\,2dy + \int_{y=1}^{y=2}(-50)\,2dy$$

$$= -50y^2\Big|_0^1 - 100y\Big|_1^2 = -150\ ft^3/s$$

or

$$\bar{V} = -150\ ft^3/s\,/4\,ft^2 = \underline{\underline{-37.5\ ft/s}}$$

Note that $\bar{V} < 0$ since it is inflow.

5.2 Newton's Second Law--The Linear Momentum and Moment-of-Momentum Equations

The Linear Momentum Equation

$$\frac{D}{Dt}\int_{\text{sys}} \mathbf{V}\rho \, d\forall = \sum \mathbf{F}_{\text{sys}} \qquad (5.19)$$

$$\frac{\partial}{\partial t}\int_{\text{cv}} \mathbf{V}\rho \, d\forall + \int_{\text{cs}} \mathbf{V}\rho\mathbf{V}\cdot\hat{\mathbf{n}} \, dA = \sum \mathbf{F}_{\substack{\text{contents of the}\\ \text{control volume}}} \qquad (5.22)$$

- Newton's second law of motion, Eq. 5.19, states that the time rate of change of the linear momentum of the system equals the net external force acting on the system.
- The *linear momentum equation*, Eq. 5.22, relates the rate of change of momentum within the control volume, the net flowrate of momentum through the control surface, and the net force acting on the contents of the control volume.
- For steady flow the net flux of momentum through the control surface equals the net force acting on the contents of the control volume.
- *Caution:* The linear momentum equation is a *vector* equation. The *x*, *y*, and *z* components of the equation must be treated separately.
- *Caution:* The term **V•n̂** is positive for outflows and negative for inflows. The sign of the term **VρV•n̂** depends on whether the flow is inflow or outflow *and* on the direction of the velocity relative to the coordinate system chosen.
- *Caution:* If the pressure is greater than atmospheric pressure, the pressure force on a control surface is directed *into* the control volume, regardless of the direction of the flow.

EXAMPLE: (Jet flow) A jet of water moving with a cross-sectional area of 0.15 m^2 is deflected through a 180 degree turn (i.e., turned back into the direction from which it came) by a vane. If the water speed remains 10 m/s throughout, determine the force required to hold the vane in place.

SOLUTION:

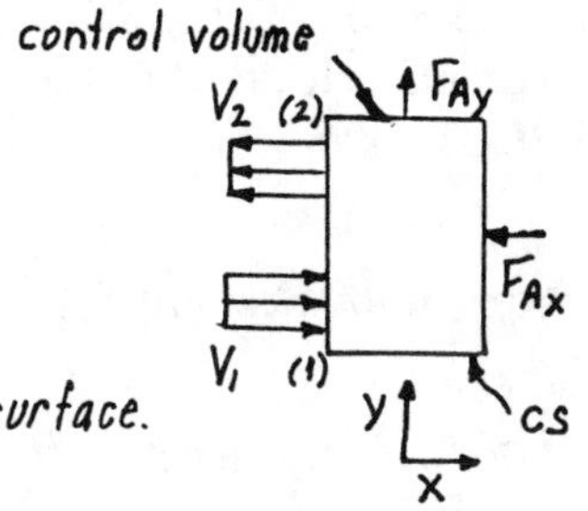

For steady flow Eq. 5.22 becomes

$$\int_{cs} \vec{V}\rho\vec{V}\cdot\hat{n}\, dA = \Sigma \vec{F}$$

<u>x-component</u>: $\int_{cs} u\rho\vec{V}\cdot\hat{n}\, dA = \Sigma F_x = -F_{Ax}$

since the pressure is uniform on the control surface.

Thus, $V_1\rho(-V_1)A_1 + (-V_2)\rho V_2 A_2 = -F_{Ax}$

or

$F_{Ax} = 2\dot{m}V_1$ since $V_1 = V_2 = 10\text{m/s}$ and $\dot{m} = \rho A_1 V_1 = \rho A_2 V_2$

Hence, $F_{Ax} = 2\,(999\,\text{kg/m}^3)(0.15\text{m}^2)(10\text{m/s})(10\text{m/s}) = 30{,}000\,\text{kg}\cdot\text{m/s}^2$
$= \underline{\underline{30.0\,\text{kN}}}$

<u>y-component</u>: $\int_{cs} v\rho\vec{V}\cdot\hat{n}\, dA = \Sigma F_y = F_{Ay}$, where $v = 0$ Thus $F_{Ay} = \underline{\underline{0}}$

EXAMPLE: (Confined flow) Water flows through a horizontal bend and discharges into the atmosphere as a free jet as shown in the figure. When the flowrate is 6 ft^3/s, the pressure at the entrance to the bend is 10 psi. Determine the x-component of the anchoring force, F_{Ax}, needed to hold the bend in place.

SOLUTION:

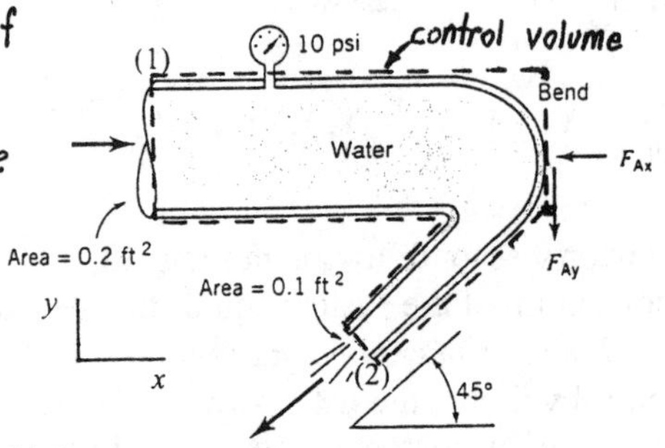

For steady flow the x-component of Eq. 5.22 becomes

$\int_{cs} u\rho\vec{V}\cdot\hat{n}\, dA = \Sigma F_x$. Thus, for the control volume indicated

(1) $V_1\rho(-V_1)A_1 + (-V_2\cos 45°)\rho V_2 A_2 = p_1 A_1 - F_{Ax}$

Since $\dot{m} = \rho V_1 A_1 = \rho V_2 A_2 = \rho Q$, Eq. (1) is

(2) $F_{Ax} = p_1 A_1 + \dot{m}(V_1 + V_2 \cos 45°)$

where $V_1 = \frac{Q}{A_1} = 6\,ft^3/s/(0.2\,ft^2) = 30\,ft/s$

and $V_2 = \frac{Q}{A_2} = 6\,ft^3/s/(0.1\,ft^2) = 60\,ft/s$

Hence, Eq. (2) gives

$F_{Ax} = 10\,lb/in.^2(144\,in.^2/ft^2)(0.2\,ft^2) + 1.94\,slugs/ft^3(6\,ft^3/s)(30 + 60\cos 45°)\,ft/s$

$= \underline{\underline{1130\,lb}}$

EXAMPLE: (Nonuniform flow) Air flowing through the variable mesh screen shown in the figure produces a nonuniform, linear, axisymmetric velocity profile as indicated. Determine the momentum flowrate through section (2) if the maximum velocity at that section is 10 ft/s.

SOLUTION:

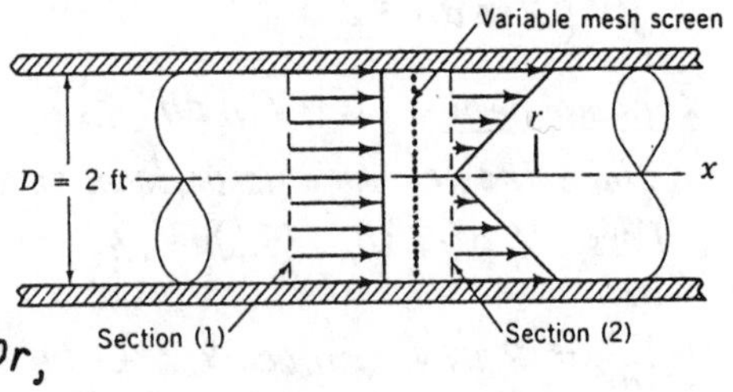

Momentum flowrate $= \int_{(2)} \vec{V}\rho\vec{V}\cdot\hat{n}\, dA$

where

$\vec{V} = 10r\hat{i}$ ft/s, with r ~ ft

(i.e. $\vec{V} = 0$ at $r = 0$; $\vec{V} = 10\hat{i}$ ft/s at $r = 1$)

Thus, with $dA = 2\pi r\, dr$ and $\vec{V}\cdot\hat{n} = 10r$,

$$\int_{(2)} \vec{V}\rho\vec{V}\cdot\hat{n}\, dA = \int_{r=0}^{r=1} (10r\hat{i}\,ft/s)(0.00238\,slug/ft^3)(10r\,ft/s)2\pi r\, dr\,(ft^2)$$

$$= 1.495\int_0^1 r^3 dr\,(slug\cdot ft/s^2) = \underline{\underline{0.374\ lb}}$$

The Moment-of-Momentum Equation

$$\frac{\partial}{\partial t}\int_{\text{cv}} (\mathbf{r} \times \mathbf{V})\rho \, d\mathcal{V} + \int_{\text{cs}} (\mathbf{r} \times \mathbf{V})\rho \mathbf{V} \cdot \hat{\mathbf{n}} \, dA = \sum (\mathbf{r} \times \mathbf{F})_{\substack{\text{contents of the}\\ \text{control volume}}} \qquad \textbf{(5.42)}$$

$$T_{\text{shaft}} = (-\dot{m}_{\text{in}})(\pm r_{\text{in}} V_{\theta \text{in}}) + \dot{m}_{\text{out}}(\pm r_{\text{out}} V_{\theta \text{out}}) \qquad \textbf{(5.50)}$$

$$\dot{W}_{\text{shaft}} = (-\dot{m}_{\text{in}})(\pm U_{\text{in}} V_{\theta \text{in}}) + \dot{m}_{\text{out}}(\pm U_{\text{out}} V_{\theta \text{out}}) \qquad \textbf{(5.53)}$$

- The *moment-of-momentum equation*, Eq. 5.42, relates the time rate of change of moment-of-momentum within the control volume, the net rate of flow of moment-of-momentum through the control surface, and the net torque acting on the contents of the control volume.
- For steady flow the net flux of moment-of-momentum through the control surface equals the net torque on the contents of the control volume.
- For rotating machines having uniform inlet and outlet flows, the moment-of-momentum equation can be written as shown in Eq. 5.50. The shaft torque, T_{shaft}, produces the change in moment-of-momentum flowrate as the fluid flows through the machine.
- *Caution:* If V_θ and U are in the same direction, the rV_θ product in Eq. 5.50 is positive; if V_θ and U are in opposite directions, the rV_θ product is negative.
- As shown by Eq. 5.53, the shaft power, $\dot{W}_{\text{shaft}}$, (torque times angular velocity) is given in terms of the mass flowrate and the inlet and outlet velocity components.

EXAMPLE: Water flows through the lawn sprinkler shown in the figure. The water speed at the exit nozzles is 20 ft/s relative to an observer riding on the rotating arm and $\theta = 30$ degrees. Determine the angular velocity of the sprinkler arm if there is no restraining torque (shaft torque) applied to it.

SOLUTION:

With the inflow in the axial direction, $V_{\theta in} = 0$ so that Eq. 5.50 becomes

$T_{shaft} = \dot{m}_{out}(\pm r_{out} V_{\theta out})$

Thus, with no restraining torque $T_{shaft} = 0$ so that $V_{\theta out} = 0$

The absolute, $\vec{V}$, relative, $\vec{W}$, and tip velocities, $\vec{U}$, are related by $\vec{V} = \vec{W} + \vec{U}$ as indicated in the figure.

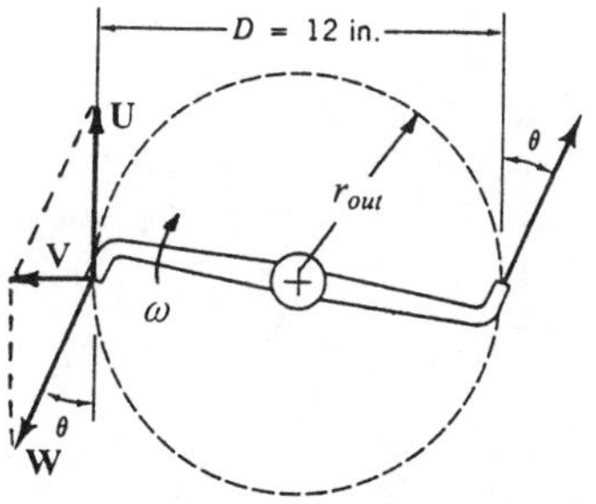

Hence, since $V_{\theta out} = 0$ and $U_{out} = r_{out}\,\omega$ it follows that $0 = -W\cos 30° + r_{out}\,\omega$,

or

$$\omega = \frac{20 \text{ ft/s} (\cos 30°)}{\frac{6}{12} \text{ ft}} = 34.6 \text{ rad/s} \,(1 \text{ rev}/2\pi \text{ rad}) = \underline{\underline{5.51 \text{ rev/s}}}$$

5.3 First Law of Thermodynamics--The Energy Equation

$$\frac{D}{Dt}\int_{\text{sys}} e\rho \, d\forall = (\dot{Q}_{\substack{\text{net}\\ \text{in}}} + \dot{W}_{\substack{\text{net}\\ \text{in}}})_{\text{sys}} \qquad \textbf{(5.55)}$$

$$\frac{\partial}{\partial t}\int_{\text{cv}} e\rho \, d\forall + \int_{\text{cs}} \left(\check{u} + \frac{p}{\rho} + \frac{V^2}{2} + gz \right) \rho \mathbf{V} \cdot \hat{\mathbf{n}} \, dA = \dot{Q}_{\substack{\text{net}\\ \text{in}}} + \dot{W}_{\substack{\text{shaft}\\ \text{net in}}} \qquad \textbf{(5.64)}$$

$$\dot{m}\left[\check{u}_{\text{out}} - \check{u}_{\text{in}} + \left(\frac{p}{\rho}\right)_{\text{out}} - \left(\frac{p}{\rho}\right)_{\text{in}} + \frac{V_{\text{out}}^2 - V_{\text{in}}^2}{2} + g(z_{\text{out}} - z_{\text{in}}) \right] = \dot{Q}_{\substack{\text{net}\\ \text{in}}} + \dot{W}_{\substack{\text{shaft}\\ \text{net in}}} \qquad \textbf{(5.67)}$$

- The first law of thermodynamics, Eq. 5.55, states that the rate of increase of stored energy of a system equals the sum of the rate of heat transfer and rate of work transfer into the system.
- The *energy equation* for a control volume, Eq. 5.64, relates the rate of change of stored energy within the control volume, the rate of energy flow across the control surface, the heat transfer rate, $\dot{Q}_{\text{net in}}$, and the shaft power, $\dot{W}_{\text{shaft net in}}$.
- For steady flow with uniform properties across the inlet and outlet, the energy equation can be written as in Eq. 5.67.
- *Caution:* Heat transfer into (out of) the control volume is positive (negative). The rate of work (i.e., power) is positive if work is done on the control volume contents; otherwise it is negative.

Comparison of the Energy Equation with the Bernoulli Equation

$$\check{u}_{out} - \check{u}_{in} - q_{\substack{net\\in}} = \text{loss} \qquad \textbf{(5.78)}$$

$$\frac{p_{out}}{\rho} + \frac{V_{out}^2}{2} + gz_{out} = \frac{p_{in}}{\rho} + \frac{V_{in}^2}{2} + gz_{in} + w_{\substack{shaft\\net\ in}} - \text{loss} \qquad \textbf{(5.82)}$$

$$\frac{p_{out}}{\gamma} + \frac{V_{out}^2}{2g} + z_{out} = \frac{p_{in}}{\gamma} + \frac{V_{in}^2}{2g} + z_{in} + h_s - h_L \qquad \textbf{(5.84)}$$

$$h_s = w_{shaft\ net\ in}/g = \frac{\dot{W}_{\substack{shaft\\net\ in}}}{\dot{m}g} = \frac{\dot{W}_{\substack{shaft\\net\ in}}}{\gamma Q} \qquad \textbf{(5.85)}$$

- The *mechanical energy equation* can be written in terms of the *work per unit mass*, $w_{shaft\ net\ in}$, and the loss as shown in Eq. 5.82.
- As shown by Eq. 5.78, the parameter *loss* is related to the change in internal energy per unit mass, $\check{u}_{out} - \check{u}_{in}$, and the heat transfer per unit mass, $q_{net\ in}$. Loss represents the reduction of useful energy that occurs because of friction.
- The mechanical energy equation can be written in terms of heads (i.e., pressure head, velocity head, elevation head, shaft head, and the head loss) as shown in Eq. 5.84.
- The relationship among the shaft head, h_s, the shaft work per unit mass, and the shaft power is as indicated in Eq. 5.85.
- The Bernoulli equation is a restricted form of the mechanical energy equation that is valid only for flows that *do not* involve pumps, turbines, or friction.
- *Caution:* The head loss and loss are never negative. The shaft head is positive for pumps; negative for turbines.

EXAMPLE: (Head loss) Water flows in a constant diameter pipe from an elevation of 20 m to an elevation of 4 m, where it exits into the atmosphere as a free jet. The velocity throughout the pipe is 5 m/s. There are no pumps or turbines involved. Determine the pressure at the 20 m elevation if the head loss through this section of pipe is 10 m.

SOLUTION:

From Eq. 5.84 with $V_{in} = V_{out}$, $h_s = 0$ (no pump or turbine), and $p_{out} = 0$ (free jet),

$$\frac{p_{in}}{\gamma} = z_{out} - z_{in} + h_L = 4m - 20m + 10m = -6m$$

Hence, $p_{in} = -6m\,(9.80\times10^3\ N/m^3) = -58.8\times10^3 N/m^2 = \underline{\underline{-58.8\ kPa}}$

EXAMPLE: (Pump work) The pressure and velocity in the inlet pipe of a water pump are 10 psi and 5 ft/s, respectively. In the smaller diameter outlet pipe the pressure and velocity are 20 psi and 30 ft/s, respectively. Determine the work per unit mass, $w_{shaft\ net\ in}$, that the pump adds to the water. Elevation changes and losses are negligible.

SOLUTION:

From Eq. 5.82 with $z_{in} = z_{out}$ and loss = 0,

$$w_{shaft\ net\ in} = \frac{p_{out} - p_{in}}{\rho} + \frac{1}{2}\left(V_{out}^2 - V_{in}^2\right)$$

$$= \left[(20-10)\ \text{lb/in.}^2(144\ \text{in.}^2/\text{ft}^2)\right]/1.94\ \text{slugs/ft}^3$$

$$+ \frac{1}{2}\left(30^2 - 5^2\right)\ \text{ft}^2/\text{s}^2 = \underline{\underline{1180\ \text{ft}\cdot\text{lb/slug}}}$$

Note that $1\ \text{lb}\cdot\text{ft/slug} = 1(\text{slug}\cdot\text{ft/s}^2)\cdot\text{ft/slug} = \text{ft}^2/\text{s}^2$

EXAMPLE: (Turbine power) Water flows from a lake at an elevation of 400 ft, through a pipe and a turbine, and exits from a large diameter pipe as a free jet with negligible velocity at an elevation of 150 ft. Determine the shaft power of the turbine if the flowrate is 50 ft^3/s and the head loss is 80 ft.

SOLUTION:

From Eq. 5.84 with $V_{in} = 0$, $p_{in} = p_{out} = 0$,

$$h_s = \frac{V_{out}^2}{2g} + z_{out} - z_{in} + h_L \quad \text{where } V_{out} = Q/A_{out} \approx 0$$

Thus, $h_s = z_{out} - z_{in} + h_L = 150\ \text{ft} - 400\ \text{ft} + 80\ \text{ft} = -170\ \text{ft}$

so that $\dot{W}_{shaft} = \gamma Q h_s = 62.4\ \text{lb/ft}^3(50\ \text{ft}^3/\text{s})(-170\ \text{ft})$

$$= -5.30\times10^5\ \text{ft}\cdot\text{lb/s}\ (1\ \text{hp}/550\ \text{ft}\cdot\text{lb/s})$$

$$= \underline{\underline{-964\ \text{hp}}}$$

Note that $\dot{W}_{shaft} < 0$ since the device removes energy from the fluid. It is a turbine.

Application of the Energy Equation to Nonuniform Flows

$$\alpha = \frac{\int_A (V^2/2)\rho \mathbf{V} \cdot \hat{\mathbf{n}}\, dA}{\dot{m}\overline{V}^2/2} \qquad \textbf{(5.86)}$$

$$\frac{p_{\text{out}}}{\rho} + \frac{\alpha_{\text{out}}\overline{V}^2_{\text{out}}}{2} + gz_{\text{out}} = \frac{p_{\text{in}}}{\rho} + \frac{\alpha_{\text{in}}\overline{V}^2_{\text{in}}}{2} + gz_{\text{in}} + w_{\substack{\text{shaft}\\ \text{net in}}} - \text{loss} \qquad \textbf{(5.87)}$$

$$\frac{p_{\text{out}}}{\gamma} + \frac{\alpha_{\text{out}}\overline{V}^2_{\text{out}}}{2g} + z_{\text{out}} = \frac{p_{\text{in}}}{\gamma} + \frac{\alpha_{\text{in}}\overline{V}^2_{\text{in}}}{2g} + z_{\text{in}} + \frac{w_{\substack{\text{shaft}\\ \text{net in}}}}{g} - h_L \qquad \textbf{(5.89)}$$

- The *kinetic energy coefficient*, α, as defined in Eq. 5.86 is used in the energy equation to account for nonuniform velocity profiles.
- If the velocity profile is uniform, $\alpha = 1$; for any nonuniform velocity profile $\alpha > 1$.
- The energy equation for flows with nonuniform velocity profiles at the inlet and/or outlet can be written as shown by Eq. 5.87 (on an energy per unit mass basis) or Eq. 5.89 (on a head basis).

EXAMPLE: The uniform water velocity at the entrance, section (1), to a horizontal, constant diameter pipe is 15 ft/s. Farther along the pipe, at section (2), the velocity is nonuniform and the kinetic energy coefficient is $\alpha_2 = 1.12$. Determine the pressure difference, $p_1 - p_2$, if the head loss between the two sections is 0.2 ft.

SOLUTION:

From Eq. 5.89 with $w_{\text{shaft net in}} = 0$ (since there is no pump or turbine in the flow),

$$p_2/\gamma + \alpha_2\overline{V}_2^2/2g + z_2 = p_1/\gamma + \alpha_1\overline{V}_1^2/2g - h_L$$

where $z_1 = z_2$ (horizontal pipe), $\overline{V}_1 = \overline{V}_2 = 15$ ft/s (constant diameter pipe so the average velocity is constant), $\alpha_1 = 1$ (uniform flow) and $h_L = 0.2$ ft. Thus,

$$p_1 - p_2 = (\alpha_2 - \alpha_1)\overline{V}_1^2\gamma/2g + \gamma h_L$$

$$= [(1.12-1)(15\text{ ft/s})^2/(2(32.2\text{ ft/s}^2)) + 0.2\text{ ft}](62.4\text{ lb/ft}^3)$$

$$= [0.419\text{ ft} + 0.2\text{ ft}](62.4\text{ lb/ft}^3) = \underline{\underline{38.6\text{ lb/ft}^2}}$$

Solutions

to Review Problems

Chapter 5

5.1R

5.1R (Continuity equation) Water flows steadily through a 2-in.-inside-diameter pipe at the rate of 200 gal/min. The 2-in. pipe branches into two 1-in.-inside-diameter pipes. If the average velocity in one of the 1-in. pipes is 30 ft/s, what is the average velocity in the other 1-in. pipe?

(ANS: 51.7 ft/s)

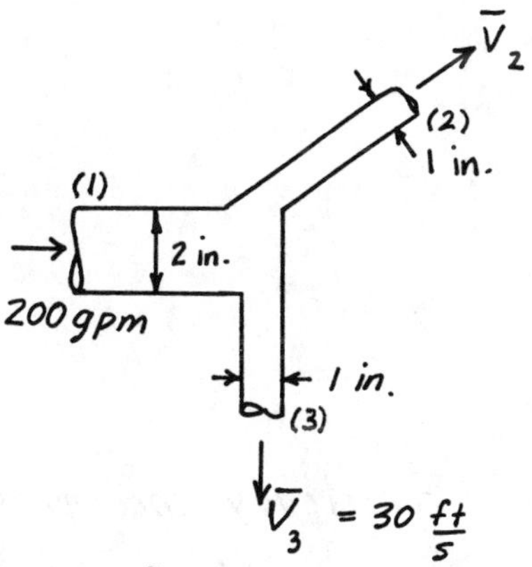

For steady incompressible flow

$$Q_1 = Q_2 + Q_3$$

or

$$Q_1 = A_2\overline{V}_2 + A_3\overline{V}_3$$

Thus

$$\overline{V}_2 = \frac{Q_1}{A_2} - \frac{A_3}{A_2}\overline{V}_3 = \frac{Q_1}{\frac{\pi D_2^2}{4}} - \frac{D_3^2}{D_2^2}\overline{V}_3$$

$$\overline{V}_2 = \frac{(200\,gpm)\left(231\ \frac{in.^3}{gal}\right)}{\frac{\pi(1\ in.)^2}{4}\left(60\ \frac{s}{min}\right)\left(12\ \frac{in.}{ft}\right)} - \frac{(1\ in.)^2\left(30\ \frac{ft}{s}\right)}{(1\ in.)^2} = 51.7\ \frac{ft}{s}$$

5.2R

5.2R (Continuity equation) Air (assumed incompressible) flows steadily into the square inlet of an air scoop with the nonuniform velocity profile indicated in Fig. P5.2R. The air exits as a uniform flow through a round pipe 1 ft in diameter. **(a)** Determine the average velocity at the exit plane. **(b)** In one minute, how many pounds of air pass through the scoop?

(ANS: 191 ft/s; 688 lb/min)

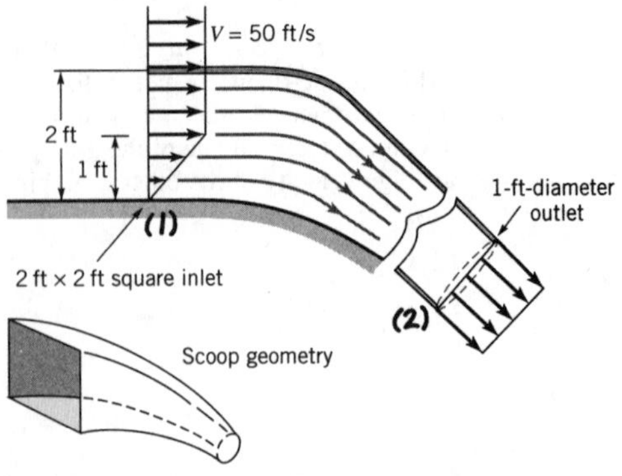

■ FIGURE P5.2R

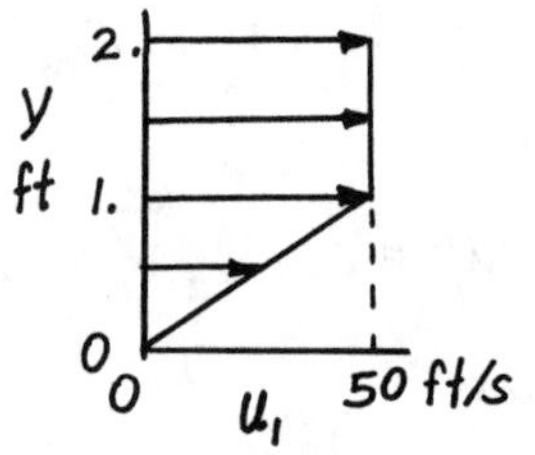

a) For steady, incompressible flow

$Q_1 = Q_2$ or

$$\int_{(1)} u_1\,dA = A_2 V_2 \quad \text{where} \quad \int_{(1)} u_1\,dA = \int_{y=0}^{y=1.} u_1(2\,dy) + \int_{y=1.}^{y=2.}(50)(2\,dy)$$

$$= (2\,ft)(1\,ft)\left(\frac{50}{2}\frac{ft}{s}\right) + (2\,ft)(1\,ft)\left(50\frac{ft}{s}\right)$$

$$= 150\frac{ft^3}{s} = Q$$

Thus,

$$V_2 = \frac{150\frac{ft^3}{s}}{\frac{\pi}{4}(1\,ft)^2} = \underline{\underline{191\frac{ft}{s}}}$$

b) $\text{Weight flowrate} = \gamma Q = \left(0.0765\frac{lb}{ft^3}\right)\left(150\frac{ft^3}{s}\right) = 11.48\frac{lb}{s}\left(\frac{60s}{min}\right) = \underline{\underline{688\frac{lb}{min}}}$

$$\text{Note: } \int_{y=0}^{y=1.} u_1(2\,dy) = \int_0^1 (50y)(2\,dy) = 100\int_0^1 y\,dy = 50y^2\Big|_0^1 = 50\frac{ft^3}{s}$$

5.3R

5.3R (Continuity equation) Water at 0.1 m³/s and alcohol ($SG = 0.8$) at 0.3 m³/s are mixed in a *y*-duct as shown in Fig. P5.3R. What is the average density of the mixture of alcohol and water?

(ANS: 849 kg/m³)

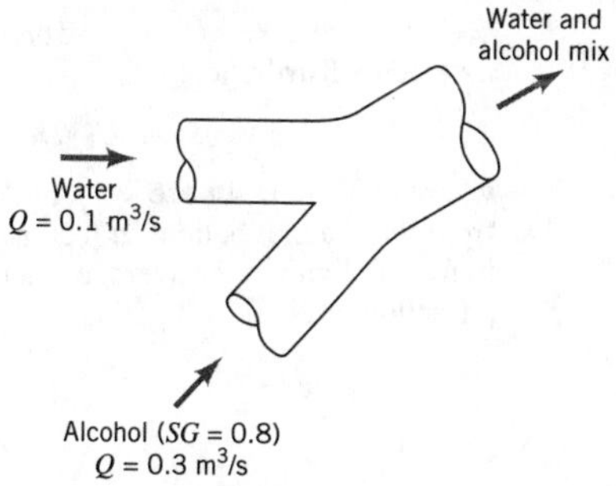

■ FIGURE P5.3R

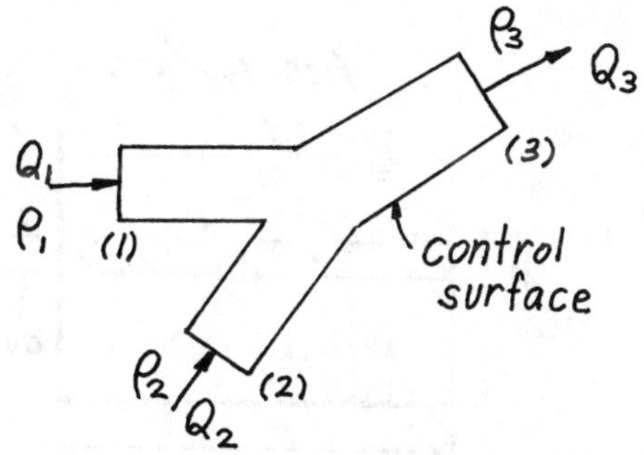

For steady flow $\int_{cs} \rho \vec{V} \cdot \hat{n}\, dA = 0$, or

$$\dot{m}_1 + \dot{m}_2 = \dot{m}_3$$

or

$$\rho_1 Q_1 + \rho_2 Q_2 = \rho_3 Q_3 \qquad (1)$$

Also, since the water and alcohol may be considered incompressible

$$Q_1 + Q_2 = Q_3 \qquad (2)$$

Combining Eqs. 1 and 2 we get

$$\rho_1 Q_1 + \rho_2 Q_2 = \rho_3 (Q_1 + Q_2)$$

or

$$\rho_3 = \frac{\rho_1 Q_1 + \rho_2 Q_2}{Q_1 + Q_2}$$

and

$$\rho_3 = \rho_1 \frac{(Q_1 + SG_2\, Q_2)}{Q_1 + Q_2}$$

Thus

$$\rho_3 = \frac{\left(999 \frac{kg}{m^3}\right)\left[0.1 \frac{m^3}{s} + (0.8)\left(0.3 \frac{m^3}{s}\right)\right]}{0.1 \frac{m^3}{s} + 0.3 \frac{m^3}{s}} = \underline{\underline{849 \frac{kg}{m^3}}}$$

5.4R

5.4R (Average velocity) The flow in an open channel has a velocity distribution

$$\mathbf{V} = U(y/h)^{1/5}\hat{\mathbf{i}} \text{ ft/s}$$

where U = free-surface velocity, y = perpendicular distance from the channel bottom in feet, and h = depth of the channel in feet. Determine the average velocity of the channel stream as a fraction of U.

(ANS: 0.833)

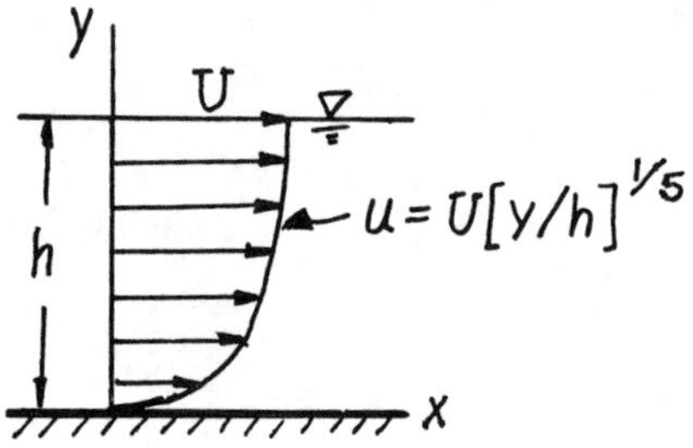

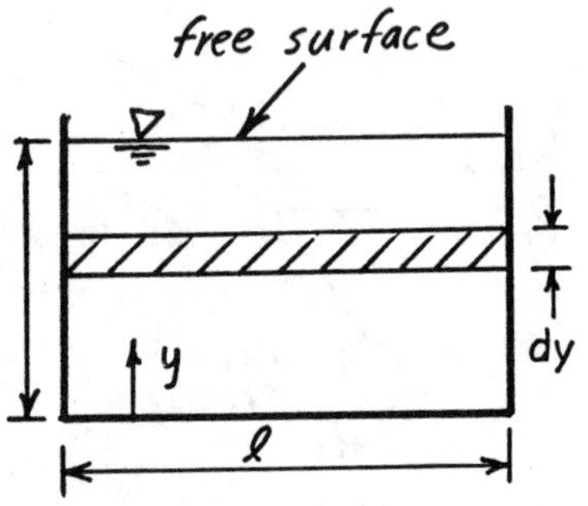

flow cross section

For any flow cross section

$$\dot{m} = \rho A \bar{u} = \int_A \rho \vec{V} \cdot \hat{n} \, dA$$

Also

$$\vec{V} \cdot \hat{n} = \vec{V} \cdot \hat{i} = U\left(\frac{y}{h}\right)^{\frac{1}{5}}$$

Thus for uniformly distributed density, ρ, over area A

$$\bar{u} = \frac{\int_0^h U\left(\frac{y}{h}\right)^{\frac{1}{5}} \ell \, dy}{\ell h}$$

and

$$\frac{\bar{u}}{U} = \int_0^1 \left(\frac{y}{h}\right)^{\frac{1}{5}} d\left(\frac{y}{h}\right) = \frac{5}{6} = \underline{\underline{0.833}}$$

5.5R

5.5R (Linear momentum) Water flows through a right angle valve at the rate of 1000 lbm/s as is shown in Fig. P5.5R. The pressure just upstream of the valve is 90 psi and the pressure drop across the valve is 50 psi. The inside diameters of the valve inlet and exit pipes are 12 and 24 in. If the flow through the valve occurs in a horizontal plane, determine the x and y components of the force exerted by the valve on the water.
(ANS: 18,200 lb; 10,800 lb)

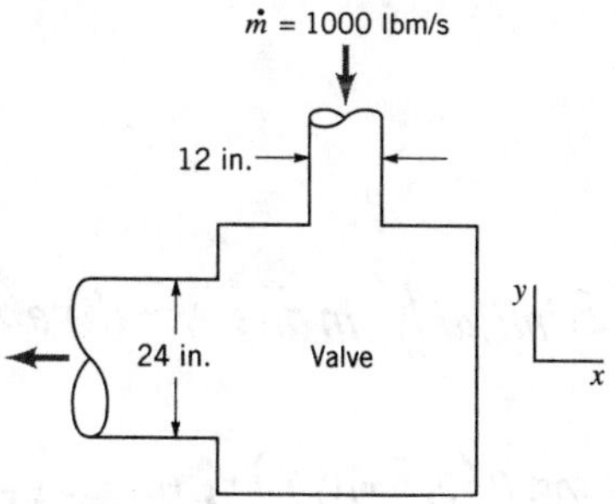

■ FIGURE P5.5R

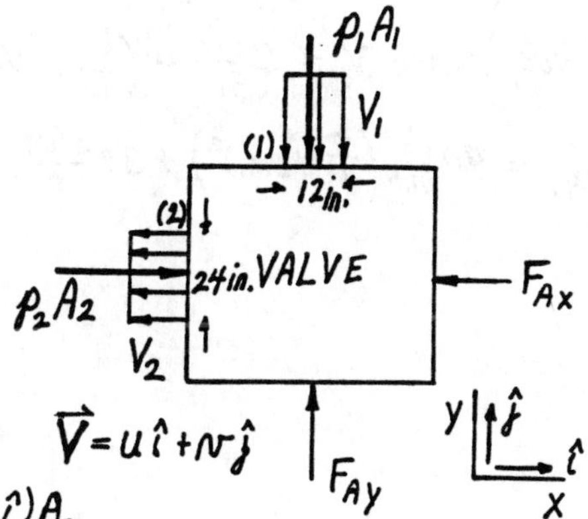

For steady flow the x-component of the momentum equation is

$$\int_{cs} u\rho \vec{V}\cdot\hat{n}\, dA = \sum F_x$$

or

$$u_1\rho(u_1\hat{i}+v_1\hat{j})\cdot\hat{j}A_1 + u_2\rho(u_2\hat{i}+v_2\hat{j})\cdot(-\hat{i})A_2$$

$$= +p_2A_2 - F_{Ax}$$

but $u_1 = 0$, $v_1 = -V_1$, $u_2 = -V_2$, and $v_2 = 0$

Thus,

$$0 + (-V_2)\rho(V_2)A_2 = +p_2A_2 - F_{Ax}$$

or

$$F_{Ax} = p_2A_2 + \dot{m}V_2, \text{ where } \dot{m} = \rho V_2A_2 = \rho V_1A_1 = 30\,\frac{slugs}{s}$$

Hence,

$$F_{Ax} = (90-50)\frac{lb}{in.^2}\left(\frac{\pi}{4}(24\,in.)^2\right) + 30\,\frac{slugs}{s}\left(4.92\,\frac{ft}{s}\right) = \underline{\underline{18{,}200\,lb}}$$

where we have used $V_2 = \frac{\dot{m}}{A_2\rho}$

or

$$V_2 = \frac{30\,\frac{slugs}{s}}{\left(\frac{\pi}{4}(2\,ft)^2\right)\left(1.94\,\frac{slugs}{ft^3}\right)} = 4.92\,\frac{ft}{s}$$

(continued)

Similarly, in the y-direction $\int_{cs} v \rho \vec{V}\cdot\hat{n}\, dA = \sum F_y$, or

$$v_1 \rho (u_1\hat{i} + v_1\hat{j})\cdot\hat{j} A_1 + v_2 \rho (u_2\hat{i} + v_2\hat{j})\cdot(-\hat{i}) A_2 = -p_1 A_1 + F_{Ay}$$

or

$$(-V_1)\rho(-V_1)A_1 + 0 = -p_1 A_1 + F_{Ay}\text{, or } F_{Ay} = p_1 A_1 + \rho A_1 V_1^2 = p_1 A_1 + \dot{m} V_1$$

Thus, since $V_1 = \frac{A_2 V_2}{A_1} = \left(\frac{D_2}{D_1}\right)^2 V_2 = \left(\frac{24\text{ in.}}{12\text{ in.}}\right)^2 \left(4.92\,\frac{\text{ft}}{\text{s}}\right) = 19.7\,\frac{\text{ft}}{\text{s}}$ we obtain

$$F_{Ay} = 90\,\frac{\text{lb}}{\text{in.}^2}\left(\frac{\pi}{4}(12\text{ in.})^2\right) + 30\,\frac{\text{slugs}}{\text{s}}\left(19.7\,\frac{\text{ft}}{\text{s}}\right) = \underline{\underline{10{,}800\text{ lb}}}$$

5.6R

5.6R (Linear momentum) A horizontal circular jet of air strikes a stationary flat plate as indicated in Fig. P5.6R. The jet velocity is 40 m/s and the jet diameter is 30 mm. If the air velocity magnitude remains constant as the air flows over the plate surface in the directions shown, determine: **(a)** the magnitude of F_A, the anchoring force required to hold the plate stationary, **(b)** the fraction of mass flow along the plate surface in each of the two directions shown, **(c)** the magnitude of F_A, the anchoring force required to allow the plate to move to the right at a constant speed of 10 m/s.

(ANS: 0.696 N; 0.933 and 0.0670; 0.391 N)

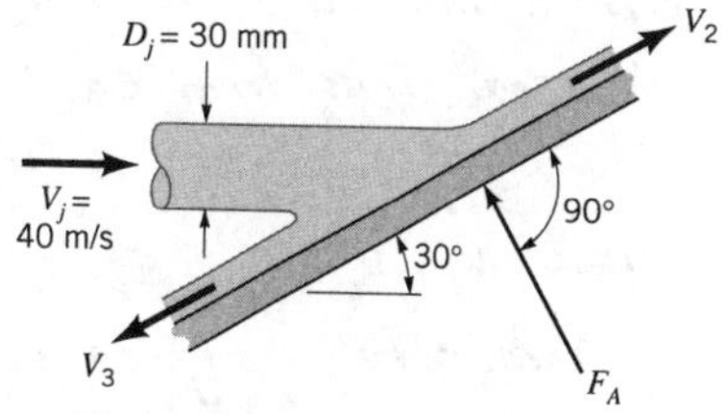

■ FIGURE P5.6R

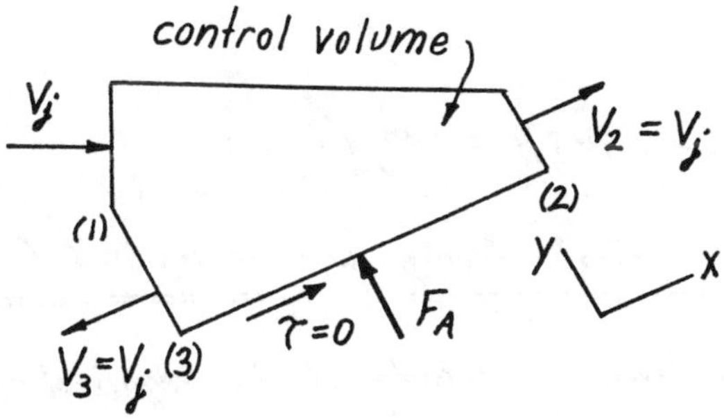

The non-deforming control volume shown in the sketch above is used.

(a) To determine the magnitude of F_A we apply the component of the linear momentum equation (Eq. 5.22) along the direction of F_A. Thus, $\int_{cs} v \rho \vec{V}\cdot\hat{n}\, dA = \sum F_y$, or

$$F_A = \dot{m} V_j \sin 30° = \rho A_j V_j V_j \sin 30° = \frac{\rho \pi D_j^2}{4} V_j^2 \sin 30°$$

or

$$F_A = \left(1.23 \frac{kg}{m^3}\right) \frac{\pi (0.030m)^2}{(4)} \left(40 \frac{m}{s}\right)^2 (\sin 30°) \left(\frac{1\, N}{kg\cdot\frac{m}{s^2}}\right) = \underline{\underline{0.696\, N}}$$

(b) To determine the fraction of mass flow along the plate surface in each of the 2 directions shown in the sketch above, we apply the component of the linear momentum equation parallel to the surface of the plate, $\int_{cs} u \rho \vec{V}\cdot\hat{n}\, dA = \sum F_x$, to obtain

$$R_{\text{along plate surface}} = \dot{m}_2 V_2 - \dot{m}_3 V_3 - \dot{m}_j V_j \cos 30° \qquad (1)$$

(continued)

Since the air velocity magnitude remains constant, the value of $R_{\text{along plate surface}}$ is zero.* Thus from Eq. 1 we obtain

$$\dot{m}_3 V_3 = \dot{m}_2 V_2 - \dot{m}_j V_j \cos 30° \quad (2)$$

Since $V_3 = V_2 = V_j$, Eq. 2 becomes

$$\dot{m}_3 = \dot{m}_2 - \dot{m}_j \cos 30° \quad (3)$$

From conservation of mass we conclude that

$$\dot{m}_j = \dot{m}_2 + \dot{m}_3 \quad (4)$$

Combining Eqs. 3 and 4 we get

$$\dot{m}_3 = \dot{m}_j - \dot{m}_3 - \dot{m}_j \cos 30°$$

or

$$\dot{m}_3 = \dot{m}_j \frac{(1-\cos 30°)}{2} = \dot{m}_j (0.0670)$$

and

$$\dot{m}_2 = \dot{m}_j (1 - 0.067) = \dot{m}_j (0.933)$$

Thus, $\dot{m}_2$ involves 93.3% of $\dot{m}_j$ and $\dot{m}_3$ involves 6.7% of $\dot{m}_j$.

(c) To determine the magnitude of F_A required to allow the plate to move to the right at a constant speed of $10 \frac{m}{s}$, we use a non-deforming control volume like the one in the sketch above that moves to the right with a speed of $10 \frac{m}{s}$. The translating control volume linear momentum equation (Eq. 5.29) leads to

$$F_A = \frac{\rho \pi D_j^2}{4} \left(V_j - 10 \frac{m}{s}\right)^2 \sin 30°$$

or

$$F_A = \left(1.23 \frac{kg}{m^3}\right) \frac{\pi (0.030\,m)^2}{4} \left(40 \frac{m}{s} - 10 \frac{m}{s}\right)^2 (\sin 30°) \left(1 \frac{N}{kg \cdot \frac{m}{s^2}}\right)$$

and

$$F_A = 0.391\ N$$

* Since $V_1 = V_2 = V_3$ and $p_1 = p_2 = p_3$ and $z_1 = z_2 = z_3$ it follows that the Bernoulli equation is valid from $1 \rightarrow 2$ and $1 \rightarrow 3$. Thus, there are no viscous effects (Bernoulli equation is valid only for inviscid flow) so that $\tau = 0$. Hence, $R_{\text{along plate}} = 0$.

5.7R

5.7R (Linear momentum) An axisymmetric device is used to partially "plug" the end of the round pipe shown in Fig. P5.7R. The air leaves in a radial direction with a speed of 50 ft/s as indicated. Gravity and viscous forces are negligible. Determine the **(a)** flowrate through the pipe, **(b)** gage pressure at point (1), **(c)** gage pressure at the tip of the plug, point (2), **(d)** force, F, needed to hold the plug in place.

(ANS: 23.6 ft³/s; 1.90 lb/ft²; 2.97 lb/ft²; 3.18 lb)

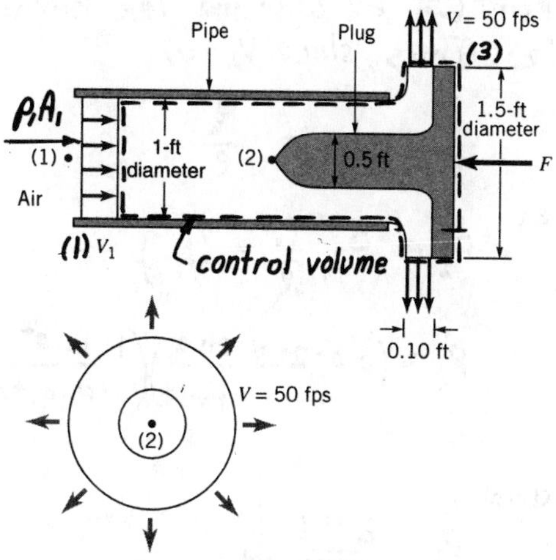

■ FIGURE P5.7R

For part (a) we determine the volume flowrate through the pipe by calculating the volume flowrate of the air leaving radially after being turned by the axisymmetric plug. Thus

$$Q = V_3 A_3 = \left(50 \frac{ft}{s}\right) \pi (1.5\ ft)(0.10\ ft) = \underline{\underline{23.6 \frac{ft^3}{s}}}$$

For part (b) we determine the gage pressure at (1) by applying the Bernoulli equation to the flow between (1) and the radial flow leaving the plug, station (3). Thus

$$\frac{p_1}{\rho} + \frac{V_1^2}{2} = \frac{p_3}{\rho} + \frac{V_3^2}{2} \qquad (1)$$

($p_3 = 0$ gage)

We get V_1 from

$$V_1 = \frac{Q}{A_1} = \frac{23.6 \frac{ft^3}{s}}{\frac{\pi (1\ ft)^2}{4}} = 30 \frac{ft}{s} \qquad (2)$$

Combining Eqs. 1 and 2 we get

$$p_1 = \rho\left(\frac{V_3^2 - V_1^2}{2}\right) = \left(0.00238 \frac{slug}{ft^3}\right)\left(\frac{1\ lb \cdot s^2}{slug \cdot ft}\right)\left[\frac{\left(50 \frac{ft}{s}\right)^2 - \left(30 \frac{ft}{s}\right)^2}{2}\right]$$

or

$$p_1 = \underline{\underline{1.90 \frac{lb}{ft^2}}}$$

(continued)

5.7R continued

For part (c) we determine the gage pressure at the tip of the plug, point (2), by applying the Bernoulli equation between points (1) and (2). Thus, since $V_2 = 0$,

$$\frac{p_1}{\rho} + \frac{V_1^2}{2} = \frac{p_2}{\rho}$$

or

$$p_2 = \rho\left(\frac{p_1}{\rho} + \frac{V_1^2}{2}\right)$$

$$p_2 = \left(0.00238\ \frac{\text{slug}}{\text{ft}^3}\right)\left(1\ \frac{\text{lb}\cdot\text{s}^2}{\text{slug}\cdot\text{ft}}\right)\left[\frac{1.90\ \frac{\text{lb}}{\text{ft}^2}}{\left(0.00238\ \frac{\text{slug}}{\text{ft}^3}\right)\left(1\ \frac{\text{lb}\cdot\text{s}^2}{\text{slug}\cdot\text{ft}}\right)} + \frac{\left(30\ \frac{\text{ft}}{\text{s}}\right)^2}{2}\right]$$

and

$$p_2 = \underline{\underline{2.97\ \frac{\text{lb}}{\text{ft}^2}}}$$

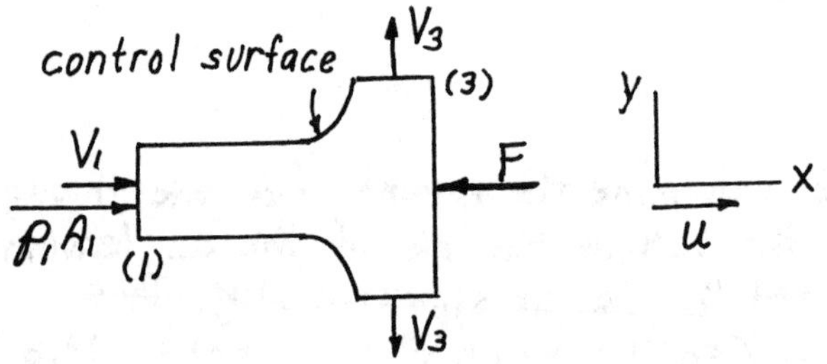

For part (d) we apply the linear momentum equation to the contents of the control volume sketched above to get

$$\int_{cs} u\rho\vec{V}\cdot\hat{n}\,dA = \sum F_x \quad \text{or} \quad \int_{(1)} u\rho\vec{V}\cdot dA = \sum F_x \quad \text{since } u_3 = 0$$

Thus,

$$-V_1\rho V_1 A_1 = p_1 A_1 - F \quad \text{or} \quad F = p_1 A_1 + V_1 \rho Q$$

Thus,

$$F = \left(1.90\ \frac{\text{lb}}{\text{ft}^2}\right)\frac{\pi(1\ \text{ft})^2}{4} + \left(30\ \frac{\text{ft}}{\text{s}}\right)\left(0.00238\ \frac{\text{slug}}{\text{ft}^3}\right)\left(1\ \frac{\text{lb}\cdot\text{s}^2}{\text{slug}\cdot\text{ft}}\right)\left(23.6\ \frac{\text{ft}^3}{\text{s}}\right)$$

$$F = \underline{\underline{3.18\ \text{lb}}}$$

5.8R

5.8R (Linear momentum) A nozzle is attached to an 80-mm inside-diameter flexible hose. The nozzle area is 500 mm². If the delivery pressure of water at the nozzle inlet is 700 kPa, could you hold the hose and nozzle stationary? Explain.

(ANS: yes, 707 N or 159 lb)

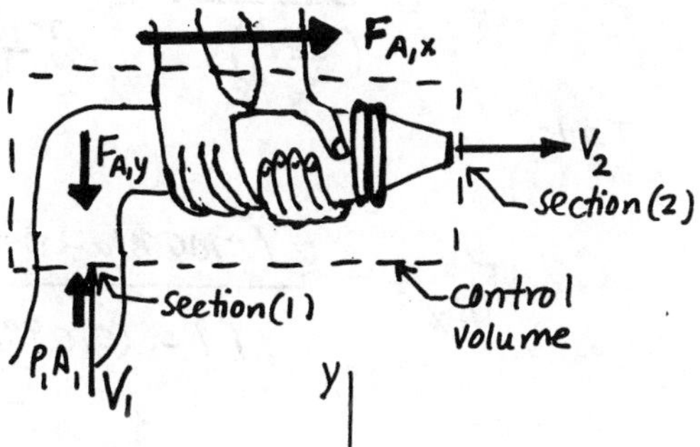

$$\int_{cs} u\rho \vec{V}\cdot\hat{n}\, dA = \sum F_x$$

The control volume shown in the sketch is used. We assume that the vertical component of the anchoring force, $F_{A,y}$, is exerted by the hose material. We further assume that the horizontal component of the anchoring force, $F_{A,x}$, must be exerted by the hands holding the hose and nozzle stationary. Application of the horizontal or x direction component of the linear momentum equation leads to

$$V_2 \rho V_2 A_2 = F_{A,x} \qquad (1)$$

Application of Bernoulli's equation between sections (1) and (2) yields

$$\frac{p_1}{\rho} + \frac{V_1^2}{2} = \frac{p_2}{\rho} + \frac{V_2^2}{2} \qquad (2)$$

From the conservation of mass equation

$$\rho V_1 A_1 = \rho V_2 A_2$$

or

$$V_1 = V_2 \frac{A_2}{A_1} \qquad (3)$$

Thus combining Eqs. 2 and 3 gives

$$V_2^2 = \frac{2(p_1 - p_2)}{\rho\left[1 - \left(\frac{A_2}{A_1}\right)^2\right]}$$

and Eq. 1 becomes

$$F_{A,x} = \frac{2(p_1 - p_2)A_2}{\left[1 - \left(\frac{A_2}{A_1}\right)^2\right]}$$

(continued)

5.8R continued

Now

$$\frac{A_2}{A_1} = \frac{(500\ mm^2)}{\left(\frac{\pi D_i^2}{4}\right)} = \frac{(500\ mm^2)}{\frac{\pi\,(80 mm)^2}{4}} = 0.0995$$

Thus,

$$F_{A,x} = \frac{2\,(700\ kPa - 0\ kPa)\,(500\ mm^2)}{[1-(0.0995)^2]\,\left(1000\ \frac{mm}{m}\right)^2}\left(1000\ \frac{N}{m^2\,kPa}\right)$$

$$F_{A,x} = 707\ N$$

or in terms of lb

$$F_{A,x} = \frac{707\ N}{4.448\ \frac{N}{lb}} = \underline{\underline{159\ lb}}$$

which is managable.

5.9R

5.9R (Linear momentum) A horizontal air jet having a velocity of 50 m/s and a diameter of 20 mm strikes the inside surface of a hollow hemisphere as indicated in Fig. P5.9R. How large is the horizontal anchoring force needed to hold the hemisphere in place? The magnitude of velocity of the air remains constant.

(ANS: 1.93 N)

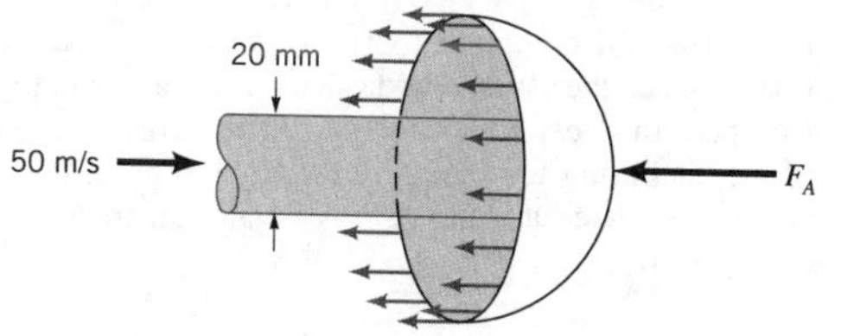

■ FIGURE P5.9R

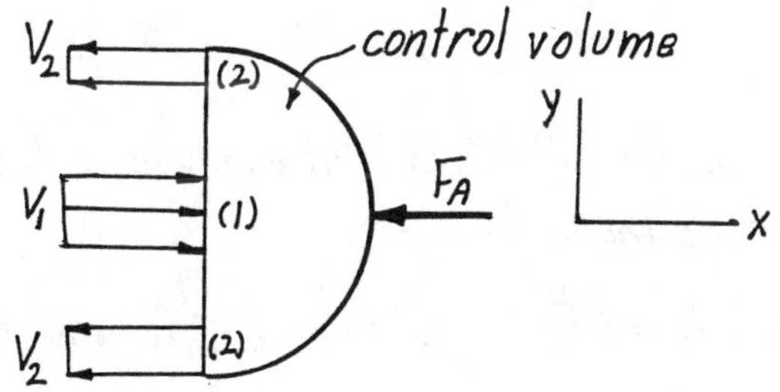

The control volume shown in the sketch is used. The x-component of the momentum equation gives

$$\int_{cs} u\rho\vec{V}\cdot\hat{n}\,dA = \sum F_x \quad \text{or}$$

$$(V_1)\rho_1(-V_1)A_1 + (-V_2)\rho_2(+V_2)A_2 = -F_A\,,$$

where for conservation of mass

$\rho_1 A_1 V_1 = \rho_2 A_2 V_2 = \dot{m}$, the mass flowrate.

Thus,

$$F_A = \dot{m}(V_1 + V_2) = 2\dot{m}V \text{ since } V_1 = V_2 = 50\,\frac{m}{s}$$

Note that $V_1 = V_2$ (i.e. the speed is constant), but $\vec{V}_1 = 50\hat{i}\,\frac{m}{s} \neq \vec{V}_2 = -50\hat{i}\,\frac{m}{s}$ (i.e. the velocity changes).

With

$$\dot{m} = \rho_1 A_1 V_1 = 1.23\,\frac{kg}{m^3}\left(\frac{\pi}{4}(0.020\,m)^2\right)\left(50\,\frac{m}{s}\right) = 0.0193\,\frac{kg}{s}$$

we obtain

$$F_A = \left(0.0193\,\frac{kg}{s}\right)\left(50\,\frac{m}{s}\right)(2) = \underline{\underline{1.93\,N}}$$

5.10R

5.10R (Linear momentum) Determine the magnitude of the horizontal component of the anchoring force required to hold in place the 10-foot-wide sluice gate shown in Fig. P5.10R. Compare this result with the size of the horizontal component of the anchoring force required to hold in place the sluice gate when it is closed and the depth of water upstream is 6 ft.

(ANS: 5310 lb; 11,200 lb)

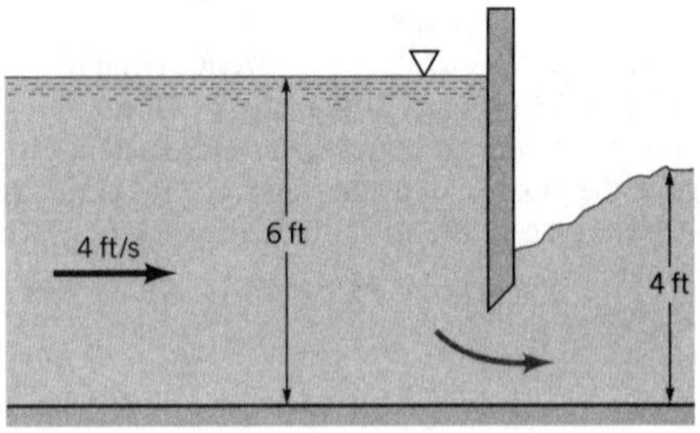

■ FIGURE P5.10R

When the gate is closed the water is stationary and the resultant water force on the gate is

$$R_{x_{closed}} = p_c A = \gamma h_c A = \gamma \frac{H^2}{2} b \quad \text{where } H = 6\,ft \text{ and } b = 10\,ft = \text{width}$$

Thus,

$$R_{x_{closed}} = \tfrac{1}{2}(62.4\,\tfrac{lb}{ft^3})(6\,ft)^2(10\,ft) = \underline{\underline{11{,}200\,lb}}$$

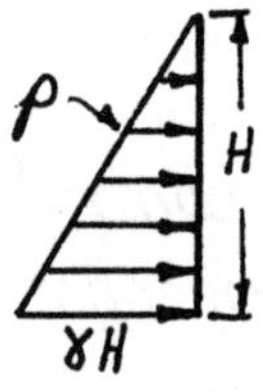

From the control volume diagram shown, the x-component of the momentum equation becomes:

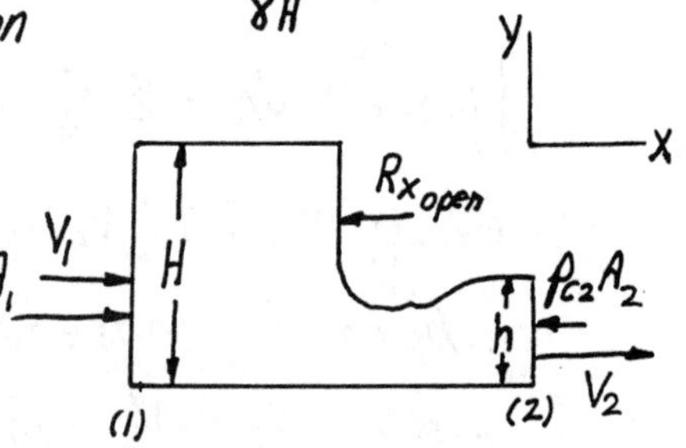

$$\int_{cs} u \rho \vec{V} \cdot \hat{n}\, dA = \Sigma F_x \quad \text{or}$$

$$V_1 \rho(-V_1)A_1 + V_2 \rho(+V_2)A_2 = p_{c1}A_1 - p_{c2}A_2 - R_{x_{open}}$$

or

$$R_{x_{open}} = p_{c1}A_1 - p_{c2}A_2 + \rho V_1^2 A_1 - \rho V_2^2 A_2$$

$$= \tfrac{1}{2}\gamma H^2 b - \tfrac{1}{2}\gamma h^2 b + \rho V_1^2 H b - \rho V_2^2 h b$$

where

$$V_1 = 4\,\tfrac{ft}{s} \text{ and } V_2 = \frac{V_1 A_1}{A_2} = V_1 \frac{H}{h} = 4\,\tfrac{ft}{s}\left(\frac{6\,ft}{4\,ft}\right) = 6\,\tfrac{ft}{s}$$

Thus,

$$R_{x_{open}} = \tfrac{1}{2}(62.4\,\tfrac{lb}{ft^3})(6\,ft)^2(10\,ft) - \tfrac{1}{2}(62.4\,\tfrac{lb}{ft^3})(4\,ft)^2(10\,ft) + 1.94\,\tfrac{slugs}{ft^3}(4\,\tfrac{ft}{s})^2(6\,ft)(10\,ft) - 1.94\,\tfrac{slugs}{ft^3}(6\,\tfrac{ft}{s})^2(4\,ft)(10\,ft)$$

or

$$R_{x_{open}} = \underline{\underline{5{,}310\,lb}} < R_{x_{closed}} = 11{,}200\,lb$$

5.11R

5.11R (Linear momentum) Two jets of liquid, one with specific gravity 1.0 and the other with specific gravity 1.3, collide and form one homogeneous jet as shown in Fig. P5.11R. Determine the speed, V, and the direction, θ, of the combined jet. Gravity is negligible.

(ANS: 6.97 ft/s; 70.3 deg)

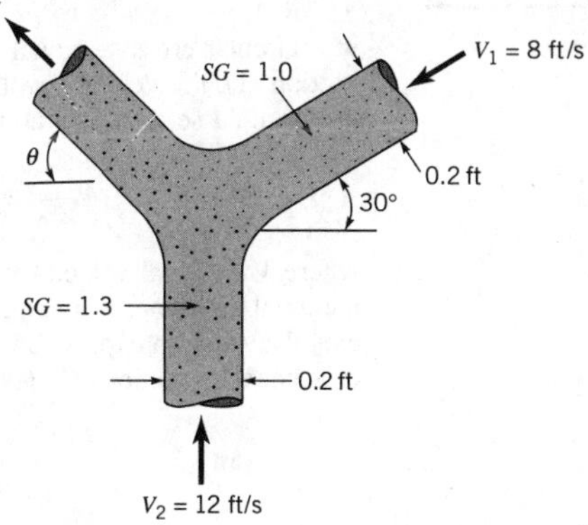

■ FIGURE P5.11R

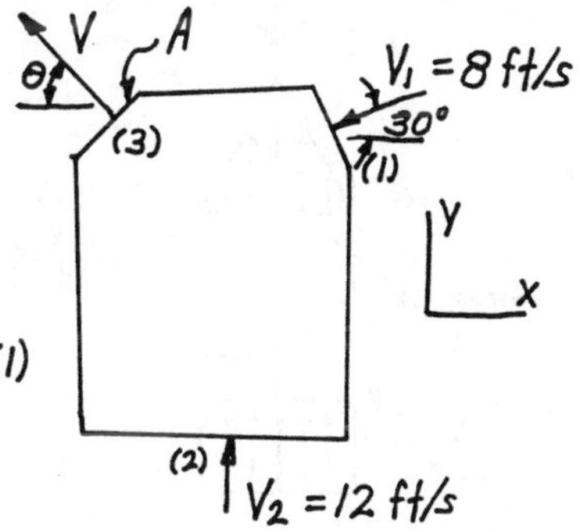

For the control volume shown, the x-component of the momentum equation becomes $\int_{cs} u\rho\vec{V}\cdot\hat{n}\,dA = \Sigma F_x$, or

$$(-V_1\cos 30^\circ)\rho_1(-V_1)A_1 - (V\cos\theta)\rho(V)A = 0 \qquad (1)$$

since there is no force acting on the control volume. Similarly, in the y-direction

$\int_{cs} v\rho\vec{V}\cdot\hat{n}\,dA = \Sigma F_y$, or

$$(-V_1\sin 30^\circ)\rho_1(-V_1)A_1 + V_2\rho_2(-V_2)A_2 + (V\sin\theta)\rho(V)A = 0 \qquad (2)$$

since $\Sigma F_y = 0$

By combining Eqs. (1) and (2) we obtain (divide (1) by (2)):

$$\cot\theta = \frac{\rho_1 V_1^2\cos 30^\circ A_1}{\rho_2 V_2^2 A_2 - \rho_1 V_1^2 \sin 30^\circ A_1} \quad \text{where } \rho_2 = 1.3\rho_1$$

Thus,

$$\cot\theta = \frac{\rho_1\left(8\frac{ft}{s}\right)^2\cos 30^\circ\,\frac{\pi}{4}(0.2ft)^2}{(1.3\rho_1)\left(12\frac{ft}{s}\right)^2\frac{\pi}{4}(0.2ft)^2 - \rho_1\left(8\frac{ft}{s}\right)^2\frac{\pi}{4}(0.2ft)^2\sin 30^\circ} = 0.357$$

so that

$$\theta = \underline{\underline{70.3^\circ}}$$

Also, for conservation of mass: $\rho VA = \rho_1 V_1 A_1 + \rho_2 V_2 A_2$ (3)

By combining Eqs. (1) and (3)

$\rho_1 V_1^2\cos 30^\circ A_1 - V\cos\theta\left[\rho_1 V_1 A_1 + \rho_2 V_2 A_2\right] = 0$ or since $A_1 = A_2$

$$V = \frac{\rho_1 V_1^2\cos 30^\circ}{\cos\theta(\rho_1 V_1 + \rho_2 V_2)} = \frac{\rho_1\left(8\frac{ft}{s}\right)^2\cos 30^\circ}{(\cos 70.3^\circ)\left[\rho_1\left(8\frac{ft}{s}\right) + 1.3\rho_1\left(12\frac{ft}{s}\right)\right]} = \underline{\underline{6.97\frac{ft}{s}}}$$

5.12R

5.12R (Linear momentum) Water flows vertically upward in a circular cross-sectional pipe as shown in Fig. P5.12R. At section (1), the velocity profile over the cross-sectional area is uniform. At section (2), the velocity profile is

$$\mathbf{V} = w_c\left(\frac{R-r}{R}\right)\hat{\mathbf{k}}$$

where $\mathbf{V}$ = local velocity vector, w_c = centerline velocity in the axial direction, R = pipe radius, and r = radius from pipe axis. Develop an expression for the fluid pressure drop that occurs between sections (1) and (2).

(ANS: $p_1 - p_2 = R_z/\pi R^2 + 0.50\,\rho w_1^2 + g\rho h$, where R_z = friction force)

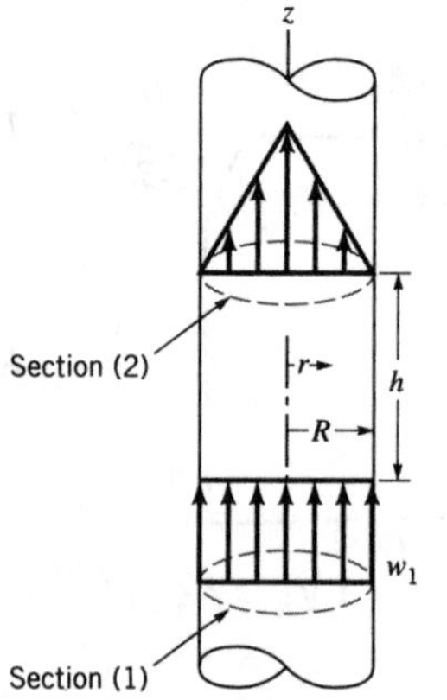

■ FIGURE P5.12R

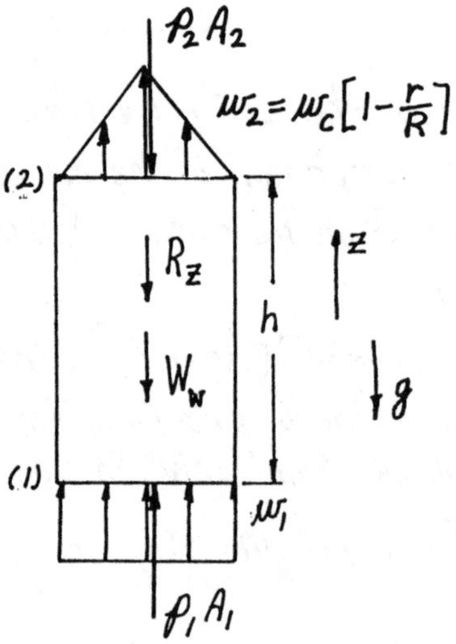

R_z = axial force of pipe wall on the fluid

W_w = weight of water

For the control volume shown in the figure the z-component of the momentum equation is

$$\int_{cs} w\rho\vec{V}\cdot\hat{n}\,dA = \sum F_z \quad \text{or}$$

$$-w_1\rho w_1 A_1 + \int_{r=0}^{R} w_2\rho w_2\,(2\pi r\,dr) = p_1A_1 - p_2A_2 - R_z - W_w$$

where $dA = 2\pi r\,dr$

Thus, with $A_1 = A_2 \equiv A$ this becomes

$$p_1 - p_2 = \frac{R_z}{A} + \frac{W_w}{A} - \rho w_1^2 + \frac{2\pi\rho}{A}\int_0^R\left[w_c\left(1-\frac{r}{R}\right)\right]^2 r\,dr \qquad (1)$$

But with $x \equiv \frac{r}{R}$,

$$\int_0^R w_c^2\left(1-\frac{r}{R}\right)^2 r\,dr = w_c^2\int_{r=0}^{R}\left(1-2\frac{r}{R}+\frac{r^2}{R^2}\right)r\,dr = w_c^2R^2\int_{x=0}^{x=1}(x-2x^2+x^3)\,dx$$

$$= w_c^2R^2\left[\frac{x^2}{2}-\frac{2}{3}x^3+\frac{1}{4}x^4\right]_0^1 = \frac{1}{12}R^2(w_c^2)$$

(continued)

Thus, since $A = \pi R^2$, Eq. (1) becomes

$$p_1 - p_2 = \frac{R_z}{\pi R^2} + \frac{W_w}{\pi R^2} - \rho w_1^2 + \frac{2\pi\rho}{\pi R^2}\left(\frac{1}{12}R^2\right) w_c^2 \qquad (2)$$

We can determine w_c in terms of w_1 by using the continuity equation:
$\int_{cs} \rho \vec{V}\cdot\hat{n}\,dA = 0$, or since $\rho \equiv$ constant:

$$A_1 w_1 = \int w_2\, dA = \int_{r=0}^{R} w_c\left[1 - \frac{r}{R}\right](2\pi r\, dr) = 2\pi w_c R^2 \int_{x=0}^{1} (x - x^2)\, dx$$

or

$$\pi R^2 w_1 = 2\pi w_c R^2 \left(\frac{1}{6}\right)$$

Thus,

$$w_c = 3 w_1 \quad \text{and Eq. (2) becomes}$$

$$p_1 - p_2 = \frac{R_z}{\pi R^2} + \frac{W_w}{\pi R^2} - \rho w_1^2 + \rho\left(\frac{1}{6}\right)(3 w_1)^2$$

or

$$p_1 - p_2 = \frac{R_z}{\pi R^2} + \frac{W_w}{\pi R^2} + \frac{1}{2}\rho w_1^2$$

or with $W_w = \gamma A h = \rho g \pi R^2 h$

$$p_1 - p_2 = \frac{R_z}{\pi R^2} + \rho g h + \frac{1}{2}\rho w_1^2$$

5.13R

5.13R (Moment-of-momentum) A lawn sprinkler is constructed from pipe with $\frac{1}{4}$-in. inside diameter as indicated in Fig. P5.13R. Each arm is 6 in. in length. Water flows through the sprinkler at the rate of 1.5 lb/s. A force of 3 lb positioned halfway along one arm holds the sprinkler stationary. Compute the angle, θ, which the exiting water stream makes with the tangential direction. The flow leaves the nozzles in the horizontal plane.

(ANS: 23.9 deg)

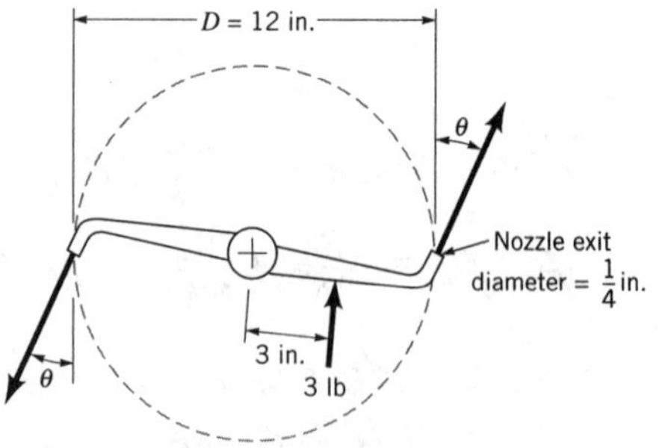

■ FIGURE P5.13R

The stationary, non-deforming control volume shown in the sketch is used. Application of the axial component of the moment-of-momentum equation (Eq. 5.50) leads to

$$T_{shaft} = \dot{m}\, r_2 V_{\theta,2} = \dot{m}\, r_2 V_2 \cos\theta \qquad (1)$$

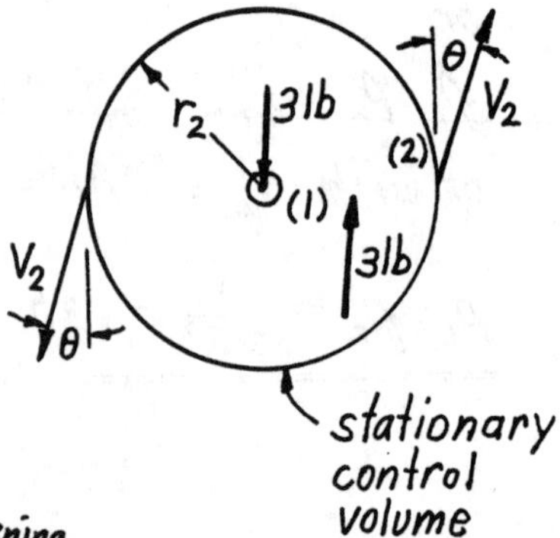

Since

$$V_2 = \frac{\dot{m}}{2\rho A_{nozzle\ opening}}$$

where

$$A_{nozzle} = \frac{\pi D_{nozzle\ opening}^2}{4}$$

Eq. 1 leads to

$$\cos\theta = \frac{T_{shaft}\, 2\rho \frac{\pi D_{nozzle\ opening}^2}{4}}{\dot{m}^2 r_2}$$

where $\dot{m} = \rho V_2 A_2 = \frac{\gamma}{g} Q = \frac{1.5\ lb/s}{32.2\ ft/s^2} = 0.0466\ \frac{slug}{s}$

$$\theta = \cos^{-1} \frac{(3\ lb)(\frac{3}{12}ft)(2)(1.94\ \frac{slugs}{ft^3})\frac{\pi}{4}(\frac{1}{48}ft)^2}{(0.0466\ \frac{slugs}{s})^2(\frac{6}{12}ft)}$$

Thus,

$$\theta = \underline{\underline{23.9^\circ}}$$

5.14R

5.14R (Moment-of-momentum) A water turbine with radial flow has the dimensions shown in Fig. P5.14R. The absolute entering velocity is 15 m/s, and it makes an angle of 30° with the tangent to the rotor. The absolute exit velocity is directed radially inward. The angular speed of the rotor is 30 rpm. Find the power delivered to the shaft of the turbine.

(ANS: −7.68 MW)

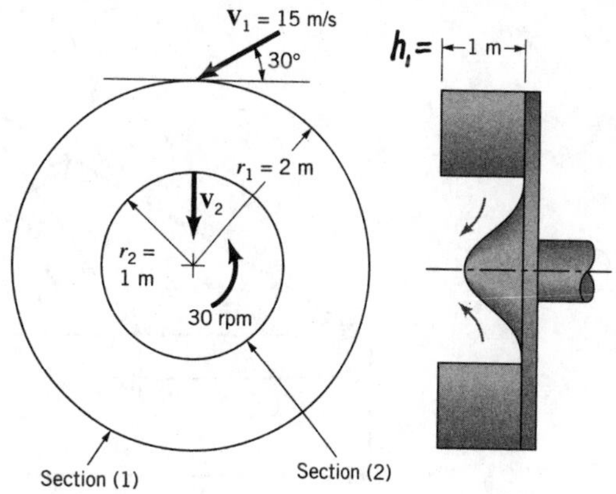

■ FIGURE P5.14R

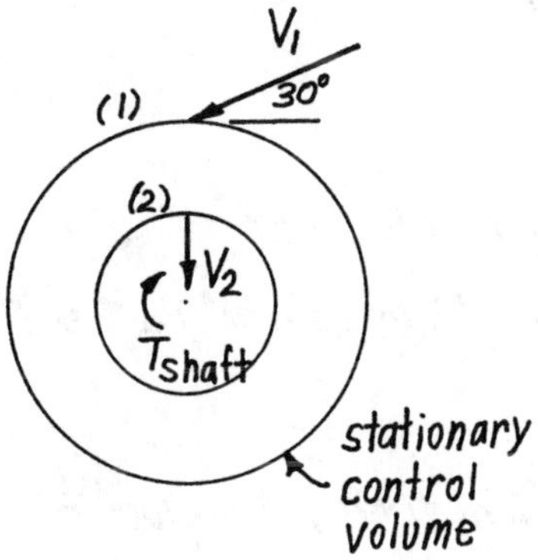

The stationary and non-deforming control volume shown in the sketch above is used. We use Eq. 5.53 to determine the shaft power involved. Thus

$$\dot{W}_{shaft} = -\dot{m}_1 U_1 V_{\theta,1} + \dot{m}_2 U_2 V_{\theta,2}\text{, where } V_{\theta,2}=0 \qquad (1)$$

The mass flowrate may be obtained from (2)

$$\dot{m}_1 = \rho V_{R,1} A_1 = \rho V_{R,1} 2\pi r_1 h_1$$

where

$V_{R,1}$ = radial component of velocity at section (1)

The blade velocity at section (1) is

$$U_1 = r_1 \omega = (2\text{ m})\left(30 \frac{\text{rev}}{\text{min}}\right)\left(2\pi \frac{\text{rad}}{\text{rev}}\right)\frac{1}{\left(60 \frac{\text{s}}{\text{min}}\right)} = 6.28 \frac{\text{m}}{\text{s}}$$

The values of $V_{\theta,1}$ and $V_{R,1}$ may be obtained with the help of a velocity triangle for the flow at section (1) as sketched below.

(continued)

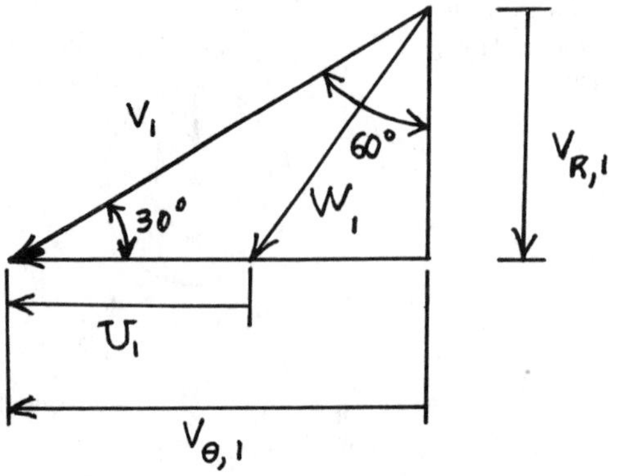

With the velocity triangle we conclude that

$$V_{R,1} = V_1 \sin 30° = V_1 \cos 60° = \left(15 \frac{m}{s}\right)(\sin 30°) = 7.5 \frac{m}{s}$$

Then from Eq. 2

$$\dot{m}_1 = \left(999 \frac{kg}{m^3}\right)\left(7.5 \frac{m}{s}\right) 2\pi (2\,m)(1\,m) = 94{,}100 \frac{kg}{s}$$

Also, with the triangle we see that

$$V_{\theta,1} = V_1 \cos 30° = V_1 \sin 60° = \left(15 \frac{m}{s}\right) \cos 30° = 13.0 \frac{m}{s}$$

Then, with Eq. 1 we obtain

$$\dot{W}_{shaft} = -\left(94{,}100 \frac{kg}{s}\right)\left(6.28 \frac{m}{s}\right)\left(13.0 \frac{m}{s}\right)\left(\frac{1\,N}{kg \cdot \frac{m}{s^2}}\right)\left(\frac{MW}{10^6 \frac{N \cdot m}{s}}\right)$$

$$\dot{W}_{shaft} = \underline{\underline{-7.68\ MW}}$$

5.15R

5.15R (Moment-of-momentum) The single stage, axial-flow turbomachine shown in Fig. P5.15R involves water flow at a volumetric flowrate of 11 m³/s. The rotor revolves at 600 rpm. The inner and outer radii of the annular flow path through the stage are 0.46 and 0.61 m, and $\beta_2 = 30°$. The flow entering the rotor row and leaving the stator row is axial viewed from the stationary casing. Is this device a turbine or a pump? Estimate the amount of power transferred to or from the fluid.

(ANS: pump; 7760 kW)

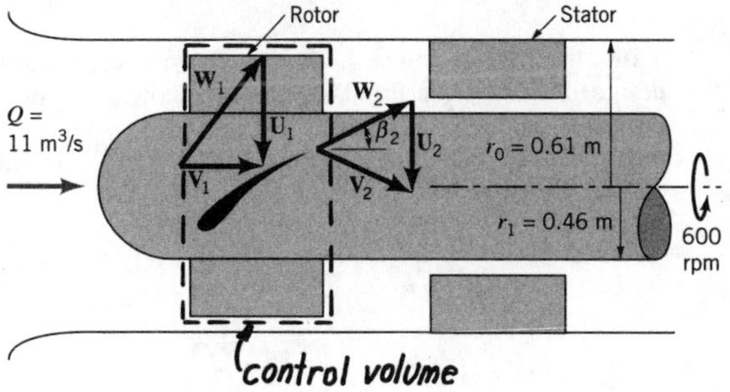

■ FIGURE P5.15R

This device is a pump because the lift force acting on each rotor blade is opposite in direction to the blade motion. The direction of the blade lift force is ascertained by noting how the blade turns the flow past the blade. The power transferred from the rotor blades to the water may be evaluated with the moment-of-momentum power equation (equation (5.53) with $V_{\theta,1}=0$):

$$\dot{W}_{\substack{shaft\\net\ in}} = \dot{m}\, U_2 V_{\theta,2} = \rho Q U_2 V_{\theta,2} \qquad (1)$$

We obtain U_2 with

$$U_2 = r_2\omega = \left(\frac{r_i + r_o}{2}\right)\omega = \left(\frac{0.46\ m + 0.61\ m}{2}\right)\frac{(600\ rpm)}{\left(60\ \frac{s}{min}\right)}\left(2\pi\ \frac{rad}{rev}\right) = 33.6\ \frac{m}{s}$$

From the rotor exit flow velocity triangle we obtain $V_{\theta,2}$ with

$$V_{\theta,2} = U_2 - V_{axial,2}\tan\beta_2 \quad \text{(see figure below)}$$

where

$$V_{axial,2} = \frac{Q}{A_2} = \frac{Q}{\pi(r_o^2 - r_i^2)} = \frac{\left(11\ \frac{m^3}{s}\right)}{\pi\left[(0.61\ m)^2 - (0.46\ m)^2\right]} = 21.8\ \frac{m}{s}$$

Thus

$$V_{\theta,2} = 33.6\ \frac{m}{s} - \left(21.8\ \frac{m}{s}\right)\tan 30° = 21.0\ \frac{m}{s}$$

and with Eq. 1 we get

$$\dot{W}_{\substack{shaft\\net\ in}} = \left(999\ \frac{kg}{m^3}\right)\left(11\ \frac{m^3}{s}\right)\left(33.6\ \frac{m}{s}\right)\left(21.0\ \frac{m}{s}\right)\left(\frac{1\ N}{kg\cdot\frac{m}{s^2}}\right) = 7.76\times10^6\ \frac{N\cdot m}{s}$$

or

$$\dot{W}_{\substack{shaft\\net\ in}} = \underline{\underline{7760\ kW}}$$

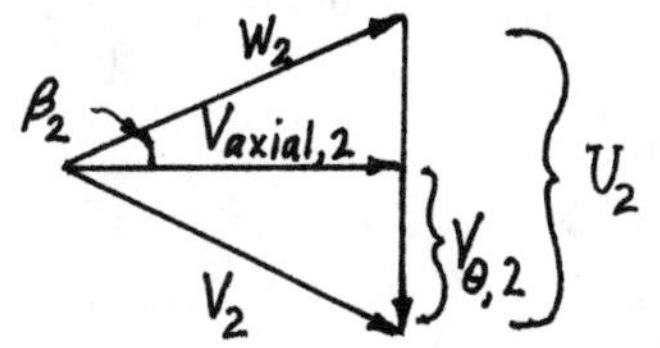

Note: Since $\dot{W}_{\substack{shaft\\net\ in}} > 0$ this device is a pump.

5.16R

5.16R (Moment-of-momentum) A small water turbine is designed as shown in Fig. P5.16R. If the flowrate through the turbine is 0.0030 slugs/s, and the rotor speed is 300 rpm, estimate the shaft torque and shaft power involved. Each nozzle exit cross-sectional area is 3.5×10^{-5} ft^2.

(ANS: -0.0107 ft·lb; -0.336 ft·lb/s)

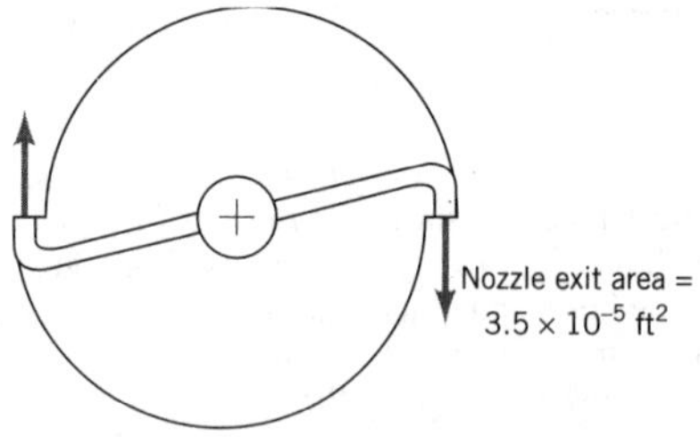

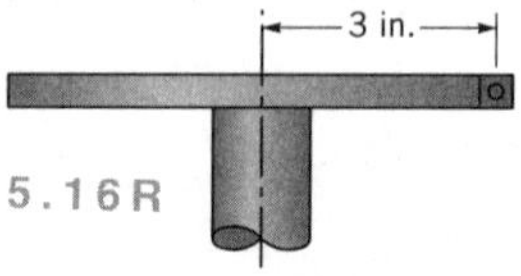

■ FIGURE P5.16R

For shaft torque we can use the axial component of the moment-of-momentum equation (Eq. 5.50). Thus, with $V_{\theta,1} = 0$:

$$T_{shaft} = -\dot{m}_2 \, r_2 \, V_{\theta,2}$$

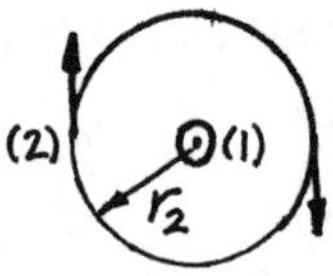

Consideration of the absolute and relative velocities of the flow out of each nozzle (see sketch below) leads to

$$V_{\theta,2} = V_2 = W_2 - U_2$$

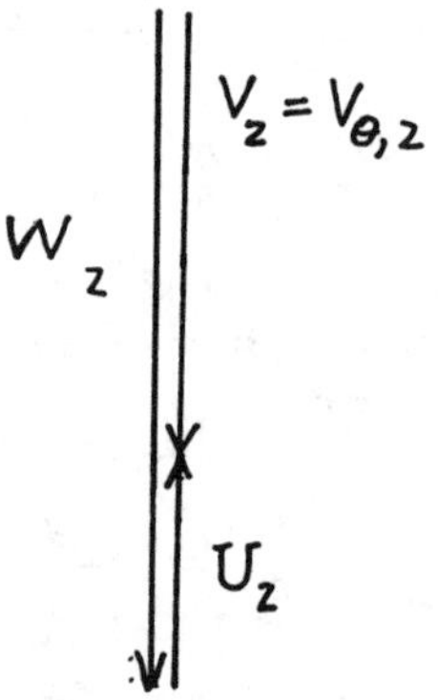

where

$$W_2 = \frac{\dot{m}}{2\rho A_{\text{nozzle exit}}}$$

and

$$U_2 = r_2 \omega$$

Thus,

$$T_{shaft} = -\dot{m}_2 r_2 \left(\frac{\dot{m}}{2\rho A_{\text{nozzle exit}}} - r_2 \omega \right)$$

$$T_{shaft} = \frac{-\left(0.003 \frac{\text{slugs}}{\text{s}}\right)(3 \text{ in.})}{\left(12 \frac{\text{in.}}{\text{ft}}\right)} \left[\frac{0.003 \frac{\text{slugs}}{\text{s}}}{2\left(1.94 \frac{\text{slugs}}{\text{ft}^3}\right)\left(3.5 \times 10^{-5} \text{ ft}^2\right)} - \frac{(3 \text{ in.})\left(300 \frac{\text{rev}}{\text{min}}\right)\left(2\pi \frac{\text{rad}}{\text{rev}}\right)}{\left(12 \frac{\text{in.}}{\text{ft}}\right)\left(60 \frac{\text{s}}{\text{min}}\right)} \right] \left(\frac{1 \text{ lb}}{\text{slug} \cdot \frac{\text{ft}}{\text{s}^2}} \right)$$

(continued)

5.16 continued

or $T_{shaft} = -0.0107$ ft·lb (minus sign mean torque opposes rotation)

Now,

$$\dot{W}_{shaft} = T_{shaft}\,\omega$$

or

$$\dot{W}_{shaft} = \frac{(-0.0107\ \text{ft}\cdot\text{lb})\left(300\ \frac{\text{rev}}{\text{min}}\right)\left(2\pi\ \frac{\text{rad}}{\text{rev}}\right)}{\left(60\ \frac{\text{s}}{\text{min}}\right)} = -0.336\ \frac{\text{ft}\cdot\text{lb}}{\text{s}}$$

(minus sign means work is out of the control volume)

5.17R

5.17R (Energy equation) Water flows steadily from one location to another in the inclined pipe shown in Fig. P5.17R. At one section, the static pressure is 12 psi. At the other section, the static pressure is 5 psi. Which way is the water flowing? Explain.

(ANS: from A to B)

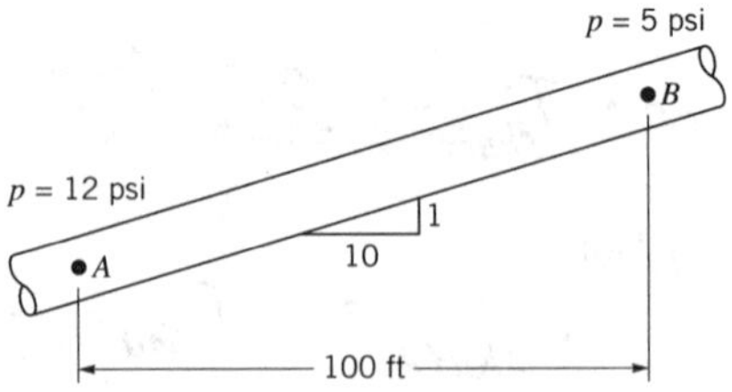

■ FIGURE P5.17R

To determine the direction of water flow we apply the energy equation (Eq. 5.82) for flow from sections (A) to (B) and flow from sections (B) to (A). The loss obtained with Eq. 5.82 is positive for the correct flow direction, but negative for the incorrect flow direction.

For flow from sections (A) to (B), Eq. 5.82 leads to

$$loss = \frac{p_A - p_B}{\rho} + \frac{V_A^2 - V_B^2}{2} + g(z_A - z_B) + w_{shaft\ net\ in}$$

(with $\frac{V_A^2 - V_B^2}{2} = 0$ and $w_{shaft\ net\ in} = 0$)

or

$$loss = \frac{(12psi - 5psi)\left(144\frac{in.^2}{ft^2}\right)}{\left(1.94\frac{slugs}{ft^3}\right)} + \left(32.2\frac{ft}{s^2}\right)(-10\,ft)\left(\frac{1\,lb}{slug \cdot \frac{ft}{s^2}}\right)$$

and

$$loss = 198\ \frac{ft \cdot lb}{slug}$$

For flow from sections (B) to (A), Eq. 5.82 leads to

$$loss = \frac{p_B - p_A}{\rho} + g(z_B - z_A)$$

or

$$loss = \frac{(5psi - 12psi)\left(144\frac{in.^2}{ft^2}\right)}{\left(1.94\frac{slugs}{ft^3}\right)} + \left(32.2\frac{ft}{s^2}\right)(10\,ft)\left(\frac{1\,lb}{slug \cdot \frac{ft}{s^2}}\right)$$

and

$$loss = -198\ \frac{ft \cdot lb}{slug}$$

The water flow is from <u>section (A) to section (B)</u> (i.e., uphill)

5.18R

5.18R (Energy equation) The pump shown in Fig. P5.18R adds 20 kW of power to the flowing water. The only loss is that which occurs across the filter at the inlet of the pump. Determine the head loss for this filter.

(ANS: 7.69 m)

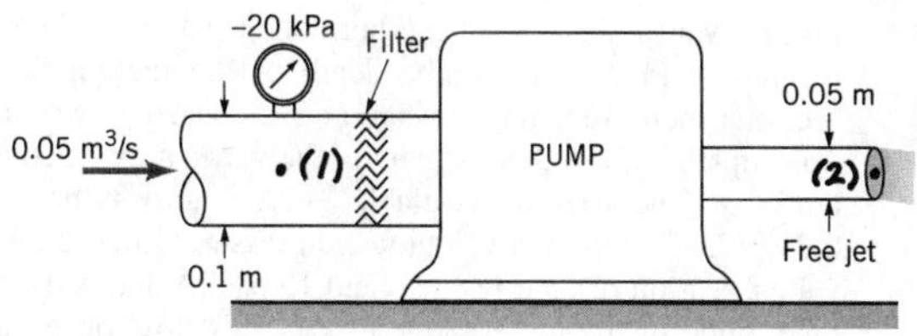

FIGURE P5.18R

The energy equation for this flow can be written as

$$\frac{p_1}{\gamma} + \frac{V_1^2}{2g} + z_1 + h_p = \frac{p_2}{\gamma} + \frac{V_2^2}{2g} + z_2 + h_L \qquad (1)$$

where

$z_1 = z_2$, $p_2 = 0$, $p_1 = -20\ kPa$, and

$$V_1 = \frac{Q}{A_1} = \frac{0.05\ \frac{m^3}{s}}{\frac{\pi}{4}(0.1m)^2} = 6.37\ \frac{m}{s}$$

$$V_2 = \frac{Q}{A_2} = \frac{0.05\ \frac{m^3}{s}}{\frac{\pi}{4}(0.05m)^2} = 25.5\ \frac{m}{s}$$

Also,

$$h_p = \frac{\dot{W}_s}{\gamma Q} = \frac{20 \times 10^3\ \frac{N\cdot m}{s}}{(9.8 \times 10^3\ \frac{N}{m^3})(0.05\ \frac{m^3}{s})} = 40.8\ m$$

Thus, Eq. (1) becomes

$$\frac{(-20 \times 10^3\ \frac{N}{m^2})}{(9.8 \times 10^3\ \frac{N}{m^3})} + \frac{(6.37\ \frac{m}{s})^2}{2(9.81\ \frac{m}{s^2})} + 40.8\ m = \frac{(25.5\ \frac{m}{s})^2}{2(9.81\ \frac{m}{s^2})} + h_L$$

or

$$h_L = \underline{\underline{7.69\ m}}$$

5.19R

5.19R (Linear momentum/energy) Eleven equally spaced turning vanes are used in the horizontal plane 90° bend as indicated in Fig. P5.19R. The depth of the rectangular cross-sectional bend remains constant at 3 in. The velocity distributions upstream and downstream of the vanes may be considered uniform. The loss in available energy across the vanes is $0.2V_1^2/2$. The required velocity and pressure downstream of the vanes, section (2), are 180 ft/s and 15 psia. What is the average magnitude of the force exerted by the air flow on each vane? Assume the force of the air on the duct walls is equivalent to the force of the air on one vane.

(ANS: 4.61 lb)

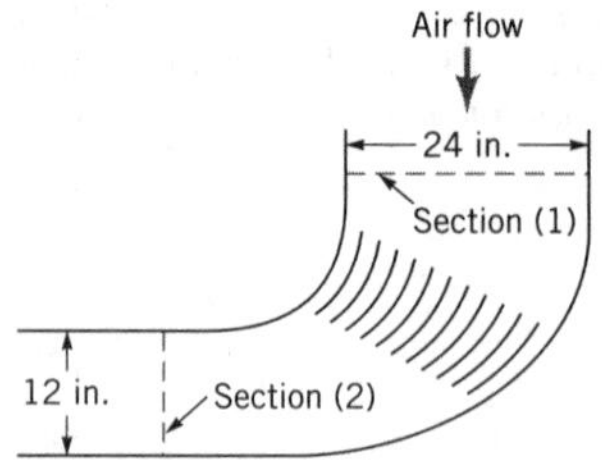

■ FIGURE P5.19R

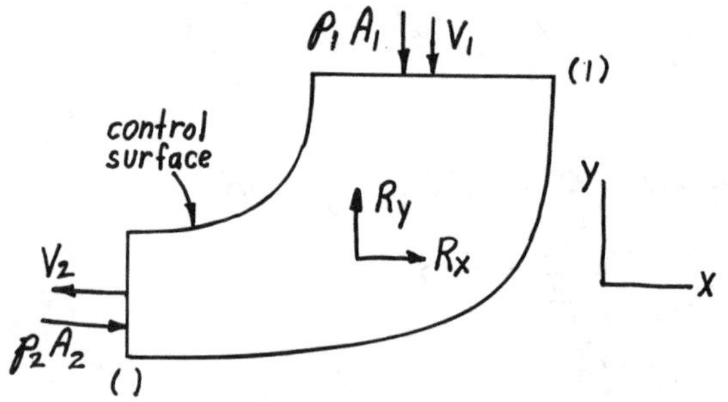

To estimate the average magnitude of the force exerted by the air flow on each vane, we determine the magnitude of the resultant force exerted by the air on the vanes and the duct walls and divide that result by 12. We assume that the duct walls act as one additional vane. The linear momentum equation (Eq. 5.22) is used to determine the x and y components of the resultant force exerted by the vanes and duct walls on the air between sections (1) and (2). Thus, $\sum \vec{F} = \int_{cs} \vec{V} \rho \vec{V} \cdot \hat{n} \, dA$, or

$$R_x = -p_2 A_2 - V_2 \rho A_2 V_2 \qquad (1)$$

and

$$R_y = p_1 A_1 + V_1 \rho A_1 V_1 \qquad (2)$$

From the conservation of mass principle (Eq. 5.13) we have

$$V_1 = V_2 \frac{A_2}{A_1} = \left(180 \frac{ft}{s}\right) \frac{(12 \text{ in.})(3 \text{ in.})}{(24 \text{ in})(3 \text{ in.})} = 90 \frac{ft}{s}$$

(continued)

5.19R continued

With the energy equation (Eq. 5.83) we obtain

$$p_1 = p_2 + \frac{\rho}{2}\left(V_2^2 - V_1^2\right) + \rho(\text{loss}) = p_2 + \frac{\rho}{2}\left(V_2^2 - V_1^2 + 0.2V_1^2\right)$$

or

$$p_1 = p_2 + \frac{\rho}{2}\left(V_2^2 - 0.8V_1^2\right) = 15\,\text{psia} + \frac{\left(2.38\times 10^{-3}\,\frac{\text{slug}}{\text{ft}^3}\right)}{2\left(144\,\frac{\text{in.}^2}{\text{ft}^2}\right)}\left[\left(180\,\frac{\text{ft}}{\text{s}}\right)^2 - 0.8\left(90\,\frac{\text{ft}}{\text{s}}\right)^2\right]\left(\frac{1\,\text{lb}}{\text{slug}\cdot\frac{\text{ft}}{\text{s}^2}}\right)$$

and

$$p_1 = 15.21\,\text{psia}$$

As suggested in Section 5.2.2, we use gage pressures at sections (1) and (2). Thus, from Eq 1 we have

$$R_x = -(15\,\text{psia} - 14.7\,\text{psia})(12\,\text{in.})(3\,\text{in.}) - \left(180\,\frac{\text{ft}}{\text{s}}\right)\left(2.38\times 10^{-3}\,\frac{\text{slug}}{\text{ft}^3}\right)\frac{(12\,\text{in.})(3\,\text{in.})\left(180\,\frac{\text{ft}}{\text{s}}\right)}{144\,\frac{\text{in.}^2}{\text{ft}^2}}\left(\frac{1\,\text{lb}}{\text{slug}\cdot\frac{\text{ft}}{\text{s}^2}}\right)$$

or

$$R_x = -30.1\,\text{lb}$$

From Eq. 2 we obtain

$$R_y = (15.21\,\text{psia} - 14.7\,\text{psia})(24\,\text{in.})(3\,\text{in.}) + \left(90\,\frac{\text{ft}}{\text{s}}\right)\left(2.38\times 10^{-3}\,\frac{\text{slug}}{\text{ft}^3}\right)\frac{(24\,\text{in.})(3\,\text{in.})\left(90\,\frac{\text{ft}}{\text{s}}\right)}{\left(144\,\frac{\text{in.}^2}{\text{ft}^2}\right)}\left(\frac{1\,\text{lb}}{\text{slug}\cdot\frac{\text{ft}}{\text{s}^2}}\right)$$

or

$$R_y = 46.4\,\text{lb}$$

Then

$$R = \sqrt{R_x^2 + R_y^2} = \sqrt{(-30.1\,\text{lb})^2 + (46.4\,\text{lb})^2} = 55.3\,\text{lb}$$

and

$$R_{\substack{\text{vane}\\ \text{average}}} = \frac{R_{\text{average}}}{12} = \frac{55.3\,\text{lb}}{12} = \underline{\underline{4.61\,\text{lb}}}$$

5.20R

5.20R (Energy equation) A hydroelectric power plant operates under the conditions illustrated in Fig. P5.20R. The head loss associated with flow from the water level upstream of the dam, section (1), to the turbine discharge at atmospheric pressure, section (2), is 20 m. How much power is transferred from the water to the turbine blades?

(ANS: 23.5 MW)

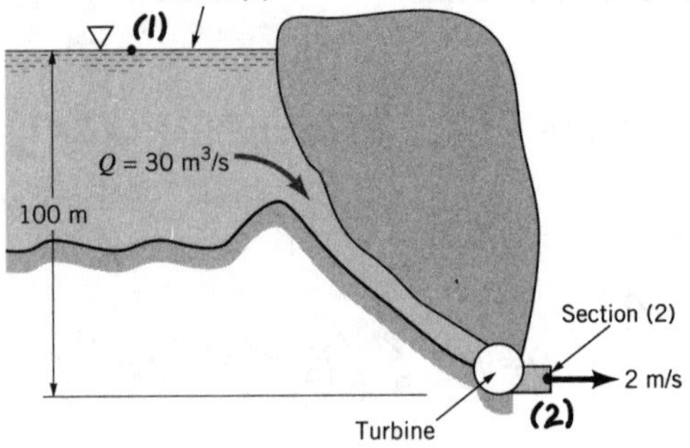

■ FIGURE P5.20R

For flow from section (1) to section (2), Eq. 5.82 leads to

$$w_{\substack{shaft\\net\ out}} = g(z_1 - z_2) - \frac{V_2^2}{2} - loss \qquad (1)$$

since

$$p_2 = p_1 = p_{atm}, \quad V_1 = 0,$$

and

$$w_{\substack{shaft\\net\ out}} = -w_{\substack{shaft\\net\ in}}$$

For power, we multiply Eq. 1 by the mass flowrate, $\dot{m}$, to get

$$\dot{W}_{\substack{shaft\\net\ out}} = \dot{m}g(z_1 - z_2) - \dot{m}\frac{V_2^2}{2} - \dot{m}\ loss$$

But

$$\dot{m} = \rho Q$$

thus

$$\dot{W}_{\substack{shaft\\net\ out}} = \rho Q g(z_1 - z_2) - \rho Q\frac{V_2^2}{2} - \rho Q\ loss$$

or

$$\dot{W}_{\substack{shaft\\net\ out}} = \left(999\frac{kg}{m^3}\right)\left(30\frac{m^3}{s}\right)\left(9.81\frac{m}{s^2}\right)(100\ m)\left(1\frac{N}{kg\cdot\frac{m}{s^2}}\right)$$
$$-\left(999\frac{kg}{m^3}\right)\left(30\frac{m^3}{s}\right)\frac{\left(2\frac{m}{s}\right)^2}{2}\left(1\frac{N}{kg\cdot\frac{m}{s^2}}\right)$$
$$-\left(999\frac{kg}{m^3}\right)\left(30\frac{m^3}{s}\right)(20\ m)\left(9.81\frac{m}{s^2}\right)\left(1\frac{N}{kg\cdot\frac{m}{s^2}}\right)$$

and

$$\dot{W}_{\substack{shaft\\net\ out}} = 23.5\times10^6\ \frac{N\cdot m}{s} = \underline{\underline{23.5\ MW}}$$

5.21R

5.21R (Energy equation) A pump transfers water from one large reservoir to another as shown in Fig. P5.21R*a*. The difference in elevation between the two reservoirs is 100 ft. The friction head loss in the piping is given by $K_L \overline{V}^2/2g$, where $\overline{V}$ is the average fluid velocity in the pipe and K_L is the loss coefficient, which is considered constant. The relation between the total head rise, H, across the pump and the flowrate, Q, through the pump is given in Fig. 5.21R*b*. If $K_L = 40$, and the pipe diameter is 4 in., what is the flowrate through the pump? (ANS: $0.653\ ft^3/s$)

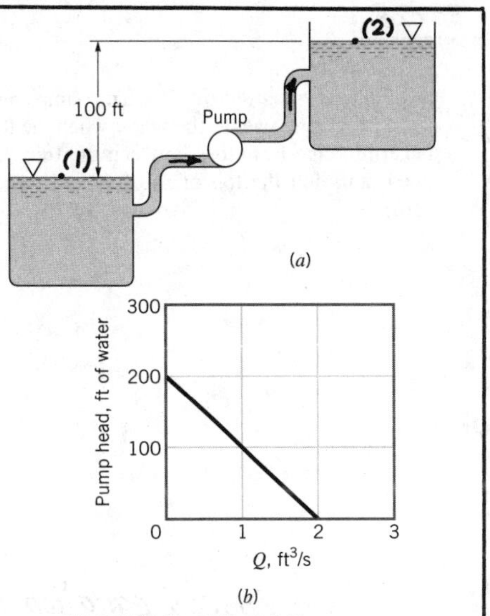

■ FIGURE P5.21R

For the flow from section (1) to section (2), Eq. 5.84 leads to

$$h_p = z_2 - z_1 + h_L \quad \text{since } V_1 = V_2 = 0 \qquad (1)$$

From Fig. P5.117b we conclude that

$$h_p = 200 - 100\,Q \qquad (2)$$

From the problem statement

$$h_L = K_L \frac{\overline{V}^2}{2g}$$

or since

$$\overline{V} = \frac{Q}{A} = \frac{Q}{\frac{\pi D^2}{4}}$$

we have

$$h_L = \frac{K_L Q^2}{2g\left(\frac{\pi D^2}{4}\right)^2} = \frac{(40)\left(Q\ \frac{ft^3}{s}\right)^2}{(2)\left(32.2\ \frac{ft}{s^2}\right)\left[\frac{\pi\,(4\ in.)^2}{\left(12\ \frac{in.}{ft}\right)^2(4)}\right]^2} = 81.6\,Q^2\ ft \qquad (3)$$

Combining Eqs. 1, 2 and 3 we obtain

$$81.6\,Q^2 + 100\,Q - 100 = 0 \qquad (4)$$

The root of Eq. 4 that makes physical sense (i.e. $Q>0$) is

$$Q = 0.653\ \frac{ft^3}{s}$$

5.22R

5.22R (Energy equation) The pump shown in Fig. P5.22R adds 1.6 horsepower to the water when the flowrate is 0.6 ft³/s. Determine the head loss bewteen the free surface in the large, open tank and the top of the fountain (where the velocity is zero).

(ANS: 7.50 ft)

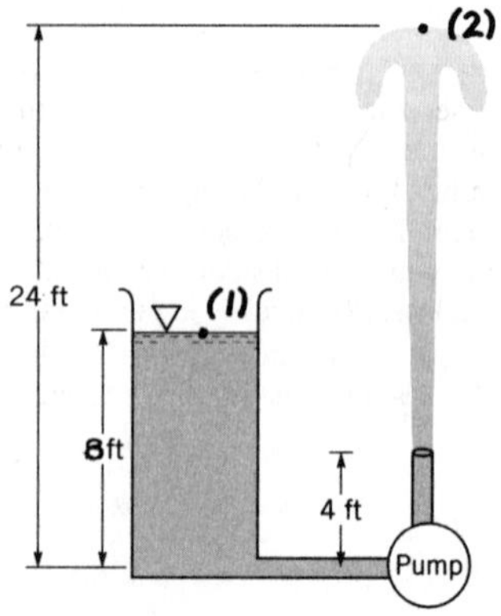

■ FIGURE P5.22R

The energy equation for this flow can be written as

$$\frac{p_1}{\gamma}+\frac{V_1^2}{2g}+z_1+h_s=\frac{p_2}{\gamma}+\frac{V_2^2}{2g}+z_2+h_L \qquad (1)$$

where

$$p_1=p_2=V_1=V_2=0$$

Also,

$$h_s=\frac{\dot{W}_s}{\gamma Q}=\frac{1.6\,hp\left(550\ \frac{ft\cdot lb}{s}/hp\right)}{\left(62.4\ \frac{lb}{ft^3}\right)\left(0.6\ \frac{ft^3}{s}\right)}=23.5\ ft$$

Thus, Eq. (1) becomes

$$h_L=z_1+h_s-z_2=8\ ft+23.5\ ft-24\ ft=\underline{\underline{7.50\ ft}}$$

Some of this head loss may occur in the pipe and some in the water jet as it interacts with the surrounding air.

6 Differential Analysis of Fluid Flow

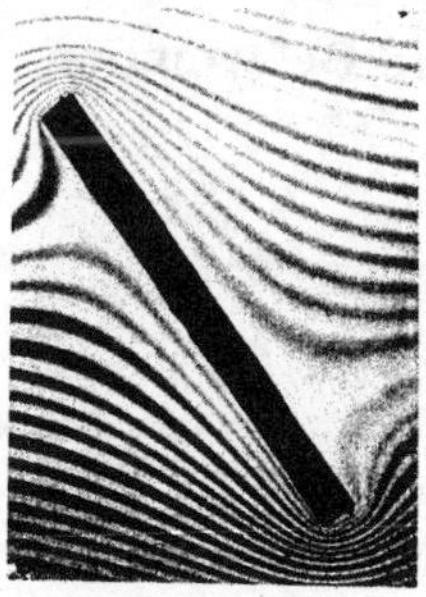

6.1 Fluid Element Kinematics

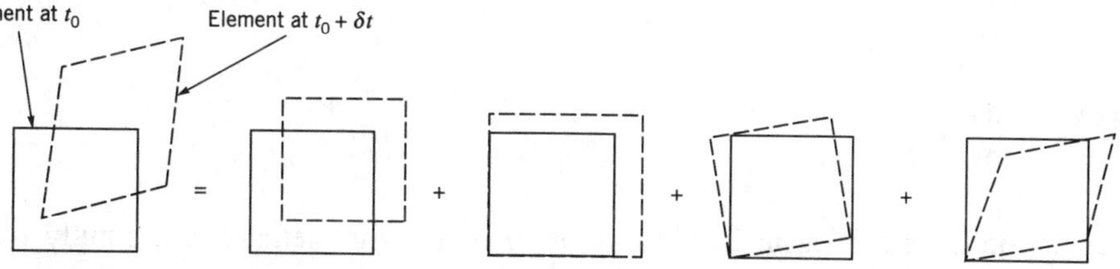

■ **FIGURE 6.1** **Types of motion and deformation for a fluid element.**

$$\mathbf{V} = u\hat{\mathbf{i}} + v\hat{\mathbf{j}} + w\hat{\mathbf{k}} \qquad \textbf{(6.1)}$$

$$\mathbf{a} = \frac{\partial \mathbf{V}}{\partial t} + u\frac{\partial \mathbf{V}}{\partial x} + v\frac{\partial \mathbf{V}}{\partial y} + w\frac{\partial \mathbf{V}}{\partial z} \qquad \textbf{(6.2)}$$

- As a fluid element moves in a flow field, it can undergo translation, linear deformation, rotation, and angular deformation as illustrated for a fluid element in the shape of a cube in Fig. 6.1.
- The velocity field can be described by specifying the velocity, **V**, at all points and at all times in the flow field of interest.
- In terms of rectangular components, the velocity, **V**, and acceleration, **a**, can be expressed as in Eqs. 6.1 and 6.2, respectively.

Linear motion and Deformation

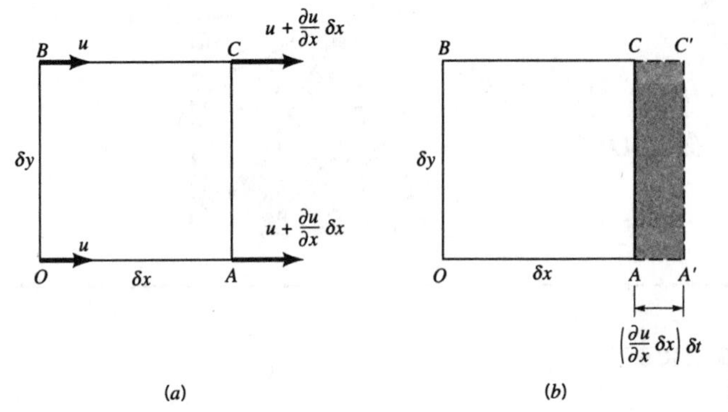

■ **FIGURE 6.3**
Linear deformation of a fluid element.

$$\frac{1}{\delta \forall}\frac{d(\delta \forall)}{dt} = \frac{\partial u}{\partial x} + \frac{\partial v}{\partial y} + \frac{\partial w}{\partial z} = \nabla \cdot \mathbf{V} \qquad \textbf{(6.9)}$$

- If all points on a fluid element have the same velocity, the element will simply *translate* from one position to another without deformation.
- If velocity gradients are present, the fluid element may be *linearly* deformed as illustrated in Fig. 6.3. The rate of change of the element volume per unit volume is called the *volumetric dilatation rate* and given by Eq. 6.9.
- For an incompressible fluid the change in the volume of the element is zero so that $\nabla \cdot \mathbf{V} = 0$.

Angular Motion and Deformation

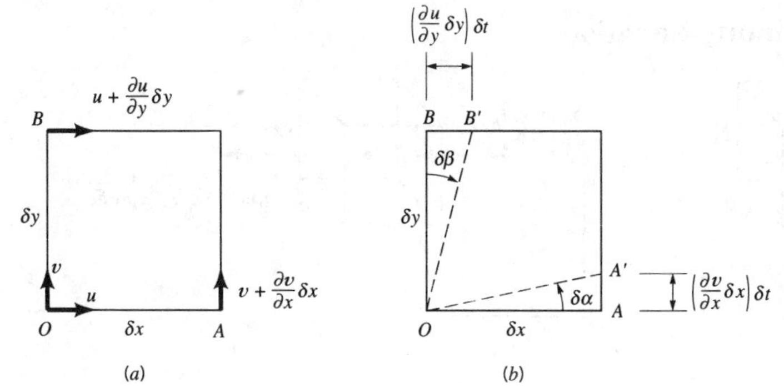

■ **FIGURE 6.4**
Angular motion and deformation of a fluid element.

$$\omega_z = \frac{1}{2}\left(\frac{\partial v}{\partial x} - \frac{\partial u}{\partial y}\right) \qquad \textbf{(6.12)}$$

$$\boldsymbol{\zeta} = 2\,\boldsymbol{\omega} = \nabla \times \mathbf{V} \qquad \textbf{(6.17)}$$

$$\dot{\gamma} = \frac{\partial v}{\partial x} + \frac{\partial u}{\partial y} \qquad \textbf{(6.18)}$$

- As shown in Fig. 6.4, if velocity gradients are present in a flow field, a fluid element will generally have *angular* motion and deformation.
- For a two-dimensional flow field (where the flow is in the *x-y* plane) the velocity components u and v are not functions of z and the velocity component w is zero. In this case the *rotation* of a fluid element is given by Eq. 6.12.
- The *vorticity*, $\boldsymbol{\zeta}$, of a fluid element is given by Eq. 6.17 and represents a measure of the rotation of the element.
- Flow fields for which the vorticity is zero are called *irrotational* flow fields.
- The *rate of shearing strain*, $\dot{\gamma}$, (also called the rate of angular deformation) is given by Eq. 6.18 for a two-dimensional flow field and is related to the shearing stress which causes the fluid element to change shape.

EXAMPLE: The velocity components for a certain two-dimensional flow field are $u = 3x^2$ and $v = -6xy$. Determine the vorticity. Is this an irrotational flow field?

SOLUTION:

For two-dimensional flow $\nabla \times \vec{V} = \left(\frac{\partial v}{\partial x} - \frac{\partial u}{\partial y}\right)\hat{k}$ so that with $\frac{\partial v}{\partial x} = -6y$ and $\frac{\partial u}{\partial y} = 0$ it follows that $\nabla \times \vec{V} = \text{vorticity} = -6y\,\hat{k}$.
Since the vorticity is not zero, this flow is not irrotational.

6.2 Conservation of Mass

Differential Form of Continuity Equation

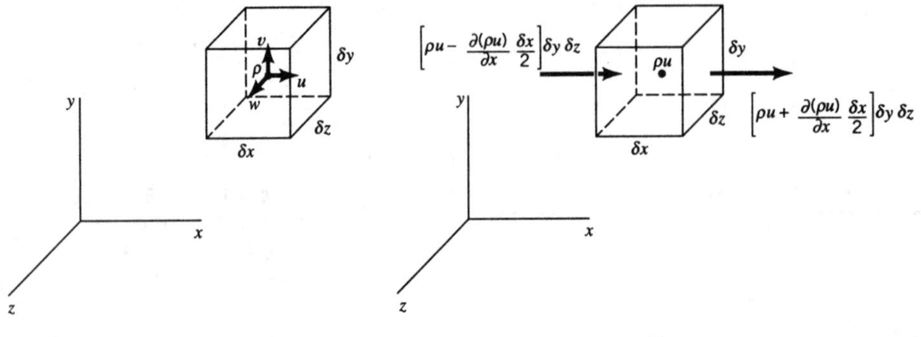

■ FIGURE 6.5 **A differential element for the development of conservation of mass equation.**

$$\frac{\partial \rho}{\partial t} + \frac{\partial(\rho u)}{\partial x} + \frac{\partial(\rho v)}{\partial y} + \frac{\partial(\rho w)}{\partial z} = 0 \qquad \textbf{(6.27)}$$

$$\frac{\partial u}{\partial x} + \frac{\partial v}{\partial y} + \frac{\partial w}{\partial z} = 0 \qquad \textbf{(6.31)}$$

- Conservation of mass applied to the small, stationary cubical element of Fig. 6.5 yields Eq. 6.27 which is the general differential equation for the conservation of mass. This equation is commonly referred to as the *continuity equation.*
- For an incompressible fluid the fluid density is constant throughout the flow field, and the continuity equation can be expressed as Eq. 6.31.
- *Caution:* Equation 6.31 is valid for both steady *and* unsteady incompressible flows.

EXAMPLE: It is suggested that the velocity components for a certain incompressible flow field are given by $u = 2xy$, $v = y^2$, and $w = -4yz$. Does this flow field satisfy conservation of mass?

SOLUTION:

To satisfy conservation of mass for incompressible flow, $\frac{\partial u}{\partial x} + \frac{\partial v}{\partial y} + \frac{\partial w}{\partial z} = 0$. Thus, with the given velocity field

$\frac{\partial u}{\partial x} = 2y$, $\frac{\partial v}{\partial y} = 2y$, and $\frac{\partial w}{\partial z} = -4y$ so that

$\frac{\partial u}{\partial x} + \frac{\partial v}{\partial y} + \frac{\partial w}{\partial z} = 2y + 2y - 4y \equiv 0$. Yes, conservation of mass is satisfied.

Cylindrical Polar Coordinates

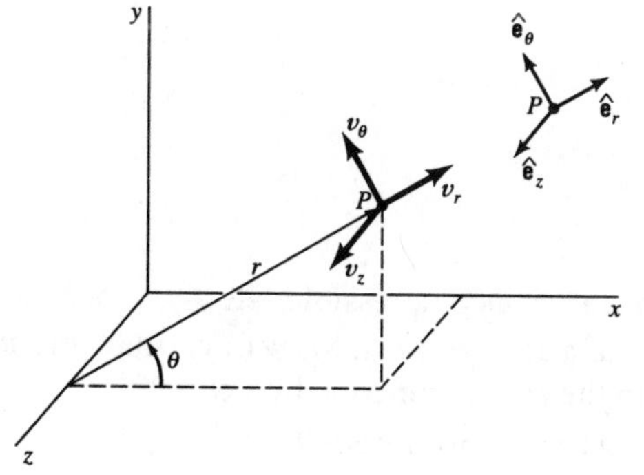

■ FIGURE 6.6 **The representation of velocity components in cylindrical polar coordinates.**

$$\mathbf{V} = v_r\hat{\mathbf{e}}_r + v_\theta\hat{\mathbf{e}}_\theta + v_z\hat{\mathbf{e}}_z \qquad (6.32)$$

$$\frac{\partial \rho}{\partial t} + \frac{1}{r}\frac{\partial(r\rho v_r)}{\partial r} + \frac{1}{r}\frac{\partial(\rho v_\theta)}{\partial \theta} + \frac{\partial(\rho v_z)}{\partial z} = 0 \qquad (6.33)$$

$$\frac{1}{r}\frac{\partial(r v_r)}{\partial r} + \frac{1}{r}\frac{\partial v_\theta}{\partial \theta} + \frac{\partial v_z}{\partial z} = 0 \qquad (6.35)$$

- For some problems it is convenient to express the velocity components in cylindrical polar coordinates as shown in Fig.6.6 and Eq. 6.32.
- The differential form of the continuity equation (conservation of mass) in cylindrical coordinates is given by Eq. 6.33.
- For incompressible fluids, Eq. 6.33 reduces to Eq. 6.35.

The Stream Function

$$u = \frac{\partial \psi}{\partial y} \qquad v = -\frac{\partial \psi}{\partial x} \qquad \textbf{(6.37)}$$

$$v_r = \frac{1}{r}\frac{\partial \psi}{\partial \theta} \qquad v_\theta = -\frac{\partial \psi}{\partial r} \qquad \textbf{(6.42)}$$

- As shown in Eq. 6.37, the velocity components in an incompressible, steady, plane, two-dimensional flow field can be expressed in terms of a *stream function*, $\psi\ (x,y)$. In cylindrical coordinates the velocity components are related to the stream function by Eq. 6.42.
- Lines along which the stream function has a constant value are *streamlines.*
- The difference in the value of the stream function from one streamline to another is related to the volume rate of flow between those streamlines.
- *Caution:* The concept of the stream function is not applicable to general three-dimensional flows.

EXAMPLE: The stream function for a certain incompressible flow field is given by the equation $\psi = x^2y^2$. Determine the corresponding velocity components, u and v, and show that the flow field represented by this stream function satisfies the continuity equation.

SOLUTION:

From Eq. 6.37, $u = \frac{\partial \psi}{\partial y} = 2x^2y$ and $v = -\frac{\partial \psi}{\partial x} = -2xy^2$

Since $\frac{\partial u}{\partial x} + \frac{\partial v}{\partial y} = 4xy - 4xy = 0$ it follows that the continuity equation is satisfied.

6.3 Conservation of Linear Momentum

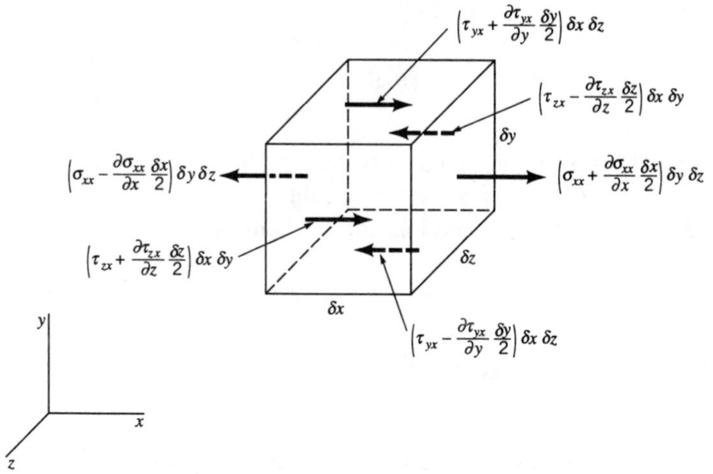

■ **FIGURE 6.11**
Surface forces in the *x* direction acting on a fluid element.

$$\rho g_x + \frac{\partial \sigma_{xx}}{\partial x} + \frac{\partial \tau_{yx}}{\partial y} + \frac{\partial \tau_{zx}}{\partial z} = \rho\left(\frac{\partial u}{\partial t} + u\frac{\partial u}{\partial x} + v\frac{\partial u}{\partial y} + w\frac{\partial u}{\partial z}\right) \quad \textbf{(6.50a)}$$

$$\rho g_y + \frac{\partial \tau_{xy}}{\partial x} + \frac{\partial \sigma_{yy}}{\partial y} + \frac{\partial \tau_{zy}}{\partial z} = \rho\left(\frac{\partial v}{\partial t} + u\frac{\partial v}{\partial x} + v\frac{\partial v}{\partial y} + w\frac{\partial v}{\partial z}\right) \quad \textbf{(6.50b)}$$

$$\rho g_z + \frac{\partial \tau_{xz}}{\partial x} + \frac{\partial \tau_{yz}}{\partial y} + \frac{\partial \sigma_{zz}}{\partial z} = \rho\left(\frac{\partial w}{\partial t} + u\frac{\partial w}{\partial x} + v\frac{\partial w}{\partial y} + w\frac{\partial w}{\partial z}\right) \quad \textbf{(6.50c)}$$

- Equations 6.50 are the general equations of motion for a fluid and are derived from Newton's second law of motion.
- The resultant force on a fluid element is due to both surface forces and body forces.
- As shown in Fig. 6.11, surface forces are described in terms of shearing stresses and normal stresses acting on the surface of the fluid element.
- Typically, the only body force of interest is the weight of the element.
- Before the equations of motion can be used to solve specific problems, additional information about the stresses must be known.

6.4 Inviscid Flow

Euler's Equations of Motion

$$\rho \mathbf{g} - \nabla p = \rho \left[\frac{\partial \mathbf{V}}{\partial t} + (\mathbf{V} \cdot \nabla)\mathbf{V} \right] \qquad \textbf{(6.52)}$$

- Flow fields in which shearing stresses are assumed to be negligible are said to be *inviscid, nonviscous, or frictionless.* These terms are used interchangeably.
- For an inviscid flow field, Euler's equations of motion, Eq. 6.52, apply.

The Bernoulli Equation

$$\frac{p_1}{\gamma} + \frac{V_1^2}{2g} + z_1 = \frac{p_2}{\gamma} + \frac{V_2^2}{2g} + z_2 \qquad \textbf{(6.58)}$$

- The Bernoulli equation, Eq. 6.58, is derived by integrating Eq. 6.52 along a streamline.
- The Bernoulli equation, as expressed by Eq. 6.58, is restricted to the following conditions: inviscid flow, steady flow, incompressible flow, and flow along a streamline.
- For irrotational flow (i.e., zero vorticity throughout), the Bernoulli equation can be applied between *any* two points in the flow field.

The Velocity Potential

$$u = \frac{\partial \phi}{\partial x} \qquad v = \frac{\partial \phi}{\partial y} \qquad w = \frac{\partial \phi}{\partial z} \qquad \textbf{(6.64)}$$

$$\nabla^2 \phi = 0 \qquad \textbf{(6.66)}$$

$$v_r = \frac{\partial \phi}{\partial r} \qquad v_\theta = \frac{1}{r}\frac{\partial \phi}{\partial \theta} \qquad v_z = \frac{\partial \phi}{\partial z} \qquad \textbf{(6.70)}$$

- For irrotational flow fields, the velocity components can be expressed in terms of a scalar function, $\phi(x, y, z, t)$, as shown in Eq. 6.64. The function ϕ is called the *velocity potential.*
- For inviscid, incompressible, irrotational flow fields, the velocity potential must satisfy Laplace's equation (Eq. 6.66). This type of flow is commonly called a *potential flow.*
- In cylindrical polar coordinates, the velocity components can be expressed in terms of the velocity potential as shown in Eq. 6.70.

EXAMPLE: The velocity potential for a certain flow field is $\phi = 2x^3 - 6xy^2$. Determine the corresponding velocity components, u and v, and show that this flow field is irrotational.

SOLUTION:

From Eq. 6.64, $u = \frac{\partial \phi}{\partial x} = \underline{6x^2 - 6y^2}$ and $v = \frac{\partial \phi}{\partial y} = \underline{-12xy}$

Since $\frac{\partial u}{\partial y} - \frac{\partial v}{\partial x} = -12y - (-12y) \equiv 0$ it follows that the flow is irrotational.

6.5 Some Basic, Plane Potential Flows

- For irrotational flow, the stream function must satisfy Laplace's equation.
- Lines along which the velocity potential, ϕ, is constant (called *equipotential* lines) are orthogonal to lines along which the stream function, ψ, is constant (*streamlines*). This network of intersecting lines is called a *flow net* and is useful in visualizing flow patterns.
- Basic velocity potentials and stream functions for plane potential flows include those for uniform flow, source and sink, vortex, and a doublet. (See Table 6.1 on the next page.)
- For steady potential flows the Bernoulli equation can be used to determine the pressure.

EXAMPLE: Water from a pipe flows through the gap between two circular disks as shown in the figure. The flow in the gap is approximated by the flow from a line sink located along the axis of the disks at $r = 0$. Assume that, the velocity potential for this flow is $\phi = -10 \ln r$, where r is the radial distance from the center of the disks and ϕ is in ft^2/s. Determine the pressure difference between point A at $r_A = 10$ ft and point B at $r_B = 2$ ft.

SOLUTION:

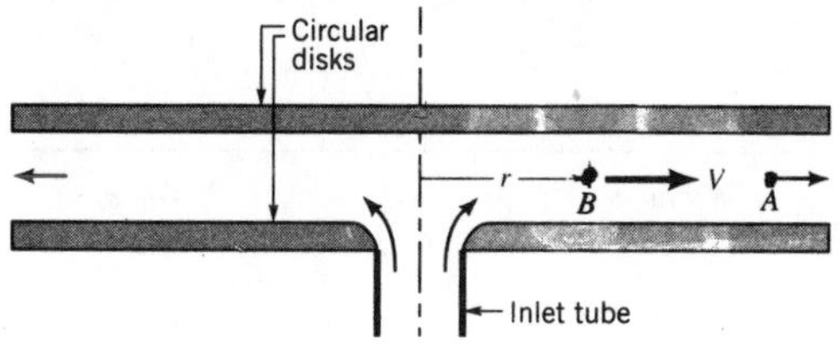

From the Bernoulli equation, $p_A + \gamma z_A + \frac{1}{2}\rho V_A^2 = p_B + \gamma z_B + \frac{1}{2}\rho V_B^2$.

Thus, with $z_A = z_B$, $p_A - p_B = \frac{1}{2}\rho(V_B^2 - V_A^2)$.

With $\phi = -10 \ln r$ it follows that $v_r = \frac{\partial \phi}{\partial r} = -10/r$ and $v_\theta = \frac{1}{r}\frac{\partial \phi}{\partial \theta} = 0$.

Hence, $V_A = -10/r_A = -(10\,ft^2/s)/10\,ft = -1\,ft/s$ and

$V_B = -10/r_B = -(10\,ft^2/s)/2\,ft = -5\,ft/s$ so that

$p_A - p_B = \frac{1}{2}(1.94\,slugs/ft^3)[(-5\,ft/s)^2 - (-1\,ft/s)^2] = \underline{\underline{23.3\,lb/ft^2}}$

EXAMPLE: A magnetic stirrer causes a liquid to rotate in a beaker. The resulting motion can be approximated as a free vortex centered along the vertical axis of the beaker with a stream function $\psi = K \ln r$. If the tangential velocity of the liquid is 1.6 m/s at $r = 75$ mm, determine the tangential velocity at $r = 25$ mm.

SOLUTION:

For a free vortex, $\psi = -K \ln r$ and $v_\theta = -\frac{\partial \psi}{\partial r} = \frac{K}{r}$, $v_r = \frac{1}{r}\frac{\partial \psi}{\partial \theta} = 0$

Thus, with $v_\theta = 1.6\,m/s$ at $r = 0.075\,m$ it follows that

$K = r v_\theta = 0.075\,m\,(1.6\,m/s) = 0.12\,m^2/s$. Thus, at $r = 0.025\,m$

$v_\theta = K/r = (0.12\,m^2/s)/0.025\,m = \underline{\underline{4.80\,m/s}}$

■ **TABLE 6.1**

Summary of Basic, Plane Potential Flows.

Description of Flow Field	Velocity Potential	Stream Function	Velocity Components[a]
Uniform flow at angle α with the x axis (see Fig. 6.16*b*)	$\phi = U(x \cos \alpha + y \sin \alpha)$	$\psi = U(y \cos \alpha - x \sin \alpha)$	$u = U \cos \alpha$ $v = U \sin \alpha$
Source or sink (see Fig. 6.17) $m > 0$ source $m < 0$ sink	$\phi = \frac{m}{2\pi} \ln r$	$\psi = \frac{m}{2\pi} \theta$	$v_r = \frac{m}{2\pi r}$ $v_\theta = 0$
Free vortex (see Fig. 6.18) $\Gamma > 0$ counterclockwise motion $\Gamma < 0$ clockwise motion	$\phi = \frac{\Gamma}{2\pi} \theta$	$\psi = -\frac{\Gamma}{2\pi} \ln r$	$v_r = 0$ $v_\theta = \frac{\Gamma}{2\pi r}$
Doublet (see Fig. 6.23)	$\phi = \frac{K \cos \theta}{r}$	$\psi = -\frac{K \sin \theta}{r}$	$v_r = -\frac{K \cos \theta}{r^2}$ $v_\theta = -\frac{K \sin \theta}{r^2}$

[a]Velocity components are related to the velocity potential and stream function through the relationships:

$$u = \frac{\partial \phi}{\partial x} = \frac{\partial \psi}{\partial y} \qquad v = \frac{\partial \phi}{\partial y} = -\frac{\partial \psi}{\partial x} \qquad v_r = \frac{\partial \phi}{\partial r} = \frac{1}{r}\frac{\partial \psi}{\partial \theta} \qquad v_\theta = \frac{1}{r}\frac{\partial \phi}{\partial \theta} = -\frac{\partial \psi}{\partial r}$$

6.6 Superposition of Basic, Plane Potential Flows

Source in a Uniform Stream – Half-Body

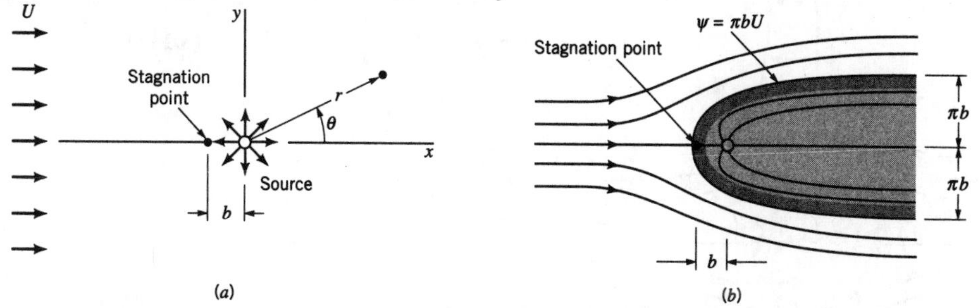

■ **FIGURE 6.24 The flow around a half-body: (*a*) superposition of a source and a uniform flow; (*b*) replacement of streamline $\psi = \pi bU$ with solid boundary to form half-body.**

$$\psi = \psi_{\text{uniform flow}} + \psi_{\text{source}}$$

$$= Ur \sin\theta + \frac{m}{2\pi}\theta \qquad \textbf{(6.97)}$$

$$b = \frac{m}{2\pi U} \qquad \textbf{(6.99)}$$

- For potential flow, basic velocity potentials or stream functions representing simple flow patterns can be added to obtain more complicated flow patterns. This method of solution is termed the *method of superposition*.
- The superposition of the stream function for a source and for a uniform flow, as shown in Eq. 6.97, provides a solution for flow around a *half-body* as illustrated in Fig. 6.24.
- The relationship between the source strength and the location of the stagnation point is given by Eq. 6.99.
- Since the flow is inviscid, incompressible, and irrotational, the pressure distribution throughout the flow field can be obtained from the Bernoulli equation.
- *Caution:* Potential flow solutions allow the fluid to "slip" at solid boundaries (since viscosity is neglected), and, therefore, do not accurately represent the velocity very near a solid boundary.

EXAMPLE: A certain body has the shape of a half-body (Fig. 6.24) with a thickness that approaches 4 ft far from the nose of the body. If this body is placed in a uniform stream of fluid having a velocity U = 20 ft/s, determine the fluid velocity at a point ahead of the body at x = – 3 ft, y = 0.

SOLUTION:

From Eq. 6.97, $\psi = Ur\sin\theta + \frac{m}{2\pi}\theta$ so that

$v_r = \frac{1}{r}\partial\psi/\partial\theta = U\cos\theta + \frac{m}{2\pi}/r$ and $v_\theta = -\frac{\partial\psi}{\partial r} = -U\sin\theta$.

From Eq. 6.99, $m = 2\pi bU$, Since the half-body thickness is $2\pi b$, it follows that $m = (4\text{ft})U$. Thus, at $x = -3$ ft ($r = 3$ ft) and $y = 0$ ($\theta = \pi$),

$v_r = U\cos\pi + (4\text{ft})U/[2\pi(3\text{ft})] = -0.788U = -0.788(20\text{ ft/s}) = \underline{\underline{-15.8\text{ ft/s}}}$

and $v_\theta = -U\sin\pi = 0$. Note: $v_r < 0$ indicates that the velocity is directed toward the nose of the half-body.

Flow Around a Circular Cylinder

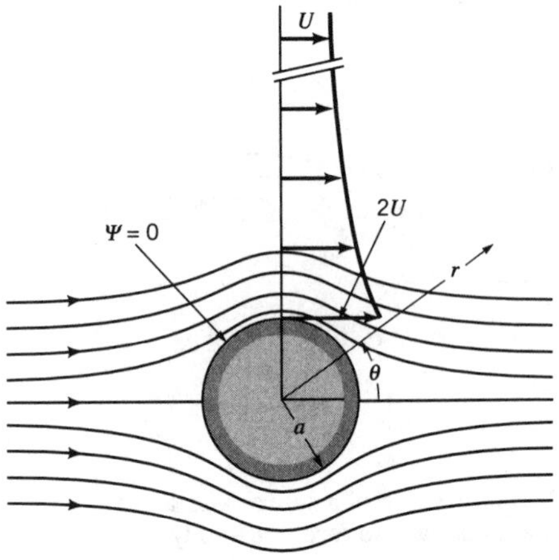

$$\psi = Ur\left(1 - \frac{a^2}{r^2}\right)\sin\theta \qquad \textbf{(6.112)}$$

■ **FIGURE 6.26 The flow around a circular cylinder.**

- Superposition of a stream function for a doublet and for a uniform flow, as shown in Eq. 6.112, provides a solution for flow around a *circular cylinder* as illustrated in Fig. 6.26.
- As shown in Fig. 6.26, the maximum velocity occurs at the top (and bottom) of the cylinder.
- The resultant fluid force acting on the cylinder, determined by integrating the pressure over the surface of the cylinder, is zero.
- *Caution:* Experience shows that for real fluids (which have viscosity), the resultant force on a circular cylinder is not zero.
- Potential flow around a rotating circular cylinder in a uniform stream can be approximated by adding a free vortex to the stream function (or velocity potential) for flow around a cylinder.
- *Caution:* Even though potential flow solutions may be "exact" mathematically, they remain approximate physically because of the fundamental assumption of a frictionless fluid.

EXAMPLE: Water flows past a circular cylinder, as illustrated in Fig. 6.26. The upstream velocity is U = 3 m/s and the cylinder radius is a = 0.5 m. Determine the difference in pressure between the stagnation point and point A located at $\theta = \pi/2$, r = 1 m. Assume frictionless flow and neglect elevation changes.

SOLUTION:

The Bernoulli equation written between the stagnation point, where the velocity is zero, and point A is (neglecting elevation changes)

$$p_{stag} = p_A + \frac{1}{2}\rho V_A^2$$

From Eq. 6.112, $\psi = Ur\left(1-\frac{a^2}{r^2}\right)\sin\theta$ so that

$$v_r = \frac{1}{r}\frac{\partial\psi}{\partial\theta} = U\left(1-\frac{a^2}{r^2}\right)\cos\theta \text{ and } v_\theta = -\frac{\partial\psi}{\partial r} = -U\left(1+\frac{a^2}{r^2}\right)\sin\theta$$

Thus, at point A where $r=1$ and $\theta=\pi/2$,

$$v_r = 0 \text{ and } v_\theta = -3\,m/s\left[1+(0.5\,m/1\,m)^2\right] = -3.75\,m/s, \text{ or } V_A = 3.75\,m/s.$$

Hence, $p_{stag} - p_A = \frac{1}{2}\rho V_A^2 = \frac{1}{2}(999\,kg/m^3)(3.75\,m/s)^2 = \underline{7.02\,kPa}$

6.8 Viscous Flow

$$\sigma_{xx} = -p + 2\mu \frac{\partial u}{\partial x} \quad \textbf{(6.125a)} \qquad \tau_{xy} = \tau_{yx} = \mu \left(\frac{\partial u}{\partial y} + \frac{\partial v}{\partial x} \right) \quad \textbf{(6.125d)}$$

$$\sigma_{yy} = -p + 2\mu \frac{\partial v}{\partial y} \quad \textbf{(6.125b)} \qquad \tau_{yz} = \tau_{zy} = \mu \left(\frac{\partial v}{\partial z} + \frac{\partial w}{\partial y} \right) \quad \textbf{(6.125e)}$$

$$\sigma_{zz} = -p + 2\mu \frac{\partial w}{\partial z} \quad \textbf{(6.125c)} \qquad \tau_{zx} = \tau_{xz} = \mu \left(\frac{\partial w}{\partial x} + \frac{\partial u}{\partial z} \right) \quad \textbf{(6.125f)}$$

(x direction)

$$\rho \left(\frac{\partial u}{\partial t} + u \frac{\partial u}{\partial x} + v \frac{\partial u}{\partial y} + w \frac{\partial u}{\partial z} \right) = -\frac{\partial p}{\partial x} + \rho g_x + \mu \left(\frac{\partial^2 u}{\partial x^2} + \frac{\partial^2 u}{\partial y^2} + \frac{\partial^2 u}{\partial z^2} \right) \quad \textbf{(6.127a)}$$

(y direction)

$$\rho \left(\frac{\partial v}{\partial t} + u \frac{\partial v}{\partial x} + v \frac{\partial v}{\partial y} + w \frac{\partial v}{\partial z} \right) = -\frac{\partial p}{\partial y} + \rho g_y + \mu \left(\frac{\partial^2 v}{\partial x^2} + \frac{\partial^2 v}{\partial y^2} + \frac{\partial^2 v}{\partial z^2} \right) \quad \textbf{(6.127b)}$$

(z direction)

$$\rho \left(\frac{\partial w}{\partial t} + u \frac{\partial w}{\partial x} + v \frac{\partial w}{\partial y} + w \frac{\partial w}{\partial z} \right) = -\frac{\partial p}{\partial z} + \rho g_z + \mu \left(\frac{\partial^2 w}{\partial x^2} + \frac{\partial^2 w}{\partial y^2} + \frac{\partial^2 w}{\partial z^2} \right) \quad \textbf{(6.127c)}$$

- To incorporate viscous effects into the differential analysis of fluid motion, it is necessary to establish a relationship between the viscous stresses and the velocities.
- For incompressible, *Newtonian fluids*, stresses are *linearly* related to the rates of deformation as shown in Eqs. 6.125.
- With a knowledge of the stress-velocity (gradients) relationship, the general equations of motion can be formulated. These differential equations, Eqs. 6.127, are called the *Navier-Stokes* equations.
- The Navier-Stokes equations, when combined with the continuity equation, provide a complete mathematical description of the flow of incompressible, Newtonian fluids.
- *Caution:* Due to the general complexity of the Navier-Stokes equations, they are not amenable to exact mathematical solutions except in a few instances.

6.9 Some Simple Solutions for Viscous, Incompressible Fluids

Steady, Laminar Flow Between Fixed Parallel Plates

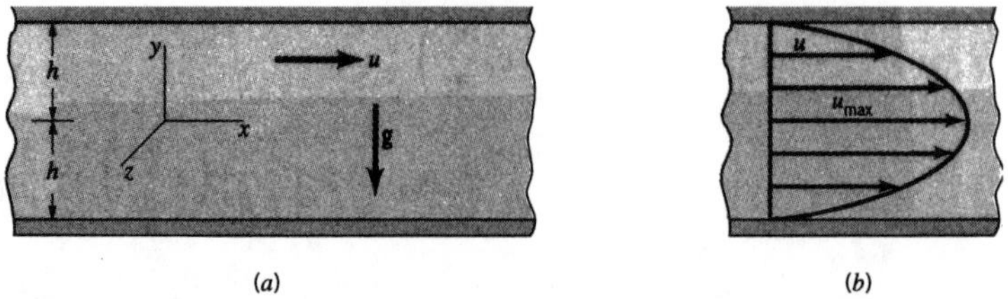

■ FIGURE 6.30 **The viscous flow between parallel plates: (*a*) coordinate system and notation used in analysis; (*b*) parabolic velocity distribution for flow between parallel fixed plates.**

$$u = \frac{1}{2\mu}\left(\frac{\partial p}{\partial x}\right)(y^2 - h^2) \qquad \textbf{(6.134)}$$

$$q = \frac{2h^3\,\Delta p}{3\mu\ell} \qquad \textbf{(6.136)}$$

- As shown by Eq. 6.134 and Fig. 6.30, the velocity distribution for steady viscous flow between two fixed parallel plates is parabolic.
- The volume rate of flow passing between the parallel plates per unit width in the z-direction, q, is given by Eq. 6.136, where Δp is the pressure drop between two points a distance ℓ apart.
- The maximum velocity occurs midway between the two plates and is equal to 3/2 times the mean velocity. That is, $u_{max} = 3V/2$.
- *Caution:* The simple, steady flow solutions to the Navier-stokes equations are only valid for *laminar* flow. It is known that as the Reynolds number, $\mathrm{Re} = \rho(2h)V/\mu$, increases beyond about 1400 the flow may become *turbulent*, with corresponding random fluctuations in velocity.

EXAMPLE: A viscous liquid (μ = 0.40 N·s/m^2) flows steadily between fixed parallel plates as shown in Fig. 6.30. Some measurements indicate that the pressure drop, Δp, is 2300 Pa over a 2-m length along the channel when the mean velocity is 0.75 m/s. Determine the spacing, $2h$, between the two plates. Assume laminar flow.

SOLUTION:

From Eq. 6.136, $q = 2h^3\Delta p/3\mu\ell$, where $q = V(2h)$. Thus, $V = h^2\Delta p/3\mu\ell$ so that

$h = [3\mu\ell V/\Delta p]^{1/2} = [3(0.40\,\mathrm{N\cdot s/m^2})(2\,\mathrm{m})(0.75\,\mathrm{m/s})/(2300\,\mathrm{N/m^2})]^{1/2}$

$= 0.0280\,\mathrm{m}$. Thus, spacing $= 2h = 0.0560\,\mathrm{m} = \underline{\underline{56.0\,\mathrm{mm}}}$.

Couette Flow

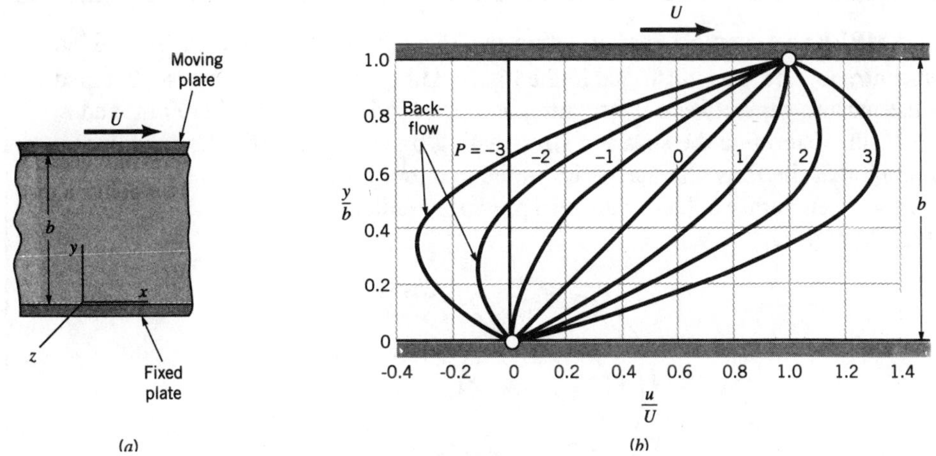

■ **FIGURE 6.31 The viscous flow between parallel plates with bottom plate fixed and upper plate moving (Couette flow): (*a*) coordinate system and notation used in analysis; (*b*) velocity distribution as a function of parameter, *P*, where $P = -(b^2/2\mu U)\partial p/\partial x$. (From Ref. 8, used by permission.)**

$$\frac{u}{U} = \frac{y}{b} - \frac{b^2}{2\mu U}\left(\frac{\partial p}{\partial x}\right)\left(\frac{y}{b}\right)\left(1 - \frac{y}{b}\right) \qquad \textbf{(6.141)}$$

$$u = U\frac{y}{b} \qquad \textbf{(6.142)}$$

- Flow between parallel plates with one plate fixed and the other plate moving, as illustrated in Fig. 6.31*a*, is called *Couette* flow.
- As shown by Eq. 6.141 and Fig. 6.31*b*, the velocity distribution for Couette flow depends on the pressure gradient parameter $P = -(b^2/2\mu U)\partial p/\partial x$.
- The simplest type of Couette flow is one for which the pressure gradient is zero, and the fluid motion is caused by the fluid being dragged along by the moving boundary. For this case, the velocity varies linearly between the plates as shown in Eq. 6.142.
- The velocity distribution in the narrow gap between closely spaced concentric cylinders, in which one cylinder is fixed and the other cylinder rotates with a constant angular velocity, can be approximated by Eq. 6.142.

EXAMPLE: A vertical shaft rotating at 70 rev/min is surrounded by a fixed concentric cylinder as illustrated in the figure. Oil ($\mu = 0.012$ lb·s/ft^2) is contained in the narrow gap between the rotating and fixed cylinders. If $r_i = 6.00$ in. and $r_o = 6.10$ in., determine the shearing stress acting on the wall of the fixed cylinder. Assume that the flow characteristics in the gap are the same as those for laminar flow between parallel plates with zero pressure gradient.

SOLUTION:

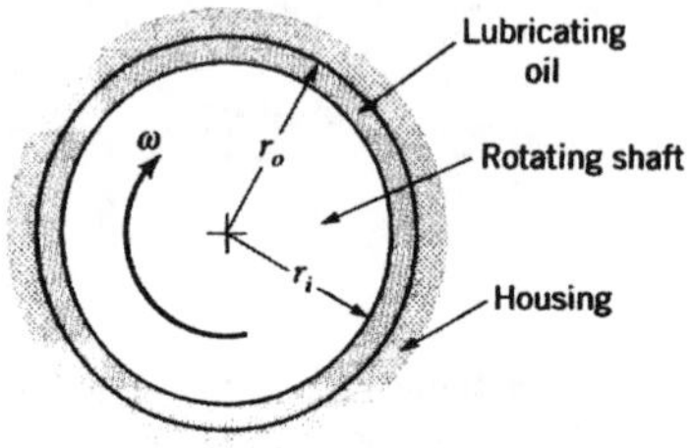

The wall shear stress is $\tau_{wall} = \mu(du/dy)_{wall}$, where $du/dy = r_i\omega/(r_o - r_i)$. Thus, $\tau_{wall} = \mu r_i \omega/(r_o - r_i)$ so that with $\omega = (70 \text{ rev/min})(2\pi \text{ rad/rev})(1 \text{ min}/60 \text{ s}) = 7.33$ rad/s,

$$\tau_{wall} = (0.012 \text{ lb·s/ft}^2)(6/12 \text{ ft})(7.33 \text{ rad/s})/(0.10/12 \text{ ft}) = \underline{\underline{5.28 \text{ lb/ft}^2}}$$

Steady, Laminar Flow in Circular Tubes

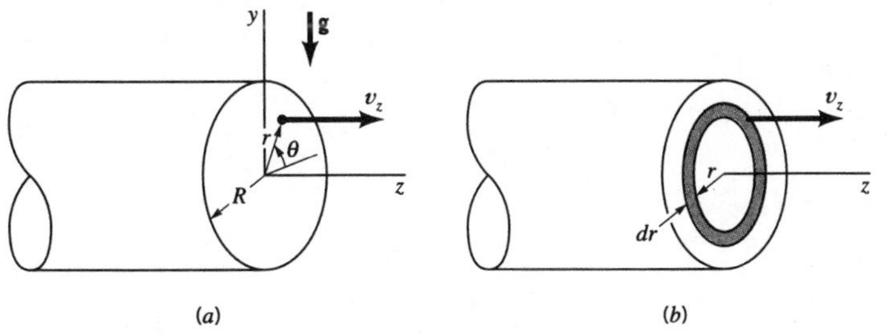

■ FIGURE 6.33 **The viscous flow in a horizontal, circular tube: (*a*) coordinate system and notation used in analysis; (*b*) flow through differential annular ring.**

$$v_z = \frac{1}{4\mu}\left(\frac{\partial p}{\partial z}\right)(r^2 - R^2) \qquad \textbf{(6.148)}$$

$$Q = \frac{\pi R^4 \, \Delta p}{8\mu \ell} \qquad \textbf{(6.151)}$$

- As shown by Eq. 6.148, the velocity distribution is parabolic for steady, laminar flow in circular tubes (Fig. 6.33).
- The volume rate of flow passing though the tube is given by Eq. 6.151, where Δp is the pressure drop which occurs over a length, ℓ, along the tube. Equation 6.151 is commonly called *Poiseuille's* law.
- The maximum velocity occurs at the center of the tube and is equal to twice the mean velocity (i.e., $v_{max} = 2\,V$).
- *Caution:* Flow in circular tubes and pipes remains laminar for Reynolds numbers, Re = $\rho(2R)V/\mu$, below 2100. For larger Reynolds numbers the flow may be turbulent and Poiseuille's law would not be correct.

EXAMPLE: Water flows with a mean velocity of 0.40 m/s through a horizontal 5-mm diameter tube. Determine: **(a)** the Reynolds number, and **(b)** the pressure drop along a 2-m length of the tube.

(a) $Re = \rho(2R)V/\mu = (999\ kg/m^3)(0.005\ m)(0.40\ m/s)/(1.12\times10^{-3}\ N\cdot s/m^2)$
$= \underline{1780}$. Since $Re < 2100$ the flow is laminar.

(b) From Eq. 6.151, $Q = \pi R^4 \Delta p/8\mu\ell$, where $Q = AV = \pi R^2 V$.
Thus, $V = R^2\Delta p/8\mu\ell$ so that
$\Delta p = 8\mu\ell V/R^2 = 8(1.12\times10^{-3}\ N\cdot s/m^2)(2m)(0.40\ m/s)/(0.005\ m/2)^2$
$= 1.15\times10^3\ N/m^2 = \underline{1.15\ kPa}$

Solutions

to Review Problems

Chapter 6

6.1R

6.1R (Acceleration) The velocity in a certain flow field is given by the equation

$$\mathbf{V} = 3yz^2\hat{\mathbf{i}} + xz\hat{\mathbf{j}} + y\hat{\mathbf{k}}$$

Determine the expressions for the three rectangular components of acceleration.

(ANS: $3xz^3 + 6y^2z$; $3yz^3 + xy$; xz)

From expression for velocity, $u = 3yz^2$, $v = xz$, and $w = y$.

Since

$$a_x = \frac{\partial u}{\partial t} + u\frac{\partial u}{\partial x} + v\frac{\partial u}{\partial y} + w\frac{\partial u}{\partial z}$$

then

$$a_x = 0 + (3yz^2)(0) + (xz)(3z^2) + (y)(6yz)$$

$$= \underline{\underline{3xz^3 + 6y^2z}}$$

Similarly,

$$a_y = \frac{\partial v}{\partial t} + u\frac{\partial v}{\partial x} + v\frac{\partial v}{\partial y} + w\frac{\partial v}{\partial z}$$

and

$$a_y = 0 + (3yz^2)(z) + (xz)(0) + (y)(x)$$

$$= \underline{\underline{3yz^3 + xy}}$$

Also,

$$a_z = \frac{\partial w}{\partial t} + u\frac{\partial w}{\partial x} + v\frac{\partial w}{\partial y} + w\frac{\partial w}{\partial z}$$

so that

$$a_z = 0 + (3yz^2)(0) + (xz)(1) + (y)(0)$$

$$= \underline{\underline{xz}}$$

6.2R

6.2R (Vorticity) Determine an expression for the vorticity of the flow field described by

$$\mathbf{V} = x^2y\hat{\mathbf{i}} - xy^2\hat{\mathbf{j}}$$

Is the flow irrotational?

(ANS: $-(x^2 + y^2)\,\hat{\mathbf{k}}$; no)

The vorticity is twice the rotation vector:

$$\vec{\zeta} = 2\vec{\omega} = \nabla \times \vec{V} \qquad \text{(Eq. 6.17)}$$

From expression for velocity, $u = x^2y$, $v = -xy^2$, and $w = 0$, and with

$$\omega_x = \frac{1}{2}\left(\frac{\partial w}{\partial y} - \frac{\partial v}{\partial z}\right) \qquad \text{(Eq. 6.13)}$$

$$\omega_y = \frac{1}{2}\left(\frac{\partial u}{\partial z} - \frac{\partial w}{\partial x}\right) \qquad \text{(Eq. 6.14)}$$

$$\omega_z = \frac{1}{2}\left(\frac{\partial v}{\partial x} - \frac{\partial u}{\partial y}\right) \qquad \text{(Eq. 6.12)}$$

it follows that

$$\omega_x = 0, \qquad \omega_y = 0, \quad \text{and} \quad \omega_z = \frac{1}{2}(-y^2 - x^2)$$

Thus,

$$\vec{\zeta} = 2(\omega_x\hat{i} + \omega_y\hat{j} + \omega_z\hat{k})$$

$$= 2\left[(0)\hat{i} + (0)\hat{j} + \frac{1}{2}(-y^2 - x^2)\hat{k}\right]$$

$$= \underline{\underline{-(x^2 + y^2)\,\hat{k}}}$$

Since $\vec{\zeta}$ is not zero everywhere, the flow is not irrotational. <u>No.</u>

6.3R

6.3R (Conservation of mass) For a certain incompressible, two-dimensional flow field the velocity component in the y direction is given by the equation

$$v = x^2 + 2xy$$

Determine the velocity component in the x direction so that the continuity equation is satisfied.
(ANS: $-x^2 + f(y)$)

To satisfy the continuity equation,

$$\frac{\partial u}{\partial x} + \frac{\partial v}{\partial y} = 0 \qquad (1)$$

Since

$$\frac{\partial v}{\partial y} = 2x$$

Then from Eq. (1)

$$\frac{\partial u}{\partial x} = -2x \qquad (2)$$

Equation (2) can be integrated with respect to x to obtain

$$\int du = -\int 2x\, dx + f(y)$$

or

$$u = \underline{\underline{-x^2 + f(y)}}$$

where $f(y)$ is an undetermined function of y.

6.4R

6.4R (Conservation of mass) For a certain incompressible flow field it is suggested that the velocity components are given by the equations

$$u = x^2y \qquad v = 4y^3z \qquad w = 2z$$

Is this a physically possible flow field? Explain.
(ANS: No)

Any physically possible incompressible flow field must satisfy conseration of mass as expressed by the relationship

$$\frac{\partial u}{\partial x} + \frac{\partial v}{\partial y} + \frac{\partial w}{\partial z} = 0 \qquad (1)$$

For the velocity distribution given,

$$\frac{\partial u}{\partial x} = 2xy, \quad \frac{\partial v}{\partial y} = 12y^2z, \text{ and } \frac{\partial w}{\partial z} = 2$$

Substitution into Eq. (1) shows that

$$2xy + 12y^2z + 2 \neq 0 \text{ for all } x, y, z.$$

Thus, this is not a physically possible flow field. No.

6.5R

6.5R (Stream function) The velocity potential for a certain flow field is

$$\phi = 4xy$$

Determine the corresponding stream function.
(ANS: $2(y^2 - x^2) + C$)

For the given velocity potential,

$$u = \frac{\partial \phi}{\partial x} = 4y \quad \text{and} \quad v = \frac{\partial \phi}{\partial y} = 4x$$

From the definition of the stream function,

$$u = \frac{\partial \psi}{\partial y} = 4y \qquad (1)$$

Integrate Eq. (1) with respect to y to obtain

$$\int d\psi = \int 4y\, dy$$

or

$$\psi = 2y^2 + f_1(x) \quad \text{where } f_1(x) \text{ is an arbitrary function of } x. \qquad (2)$$

Similarly,

$$v = -\frac{\partial \psi}{\partial x} = 4x$$

and

$$\int d\psi = -\int 4x\, dx$$

or

$$\psi = -2x^2 + f_2(y) \quad \text{where } f_2(y) \text{ is an arbitrary function of } y. \qquad (3)$$

To satisfy both Eqs. (2) and (3) $f_1(x) = f_2(y)$ for all x and y. Thus, $f_1 = f_2 =$ constant.

$$\underline{\underline{\psi = 2(y^2 - x^2) + C}}$$

Where C is a constant.

6.6R

6.6R (Velocity potential) A two-dimensional flow field is formed by adding a source at the origin of the coordinate system to the velocity potential

$$\phi = r^2 \cos 2\theta$$

Locate any stagnation points in the upper half of the coordinate plane $(0 \le \theta \le \pi)$.

(ANS: $\theta_s = \pi/2$; $r_s = (m/4\pi)^{1/2}$)

$$\phi = \frac{m}{2\pi}\ln r + r^2\cos 2\theta, \text{ where } \phi_{source} = \frac{m}{2\pi}\ln r$$

Thus,

$$v_\theta = \frac{1}{r}\frac{\partial \phi}{\partial \theta} = -2r\sin 2\theta$$

and

$$v_r = \frac{\partial \phi}{\partial r} = \frac{m}{2\pi r} + 2r\cos 2\theta$$

Stagnation points will occur where $v_\theta = 0$, $v_r = 0$, for $0 \le \theta \le \pi$.

Thus,

$$0 = -2r_s \sin 2\theta_s \qquad (1)$$

$$0 = \frac{m}{2\pi r_s} + 2r_s\cos 2\theta_s \qquad (2)$$

Equation (1) is satisfied at $r_s = 0$ or $\theta_s = 0, \frac{\pi}{2}, \pi$.

From Eq. (2)

$$\cos 2\theta_s = -\frac{m}{4\pi r_s^2} \qquad (3)$$

and for the possible values of θ_s, only $\theta_s = \frac{\pi}{2}$ will satisfy Eq. (3). Recall that $m > 0$ for a source.

Thus,

$$r_s = \sqrt{\frac{m}{4\pi}}$$

Thus, the stagnation point is located at

$$\underline{\underline{\theta_s = \frac{\pi}{2}}}, \quad \underline{\underline{r_s = \sqrt{\frac{m}{4\pi}}}}$$

6.7R

6.7R (Potential flow) The stream function for a two-dimensional, incompressible flow field is given by the equation

$$\psi = 2x - 2y$$

where the stream function has the units of ft^2/s with x and y in feet. **(a)** Sketch the streamlines for this flow field. Indicate the direction of flow along the streamlines. **(b)** Is this an irrotational flow field? **(c)** Determine the acceleration of a fluid particle at the point $x = 1$ ft, $y = 2$ ft.

(ANS: yes; no acceleration)

(a) Lines of constant ψ are streamlines. Thus, with $\psi = 2x - 2y$ the equation of a given streamline, ψ_1, (where ψ_1 is some constant) is of the form

$$\psi_1 = 2x - 2y$$

or

$$y = x - \frac{\psi_1}{2}$$

Thus, streamlines are straight lines as illustrated in the figure for three particular streamlines.

Since

$$u = \frac{\partial \psi}{\partial y} = -2 \qquad v = -\frac{\partial \psi}{\partial x} = -2$$

the direction of flow is as shown on the figure

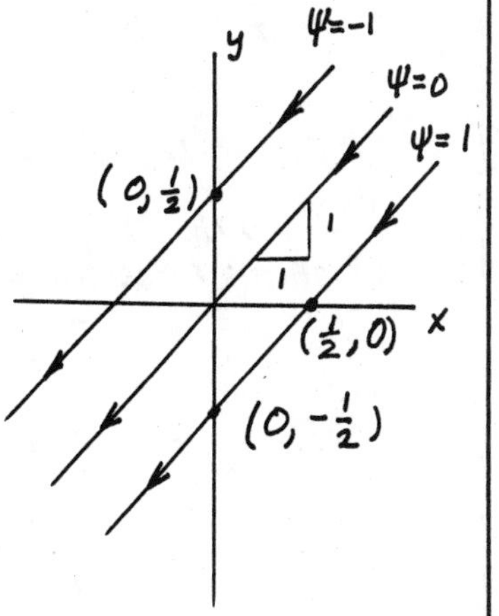

(b) The flow field is irrotational if $\omega_z = 0$ where

$$\omega_z = \frac{1}{2}\left(\frac{\partial v}{\partial x} - \frac{\partial u}{\partial y}\right) \qquad \text{(Eq. 6.12)}$$

For the stream function given

$$\frac{\partial v}{\partial x} = 0 \qquad \frac{\partial u}{\partial y} = 0$$

so that $\omega_z = 0$ and the flow field is irrotational. <u>Yes</u>.

(c) Since the velocity is constant throughout the flow field, the acceleration of <u>all</u> fluid particles is <u>zero</u>.

$$\vec{a} = \frac{\partial \vec{V}}{\partial t} + \vec{V}\cdot\nabla\vec{V} \equiv 0 \quad \text{since } \vec{V} = -2\hat{i} - 2\hat{j}$$

6.8R

6.8R (Inviscid flow) In a certain steady, incompressible, inviscid, two-dimensional flow field ($w = 0$, and all variables independent of z) the x component of velocity is given by the equation:

$$u = x^2 - y$$

Will the corresponding pressure gradient in the horizontal x direction be a function only of x, only of y, or of both x and y? Justify your answer.

(ANS: only of x)

Since the flow field must satisfy the continuity equation

$$\frac{\partial u}{\partial x} + \frac{\partial v}{\partial y} = 0$$

and with $u = x^2 - y$ it follows that

$$\frac{\partial v}{\partial y} = -\frac{\partial u}{\partial x} = -2x$$

and therefore

$$v = -2xy + f_1(x)$$

For steady, two-dimensional flow of an inviscid fluid (with the x-axis horizontal so that $g_x = 0$) the x-component of the momentum equation is

$$-\frac{\partial p}{\partial x} = \rho\left(u\frac{\partial u}{\partial x} + v\frac{\partial u}{\partial y}\right) \qquad \text{(Eq. 6.51a)}$$

Thus, for the u and v given above

$$\frac{\partial p}{\partial x} = -\rho\left[(x^2 - y)(2x) + (-2xy + f_1(x))(-1)\right]$$

$$= \rho\left[f_1(x) - 2x^3\right] = F(x)$$

The pressure gradient $\frac{\partial p}{\partial x}$ is a function only of x.

6.9R

6.9R (Inviscid flow) The stream function for the flow of a nonviscous, incompressible fluid in the vicinity of a corner (Fig. P6.9R) is

$$\psi = 2r^{4/3} \sin \tfrac{4}{3}\theta$$

Determine an expression for the pressure *gradient* along the boundary $\theta = 3\pi/4$.
(ANS: $-64\,\rho/27\,r^{1/3}$)

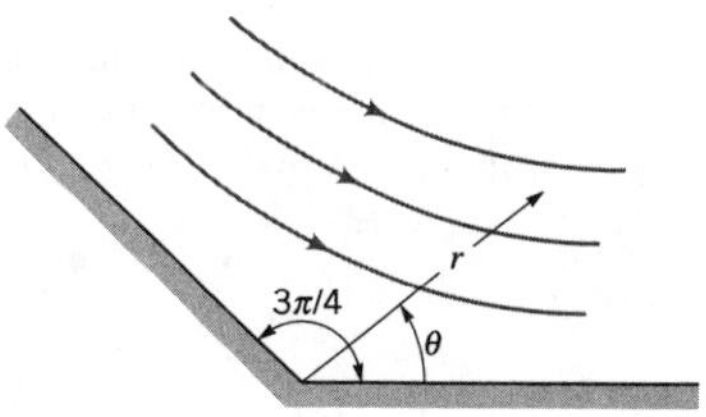

■ FIGURE P6.9R

Along the $\theta = 3\pi/4$ boundary, which is a streamline (i.e., $\psi = 0$ on $\theta = 3\pi/4$),

$$\frac{p}{\rho} + \frac{V^2}{2} = \text{constant}$$

or

$$\frac{\partial p}{\partial r} = -\rho V \frac{\partial V}{\partial r} \qquad (1)$$

For the stream function given,

$$v_r = \frac{1}{r}\frac{\partial \psi}{\partial \theta} = \frac{8}{3} r^{1/3} \cos \frac{4}{3}\theta$$

and along the $\theta = 3\pi/4$ boundary, $v_\theta = 0$, so that

$$V = v_r\left(\theta = \tfrac{3\pi}{4}\right) = -\frac{8}{3} r^{1/3}$$

Since

$$\frac{\partial V}{\partial r} = \frac{-8}{9} r^{-2/3}$$

it follows from Eq. (1) that

$$\frac{\partial p}{\partial r} = -\rho\left(-\frac{8}{3} r^{1/3}\right)\left(-\frac{8}{9} r^{-2/3}\right) = -\frac{64\,\rho}{27\, r^{1/3}}$$

6.10R

6.10R (Potential flow) A certain body has the shape of a half-body with a thickness of 0.5 m. If this body is to be placed in an airstream moving at 20 m/s, what source strength is required to simulate flow around the body?

(ANS: 10.0 m^2/s)

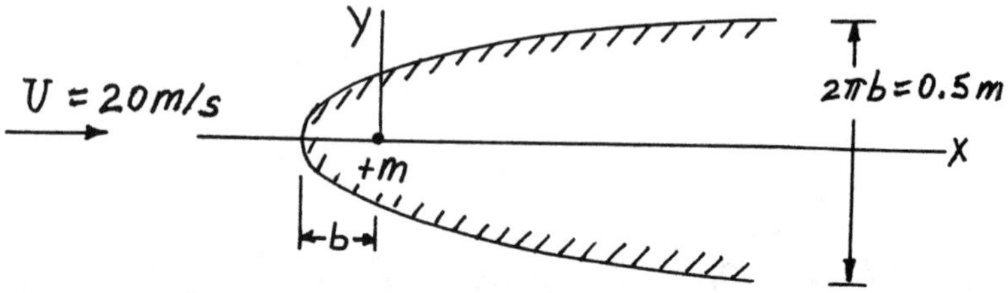

The width of half-body $= 2\pi b$ (see Fig. 6.24)

so that

$$b = \frac{(0.5\,m)}{2\pi}$$

From Eq. 6.99, the distance between the source and the nose of the body is

$$b = \frac{m}{2\pi U}$$

where m is the source strength, and therefore

$$m = 2\pi U b = 2\pi \left(20\frac{m}{s}\right)\left(\frac{0.5\,m}{2\pi}\right) = \underline{\underline{10.0\ \frac{m^2}{s}}}$$

6.11R

6.11R (Potential flow) A source and a sink are located along the x axis with the source at $x = -1$ ft and the sink at $x = 1$ ft. Both the source and the sink have a strength of 10 ft²/s. Determine the location of the stagnation points along the x axis when this source-sink pair is combined with a uniform velocity of 20 ft/s in the positive x direction.

(ANS: ± 1.08 ft)

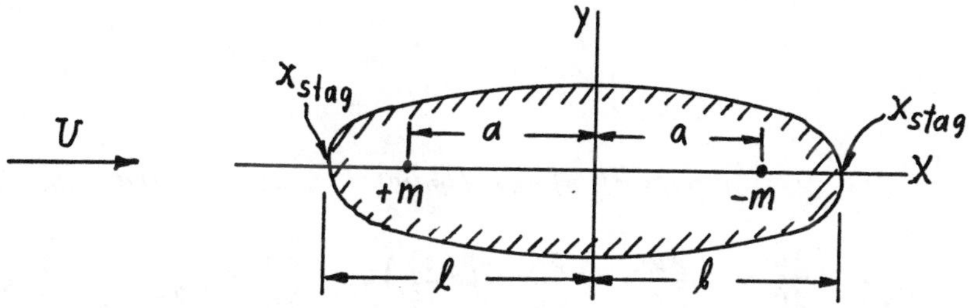

From Eq. 6.106

$$x_{stag} = \pm \ell = \pm \sqrt{\frac{ma}{\pi U} + a^2}$$

Thus, for $a = 1\ ft$, $m = 10\ \frac{ft^2}{s}$, and $U = 20\ \frac{ft}{s}$

$$x_{stag} = \pm \sqrt{\frac{(10\ \frac{ft^2}{s})(1\ ft)}{\pi\ (20\ \frac{ft}{s})} + (1\ ft)^2}$$

$$= \underline{\underline{\pm 1.08\ ft}}$$

6.12R

6.12R (Viscous flow) In a certain viscous, incompressible flow field with zero body forces the velocity components are

$$u = ay - b(cy - y^2)$$

$$v = w = 0$$

where a, b, and c are constant. **(a)** Use the Navier–Stokes equations to determine an expression for the pressure gradient in the x direction. **(b)** For what combination of the constants a, b, and c (if any) will the shearing stress, τ_{yx}, be zero at $y = 0$ where the velocity is zero?

(ANS: $2b\mu$; $a = bc$)

The x-component of the Navier-Stokes equation is (Eq. 6.127a)

(a) $$\rho\left(\frac{\partial u}{\partial t} + u\frac{\partial u}{\partial x} + v\frac{\partial u}{\partial y} + w\frac{\partial u}{\partial z}\right) = -\frac{\partial p}{\partial x} + \rho g_x + \mu\left(\frac{\partial^2 u}{\partial x^2} + \frac{\partial^2 u}{\partial y^2} + \frac{\partial^2 u}{\partial z^2}\right)$$

For the conditions specified, Eq. 6.127a above reduces to

$$0 = -\frac{\partial p}{\partial x} + 0 + \mu(2b)$$

so that

$$\frac{\partial p}{\partial x} = \underline{\underline{2b\mu}}$$

(b) $$\tau_{yx} = \mu\left(\frac{\partial u}{\partial y} + \frac{\partial v}{\partial x}\right) \qquad \text{(Eq. 6.125d)}$$

For the given velocity distribution,

$$\frac{\partial u}{\partial y} = a - bc + 2by$$

and

$$\frac{\partial v}{\partial x} = 0$$

Thus, $$\tau_{yx} = \mu(a - bc + 2by)$$

At $y = 0$

$$\tau_{yx} = \mu(a - bc)$$

and τ_{yx} will be zero if

$$\underline{\underline{a = bc}}$$

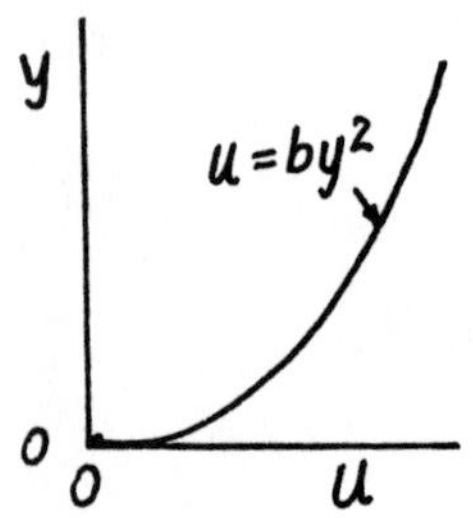

Note: With $v \equiv 0$ then $\tau_{yx} = \mu\frac{\partial u}{\partial y}$, where if $a = bc$, then $u = by^2$.
Thus, $\frac{\partial u}{\partial y} = 2by = 0$ when $y = 0$.

6.13R (Viscous flow) A viscous fluid is contained between two infinite, horizontal parallel plates that are spaced 0.5 in. apart. The bottom plate is fixed, and the upper plate moves with a constant velocity, U. The fluid motion is caused by the movement of the upper plate, and there is no pressure gradient in the direction of flow. The flow is laminar. If the velocity of the upper plate is 2 ft/s and the fluid has a viscosity of 0.03 lb·s/ft^2 and a specific weight of 70 lb/ft^3, what is the required horizontal force per square foot on the upper plate to maintain the 2 ft/s velocity? What is the pressure differential in the fluid between the top and bottom plates?

(ANS: 1.44 lb/ft^2; 2.92 lb/ft^2)

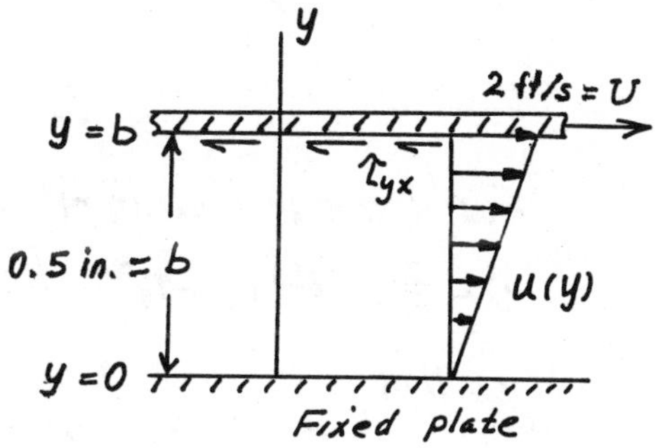

For simple Couette flow (see figure)

$$u = U\frac{y}{b} \qquad \text{(Eq. 6.142)}$$

Since,

$$\tau_{yx} = \mu\left(\frac{\partial u}{\partial y} + \frac{\partial v}{\partial x}\right) \qquad \text{(Eq. 6.125d)}$$

it follows that (with $v = 0$)

$$\tau_{yx} = \frac{\mu U}{b}$$

Thus,

$$\text{Force} = \tau_{yx} \times \text{area} = \frac{\left(0.03\ \frac{\text{lb}\cdot\text{s}}{\text{ft}^2}\right)\left(2\ \frac{\text{ft}}{\text{s}}\right)}{\left(\frac{0.5\ \text{in.}}{12\ \frac{\text{in.}}{\text{ft}}}\right)}\left(1\ \text{ft}^2\right)$$

$$= \underline{\underline{1.44\ \frac{\text{lb}}{\text{ft}^2}}}$$

Since there is no fluid motion in the y-direction, the pressure variation in the y-direction is hydrostatic so that

$$p_{bottom} - p_{top} = \gamma b = \left(70\ \frac{\text{lb}}{\text{ft}^3}\right)\left(\frac{0.5\ \text{in.}}{12\ \frac{\text{in.}}{\text{ft}}}\right) = \underline{\underline{2.92\ \frac{\text{lb}}{\text{ft}^2}}}$$

6.14R

6.14R (Viscous flow) A viscous liquid ($\mu = 0.016\ \text{lb}\cdot\text{s/ft}^2$, $\rho = 1.79\ \text{slugs/ft}^3$) flows through the annular space between two horizontal, fixed, concentric cylinders. If the radius of the inner cylinder is 1.5 in. and the radius of the outer cylinder is 2.5 in., what is the volume flowrate when the pressure drop along the axis of the annulus is 100 lb/ft² per ft?

(ANS: 0.317 ft³/s)

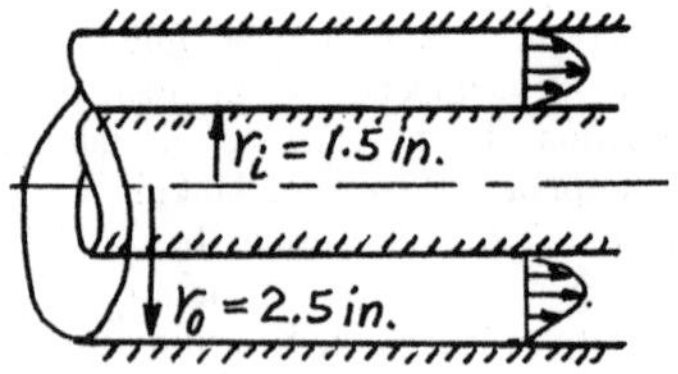

For laminar flow in an annulus (see Eq. 6.156):

$$Q = \frac{\pi}{8\mu}\frac{\Delta p}{\ell}\left[r_o^4 - r_i^4 - \frac{(r_o^2 - r_i^2)^2}{\ln(r_o/r_i)}\right] \quad \text{where } \frac{\Delta p}{\ell} = 100\ \frac{\text{lb}}{\text{ft}^3}$$

so that

$$Q = \frac{\pi}{8\left(0.016\ \frac{\text{lb}\cdot\text{s}}{\text{ft}^2}\right)}\left(100\ \frac{\text{lb}}{\text{ft}^3}\right)\left[\left(\frac{2.5}{12}\text{ft}\right)^4 - \left(\frac{1.5}{12}\text{ft}\right)^4 - \frac{\left[\left(\frac{2.5}{12}\text{ft}\right)^2 - \left(\frac{1.5}{12}\text{ft}\right)^2\right]^2}{\ln\left(\frac{2.5\text{ in.}}{1.5\text{ in.}}\right)}\right]$$

$$= 0.317\ \frac{\text{ft}^3}{\text{s}}$$

We must check to see if the flow is laminar. If not, the above result is not correct.

Now,

$$Re = \frac{\rho V D_h}{\mu}, \quad \text{where } D_h = 2(r_o - r_i)$$

or

$$D_h = 2\left(\frac{2.5\text{ in.} - 1.5\text{ in.}}{12\text{ in./ft}}\right) = 0.167\text{ ft}$$

Also,

$$V = \frac{Q}{A} = \frac{Q}{\pi(r_o^2 - r_i^2)} = \frac{0.317\ \frac{\text{ft}^3}{\text{s}}}{\pi(2.5^2 - 1.5^2)\text{in.}^2}\ \frac{144\text{ in.}^2}{\text{ft}^2} = 3.63\ \frac{\text{ft}}{\text{s}}$$

Thus,

$$Re = \frac{\left(1.79\ \frac{\text{slugs}}{\text{ft}^3}\right)\left(3.63\ \frac{\text{ft}}{\text{s}}\right)(0.167\text{ ft})}{0.016\ \frac{\text{lb}\cdot\text{s}}{\text{ft}^2}} = 67.8$$

Since $Re = 67.8 \ll 2100$ the flow is laminar and

$$Q = \underline{\underline{0.317\ \frac{\text{ft}^3}{\text{s}}}}$$

6.15R

6.15R (Viscous flow) Consider the steady, laminar flow of an incompressible fluid through the horizontal rectangular channel of Fig. P6.15R. Assume that the velocity components in the x and y directions are zero and the only body force is the weight. Start with the Navier–Stokes equations. **(a)** Determine the appropriate set of differential equations and boundary conditions for this problem. You need not solve the equations. **(b)** Show that the pressure distribution is hydrostatic at any particular cross section.

(ANS: $\partial p/\partial x = 0$; $\partial p/\partial y = -\rho g$; $\partial p/\partial z = \mu(\partial^2 w/\partial x^2 + \partial^2 w/\partial y^2)$ with $w = 0$ for $x = \pm b/2$ and $y = \pm a/2$)

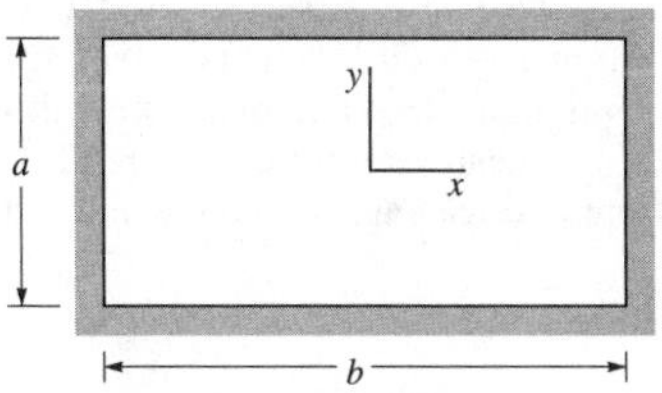

■ FIGURE P6.15R

(a) From the description of the problem, $u=0$, $v=0$, $g_x=0$, $g_y=-g$, $g_z=0$, $w \neq f(t)$, and the continuity equation indicates that $\frac{\partial w}{\partial z}=0$. With these conditions the Navier-Stokes equations (Eq. 6.127) reduce to

(x-direction) $$\frac{\partial p}{\partial x} = 0 \qquad (1)$$

(y-direction) $$\frac{\partial p}{\partial y} = -\rho g \qquad (2)$$

(z-direction) $$\frac{\partial p}{\partial z} = \mu\left(\frac{\partial^2 w}{\partial x^2} + \frac{\partial^2 w}{\partial y^2}\right) \qquad (3)$$

Boundary conditions are: $w=0$ for $x = \pm \frac{b}{2}$, $y = \pm \frac{a}{2}$

(b) Integration of Eq. (2) yields

$$\int dp = -\int \rho g \, dy$$

$$p = -\rho g y + f_1(x,z) \qquad (4)$$

However, from Eq. (1), p is not a function of x so that Eq. (4) becomes

$$p = -\rho g y + f_1(z)$$

Thus, at a given cross section (z = constant)

$$p = -\rho g y + C \qquad (5)$$

where C is a constant. Equation (5) indicates that the pressure distribution is hydrostatic at a given cross section.

6.16R

6.16R (Viscous flow) A viscous liquid, having a viscosity of 10^{-4} lb·s/ft² and a specific weight of 50 lb/ft³, flows steadily through the 2-in.-diameter, horizontal, smooth pipe shown in Fig. P6.16R. The mean velocity in the pipe is 0.5 ft/s. Determine the differential reading, Δh, on the inclined-tube manometer.

(ANS: 0.0640 ft)

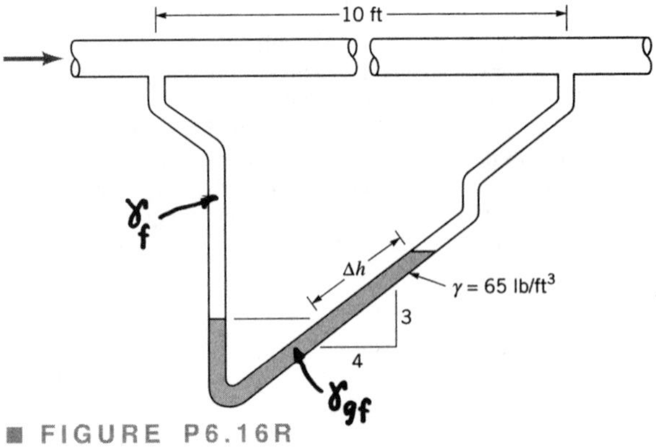

■ FIGURE P6.16R

Check Reynolds number to determine if flow is laminar:

$$Re = \frac{\rho V(2R)}{\mu} = \frac{\left(\frac{50\ \frac{lb}{ft^3}}{32.2\ ft/s^2}\right)\left(0.5\ \frac{ft}{s}\right)\left(\frac{2\ in.}{12\ in./ft}\right)}{10^{-4}\ \frac{lb\cdot s}{ft^2}} = 1290 < 2100$$

Thus, flow is laminar and

$$V = \frac{R^2}{8\mu}\frac{\Delta p}{\ell} \qquad \text{(Eq. 6.152)}$$

so that

$$\Delta p = \frac{8\mu\ell V}{R^2} = \frac{8\left(10^{-4}\ \frac{lb\cdot s}{ft^2}\right)(10\ ft)\left(0.5\ \frac{ft}{s}\right)}{\left(\frac{1\ in.}{12\ \frac{in.}{ft}}\right)^2} = 0.576\ \frac{lb}{ft^2}$$

For manometer (see figure)

$$p_1 + \gamma_f\left(\frac{3}{5}\Delta h\right) - \gamma_{gf}\left(\frac{3}{5}\Delta h\right) = p_2$$

or

$$p_1 - p_2 = \Delta p = \frac{3}{5}\Delta h\,(\gamma_{gf} - \gamma_f)$$

Thus,

$$\Delta h = \frac{\frac{5}{3}\Delta p}{\gamma_{gf} - \gamma_f} = \frac{\frac{5}{3}\left(0.576\ \frac{lb}{ft^2}\right)}{65\ \frac{lb}{ft^3} - 50\ \frac{lb}{ft^3}} = \underline{\underline{0.0640\ ft}}$$

7

Similitude, Dimensional Analysis, and Modeling

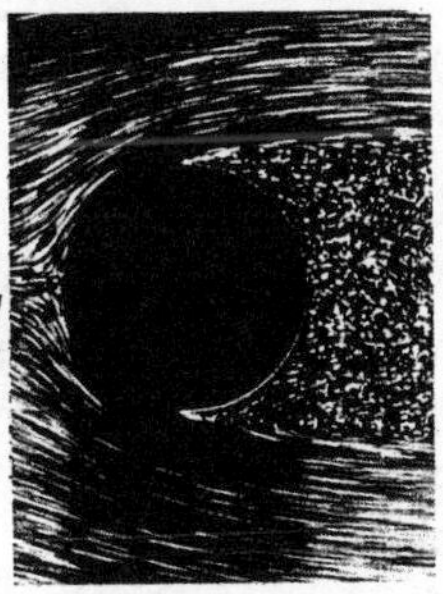

7.1 Dimensional Analysis

- Laboratory studies and experimentation play a significant role in the field of fluid mechanics. The first step in the planning of an experiment should be to decide on the physical factors, or variables, that will be important in the experiment.
- *Dimensional analysis* is a useful tool that simplifies a given problem by reducing the number of variables that need to be considered.
- As the name implies, dimensional analysis is based on a consideration of the *dimensions* required to describe the variables in a given problem.
- The dimensions of the variables are expressed in terms of the basic dimensions of mass, M, length, L, and time, T. Alternatively, force, F, length, L, and time, T, could be used as basic dimensions.
- Application of dimensional analysis will yield a new set of variables that will be *dimensionless* products of the original variables.

7.2 Buckingham Pi Theorem

- The Buckingham pi theorem which forms the theoretical basis for dimensional analysis states the following: If an equation involving k variables is dimensionally homogeneous, it can be reduced to a relationship among $k - r$ independent dimensionless products, where r is the minimum number of reference dimensions required to describe the variables.
- Dimensionless products are frequently referred to as *pi terms*, and the symbol for a pi term written as a capital Π.
- Thus, if $u_1 = f(u_2, u_3, \ldots, u_k)$, where u_i are the physical variables, then $\Pi_1 = \phi(\Pi_1, \Pi_2, \ldots, \Pi_{k-r})$, where Π_i are the dimensionless pi terms.
- Usually, the reference dimensions required to describe the variables are the basic dimensions, M, L, and T, or F, L, and T.
- *Caution:* Use *either* the dimensions M, L, and T, *or* F, L, and T, but not all four (M, L, T, and F) since M and F are related by $F \doteq MLT^{-2}$

EXAMPLE: The pressure drop, Δp, across a valve is known to be a function of the diameter of the valve opening, d, the pipe diameter, D, the fluid velocity, V, and the density, ρ, of the fluid. How many pi terms are required to describe this problem?

SOLUTION:

With $\Delta p = f(d, D, V, \rho)$ and $\Delta p \doteq FL^{-2}$, $d \doteq L$, $D \doteq L$, $V \doteq LT^{-1}$, and $\rho \doteq FL^{-4}T^{2}$, it follows that $k = 5$ (i.e., Δp, d, D, V, and ρ) and $r = 3$ (i.e., F, L, and T). Thus, from the Buckingham pi theorem, it takes $k - r = 5 - 3 = 2$ pi terms (i.e., $\Pi_1 = \phi(\Pi_2)$).

7.3 Determination of Pi Terms

- Several methods can be used to form the pi terms that arise from dimensional analysis.
- The *method of repeating variables* provides a simple, systematic technique for determining the pi terms for a given problem. The essential steps for this procedure are the following:
  Step 1. List all the variables that are involved in the problem.
  Step 2. Express each of the variables in terms of basic dimensions.
  Step 3. Determine the required number of pi terms.
  Step 4. Select a number of repeating variables, where the number is equal to the number of reference dimensions (usually the same as the number of basic dimensions). Do not use the dependent variable as one of the repeating variables.
  Step 5. Form a pi term by multiplying one of the nonrepeating variables by the product of repeating variables each raised to an exponent that will make the combination dimensionless.
  Step 6. Repeat Step 5 for each of the remaining nonrepeating variables.
  Step 7. Check all the resulting pi terms to make sure they are dimensionless.
  Step 8. Express the final form as a relationship among the pi terms.

EXAMPLE: The force (drag), $\mathcal{D}$, acting on a smooth sphere in a flowing fluid is a function of the sphere diameter, d, the fluid velocity, V, the fluid density, ρ, and the fluid viscosity, μ. Develop a suitable set of dimensionless parameters for this problem using d, V, and ρ as repeating variables.

SOLUTION:

Step 1: $\mathcal{D} = f(d, V, \rho, \mu)$

Step 2: $\mathcal{D} \doteq F$, $d \doteq L$, $V \doteq LT^{-1}$, $\rho \doteq FL^{-4}T^2$, $\mu \doteq FL^{-2}T$

Step 3: With $k=5$ and $r=3$, $k-r=2$ pi terms are required.

Step 4: Select d, V, ρ as repeating variables.

Step 5: $\Pi_1 = \mathcal{D} d^a V^b \rho^c \doteq (F)(L)^a(LT^{-1})^b(FT^{-4}T^2)^c = F^0L^0T^0$

Thus, for F, $1+c=0$; for L, $a+b-4c=0$; for T, $-b+2c=0$ which gives $a=-2, b=-2, c=-1$ so that $\Pi_1 = \mathcal{D}d^{-2}V^{-2}\rho^{-1} = \mathcal{D}/\rho V^2 d^2$

Step 6: $\Pi_2 = \mu d^a V^b \rho^c \doteq (FL^{-2}T)(L)^a(LT^{-1})^b(FL^{-4}T^2)^c = F^0L^0T^0$

Thus, for F, $1+c=0$; for L, $-2+a+b-4c=0$; for T, $1-b+2c=0$ which gives $a=-1$, $b=-1$, $c=-1$ so that $\Pi_2 = \mu d^{-1}V^{-1}\rho^{-1} = \mu/\rho V d$

Step 7: Note that $\Pi_1 \doteq (F)(FL^{-4}T^2)^{-1}(LT^{-1})^{-2}(L)^{-2} = F^0L^0T^0$ and $\Pi_2 \doteq (FL^{-2}T)(FL^{-4}T^2)^{-1}(LT^{-1})^{-1}(L)^{-1} = F^0L^0T^0$

Step 8: Thus, $\mathcal{D}/\rho V^2 d^2 = \tilde{\phi}(\mu/\rho V d)$, or since $1/\Pi_2$ is also dimensionless, an alternate form is $\mathcal{D}/\rho V^2 d^2 = \phi(\rho V d/\mu)$.

7.4 Some Additional Comments About Dimensional Analysis.

Selection of Variables

- For most fluid mechanics problems pertinent variables can be classified into three general groups – geometry, fluid properties, and external effects such as pressure, velocity, or gravity.
- One of the most important and difficult steps in dimensional analysis is the selection of the variables needed for a given problem. There is no simple procedure whereby the variables can be easily identified, and one must rely on a good understanding of the phenomenon involved and the governing physical laws.
- *Caution:* Make sure all the variables are independent, i.e., there is not a relationship among specific variables.

Determination of Reference Dimensions

- Typically, in fluid mechanics problems, the required number of reference dimensions will be three (i.e., M, L, and T, or F, L, and T), but in some problems only one or two are required.
- *Caution:* Carefully define the dimensions of the variables so that the number of reference dimensions can be correctly obtained.

Uniqueness of Pi Terms

- There is not a unique set of pi terms for a given problem, although the required *number* is fixed by the Buckingham pi theorem.
- Once a correct set of pi terms is obtained, any other set can be obtained by manipulation of the original set.

EXAMPLE: In the previous example involving the drag on a sphere, two pi terms were obtained using the repeating variable method. Assume that it is desired that Π_1 contain the viscosity, μ, rather than the density, ρ. Show how a new set of pi terms can be obtained by a manipulation of the pi terms from the previous example.

SOLUTION:

From the previous example, $\Pi_1 = \mathscr{D}/\rho V^2 d^2$ and $\Pi_2 = \rho V d/\mu$. To form a new pi term consider

$$\Pi_1' = \Pi_1 \Pi_2 = (\mathscr{D}/\rho V^2 d^2)(\rho V d/\mu) = \mathscr{D}/\mu V d$$

Thus, an alternate set of pi terms is $\underline{\underline{\mathscr{D}/\mu V d = \phi_1(\rho V d/\mu)}}$

Note: This is the set of pi terms that would result using d, V, and μ as repeating variables.

7.5 Determination of Pi Term by Inspection

- Pi terms can be formed by inspection since the only restrictions on the pi terms are that they be (1) correct in number, (2) dimensionless, and (3) independent.
- To form a pi term by inspection, pick a variable and simply cancel its dimensions by multiplying or dividing by other variables raised to appropriate powers.
- *Caution:* After forming a pi term always check to make sure it is dimensionless.

EXAMPLE: Pressure pulses travel through arteries with velocity c which is a function of the artery diameter, D, the wall thickness, h, the density of the blood, ρ, and a material property of the artery wall, E, which has the dimensions, FL^{-2}. Develop a suitable set of pi terms for this problem. Form the pi terms by inspection.

SOLUTION:

$c = f(D, h, \rho, E)$ where $c \doteq LT^{-1}$, $D \doteq L$, $h \doteq L$, $\rho \doteq FL^{-4}T^{2}$, $E \doteq FL^{-2}$

Thus, $k = 5$ and $r = 3$ so that from the Buckingham pi theorem, $k - r = 5 - 3 = 2$ pi terms are required.

Let Π_1 contain the dependent variable, c. The only variables involving T are c and ρ. Thus, to eliminate T in Π_1 we can form $c\rho^{1/2} \doteq LT^{-1}(FL^{-4}T^{2})^{1/2} = F^{1/2}L^{-1}$. But $E^{1/2} \doteq (FL^{-2})^{1/2}$ so that the quantity $c\rho^{1/2}E^{-1/2} \doteq (F^{1/2}L^{-1})(FL^{-2})^{-1/2} = F^{0}L^{0}$

Thus, $c\sqrt{\rho/E} = \Pi_1$ is dimensionless.

Check: $c\sqrt{\rho/E} \doteq (LT^{-1})[(FL^{-4}T^{2})(FL^{-2})^{-1}]^{1/2} = F^{0}L^{0}T^{0}$

Clearly the ratio $\Pi_2 = h/D \doteq L\,L^{-1} = L^{0}$ is dimensionless.

Thus, $c\sqrt{\rho/E} = \phi(h/D)$

7.6 Common Dimensionless Groups in Fluid Mechanics

■ **TABLE 7.1**

Some Common Variables and Dimensionless Groups in Fluid Mechanics

Variables: Acceleration of gravity, g; Bulk modulus, E_v; Characteristic length, ℓ; Density, ρ; Frequency of oscillating flow, ω; Pressure, p (or Δp); Speed of sound, c; Surface tension, σ; Velocity, V; Viscosity, μ

Dimensionless Groups	Name	Interpretation (Index of Force Ratio Indicated)	Types of Applications
$\frac{\rho V \ell}{\mu}$	Reynolds number, Re	$\frac{\text{inertia force}}{\text{viscous force}}$	Generally of importance in all types of fluid dynamics problems
$\frac{V}{\sqrt{g\ell}}$	Froude number, Fr	$\frac{\text{inertia force}}{\text{gravitational force}}$	Flow with a free surface
$\frac{p}{\rho V^2}$	Euler number, Eu	$\frac{\text{pressure force}}{\text{inertia force}}$	Problems in which pressure, or pressure differences, are of interest
$\frac{\rho V^2}{E_v}$	Cauchy number,[a] Ca	$\frac{\text{inertia force}}{\text{compressibility force}}$	Flows in which the compressibility of the fluid is important
$\frac{V}{c}$	Mach number,[a] Ma	$\frac{\text{inertia force}}{\text{compressibility force}}$	Flows in which the compressibility of the fluid is important
$\frac{\omega \ell}{V}$	Strouhal number, St	$\frac{\text{inertia (local) force}}{\text{inertia (convective) force}}$	Unsteady flow with a characteristic frequency of oscillation
$\frac{\rho V^2 \ell}{\sigma}$	Weber number, We	$\frac{\text{inertia force}}{\text{surface tension force}}$	Problems in which surface tension is important

[a]The Cauchy number and the Mach number are related and either can be used as an index of the relative effects of inertia and compressibility. See accompanying discussion.

- There are a number of variables that commonly arise in fluid mechanics problems, and it is standard practice to combine them into some standard dimensionless groups (pi terms) with special names.
- Some well-known pi terms are the Reynolds number, Re, the Froude number, Fr, and the Mach number, Ma (see Table 7.1).
- A physical interpretation can often be given to a dimensionless group which may be helpful in assessing its influence in a particular problem. For example, the Reynolds number can be thought of as an index for the ratio of inertia forces to viscous forces in a flowing fluid (see Table 7.1).

7.7 Correlation of Experimental Data

- One of the most important uses of dimensional analysis is as an aid in the efficient handling, interpretation, and correlation of experimental data.
- Dimensional analysis cannot provide a complete answer to a given problem since suitable experimental data are required to determine the specific relationship among the pi terms.
- The difficulty in obtaining empirically the relationship among the pi terms depends on the number of pi terms required to describe the problem.

Problems with One Pi Term

- The simplest problems involve only *one* pi term. For this case the number of variables minus the number of reference dimensions is equal to one.
- The functional relationship that must exist for one pi term is $\Pi_1 = C$ where C is a constant to be determined by experiment.

EXAMPLE: The velocity, V, of a jet of liquid as it exits from a hole in the bottom of a tank is thought to be a function of the depth, h, of the liquid in the tank and the acceleration of gravity, g. Determine, with the aid of dimensional analysis, how the velocity is related to the liquid depth.

SOLUTION:

$V = f(h, g)$, where $V \doteq LT^{-1}$, $h \doteq L$, $g \doteq LT^{-2}$ so $k = 3$ and $r = 2$.
From the Buckingham pi theorem $k - r = 3 - 2 = 1$, so only one pi term is needed. By inspection,
$\Pi_1 = V/\sqrt{gh} \doteq (LT^{-1})/[(LT^{-2})L]^{1/2} = L^0T^0$ is dimensionless.
Since there is only one pi term, $\Pi_1 = C$, where C is a constant.
Thus, $V/\sqrt{gh} = C$, or $V = C\sqrt{gh}$ Hence $V \sim \sqrt{h}$

Problem with Two or More Pi terms

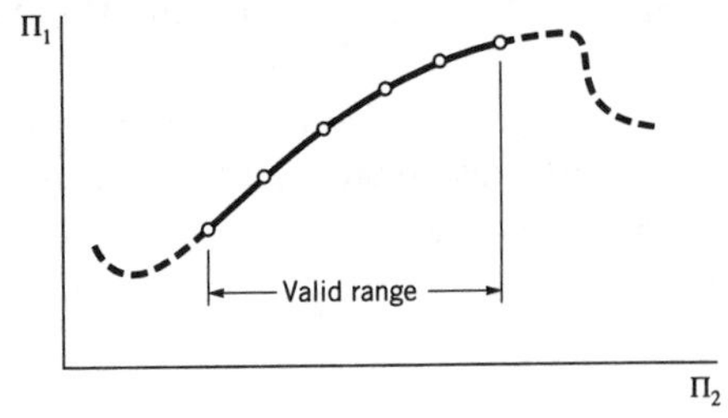

■ FIGURE 7.4 **The graphical presentation of data for problems involving two pi terms, with an illustration of the potential danger of extrapolation of data.**

- For problems involving only two pi terms, $\Pi_1 = \phi(\Pi_2)$, the results of an experiment can be conveniently presented in a simple graph of the type shown in Fig. 7.4.
- In addition to presenting the data graphically, it may be possible to obtain an empirical equation relating Π_1 and Π_2 by using a standard curve-fitting technique.
- *Caution:* The empirical relationship may only be valid over the range covered by Π_2. As illustrated in Fig. 7.4, it can be dangerous to extrapolate beyond the range covered by the experimental data.
- As the required number of pi terms increases beyond two, it becomes more difficult to display the data in a convenient graphical form and to determine a specific empirical relationship that describes the phenomenon.
- For these more complicated problems it is often more feasible to use physical *models* to predict specific characteristics of the system rather than attempting to develop a general empirical relationship.

EXAMPLE: The volume flowrate, Q, over a low head dam is a function of the depth, H, of the water flowing over the dam, the width, b, of the dam, and the acceleration of gravity, g. A dimensional analysis indicates that $Q/(gb^5)^{1/2} = \phi(H/b)$ and some experimental data are given below for a dam with b = 20 m. Plot the data in dimensionless form and determine a power-law equation relating the dimensionless parameters.

H (m)	0.1	0.2	0.3	0.4
Q (m^3/s)	0.7	2.0	3.6	5.6

SOLUTION:

With $\Pi_1 = Q/(g b^5)^{1/2} \doteq (L^3T^{-1})/[(LT^{-2})(L)^5]^{1/2} = L^0T^0$ and $\Pi_2 = H/b \doteq L/L = L^0T^0$ it follows that $Q/(gb^5)^{1/2} = \phi(H/b)$.
For the data given

Π_2	0.005	0.010	0.015	0.020
Π_1	1.25×10^{-4}	3.56×10^{-4}	6.43×10^{-4}	10.0×10^{-4}

A plot of the data is shown below. A power-law fit of the data (i.e., $\Pi_1 = C\Pi_2^{\,n}$) gives $\Pi_1 = 0.35\,\Pi_2^{\,1.50}$. That is,

$$Q/\sqrt{gb^5} = 0.35\,(H/b)^{3/2}$$

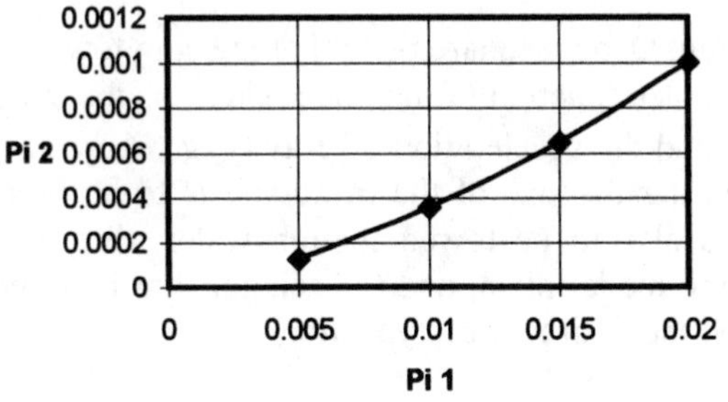

7.8 Modeling and Similitude

- An engineering *model* is a representation of a physical system that may be used to predict the behavior of the system in some desired respect.
- The physical system for which the prediction is made is called the *prototype.*

Theory of Models

$$\Pi_1 = \phi(\Pi_2, \Pi_3, \ldots, \Pi_n) \quad \textbf{(7.7)}$$

$$\Pi_{1m} = \phi(\Pi_{2m}, \Pi_{3m}, \ldots, \Pi_{nm}) \quad \textbf{(7.8)}$$

- For a given problem the phenomenon of interest can be described in terms of a set of pi terms as in Eq. 7.7.
- The same relationship, Eq. 7.8, applies to a model of the same phenomenon, although the size of the model components, fluid properties, etc., may be different from those of the prototype.
- The pi terms can be determined so that Π_1 contains the variable to be predicted from observations on the model, and the desired *prediction equation* is $\Pi_1 = \Pi_{1m}$.
- For the prediction equation to be valid all of the other pi terms must be equal, i.e., $\Pi_{2m} = \Pi_2$, $\Pi_{3m} = \Pi_3$, etc. These conditions are called the *model design conditions, similarity requirements, or modeling laws.*
- *Caution:* If important variables are omitted from the original list of variables, the model design will not be correct since important pi terms will be missing.

EXAMPLE: The drag, $\mathfrak{D}$, on a square plate that is being towed near the surface through water at 3 m/s is a function of the plate width, w, the towing velocity, V, the water density, ρ, and the acceleration of gravity, g. A geometrically similar model that is one quarter the size of the prototype is to be used to predict the drag. If the model is also to be towed though water, determine the required towing velocity for the model plate, and the relationship between the measured drag on the model and the desired prototype drag.

SOLUTION:

$\mathfrak{D} = f(w, V, \rho, g)$, where $\mathfrak{D} \doteq F$, $w \doteq L$, $V \doteq LT^{-1}$, $\rho \doteq FL^{-4}T^2$, $g \doteq LT^{-2}$
From the Buckingham pi theorem, the number of pi terms required is $k - r = 5 - 3 = 2$. Dimensional analysis yields the following pi terms:

$\mathfrak{D}/(w^2\rho V^2) = \phi(V^2/gw)$. Thus, for similarity

$V_m^2/g_m w_m = V^2/gw$, where $g_m = g$, so that

$V_m = (w_m/w)^{1/2} V = (1/4)^{1/2}(3\text{ m/s}) = \underline{\underline{1.5\text{ m/s}}}$ Also,

$\mathfrak{D}_m/(w_m^2 \rho_m V_m^2) = \mathfrak{D}/(w^2\rho V^2)$, where $\rho_m = \rho$, so that

$\mathfrak{D} = (w/w_m)^2 (V/V_m)^2 \mathfrak{D}_m = (4)^2[(3\text{ m/s})/(1.5\text{ m/s})]^2 \mathfrak{D}_m$

or $\underline{\underline{\mathfrak{D} = 64\mathfrak{D}_m}}$

Model Scales

- The ratio of a model variable to the corresponding prototype variable is called the *scale* for that variable.
- The *length* scale, ℓ_m/ℓ, is the ratio of a model characteristic length to the corresponding length of the prototype.
- *Caution:* Although the length scale is important, a different scale will exist for other variables in the problem, such as a velocity scale, V_m/V, and a time scale, t_m/t.

EXAMPLE: Flow over the spillway of a dam is governed by Froude number similarity. Determine the relationship between the velocity scale and the length scale for this type of similarity.

SOLUTION:

For Froude number similarity, $Fr_m = Fr$, or $V_m/\sqrt{g_m \ell_m} = V/\sqrt{g\ell}$.

Thus, with $g_m = g$, $V_m/V = (\ell_m/\ell)^{1/2}$

That is, velocity scale = square root of length scale

Practical Aspects of Using Models

- Whenever possible, a model design should be checked, using any available experimental data, since models typically involve simplifying assumptions with regard to the variables to be considered.
- If one or more of the similarity requirements is not met, for example, $\Pi_{2m} \neq \Pi_2$, then it follows that the prediction equation is not valid and $\Pi_1 \neq \Pi_{1m}$. Models for which this occurs are called *distorted* models.
- Distorted models can be successfully used, but the interpretation of results is obviously more difficult than the interpretation of results obtained from *true* models for which all similarity requirements are met.

7.9 Some Typical Model Studies

Flow Through Closed Conduits

$$\text{Dependent pi term} = \phi\left(\frac{\ell_i}{\ell}, \frac{\varepsilon}{\ell}, \frac{\rho V \ell}{\mu}\right) \qquad \textbf{(7.16)}$$

- Typical problems involving flow through closed conduits include flow though pipes, valves, pipe fittings, and metering devices.
- Generally, for the purpose of dimensional analysis, geometric characteristics can be described by a series of length terms, $\ell_1, \ell_2, \ell_3, \ldots \ell_i$, and ℓ, where ℓ is some particular characteristic length of the system.
- For flow in closed conduits at low Mach numbers, the dependent pi term (the one that contains the variable of interest, such as pressure drop) will generally be a function of the pi terms shown in Eq. 7.16.
- Since one of the pi terms in Eq. 7.16 is the Reynolds number, the modeling of flow through closed conduits requires *Reynolds number similarity*, $\text{Re}_m = \text{Re}$, as well as *geometric similarity*.

EXAMPLE: The pressure variation along a pipeline carrying SAE 30 oil at a velocity of 5 ft/s is to be studied using a 1:10 scale model. If the model fluid is water, what water velocity is required in the model pipeline for Reynolds number similarity?

SOLUTION:

For Reynolds number similarity, $Re_m = Re$, or $V_m D_m / \nu_m = VD/\nu$, where $\nu = \mu/\rho$. Thus,

$$V_m = (D/D_m)(\nu_m/\nu)V = (10)\left[1.21\times10^{-5}\ \text{ft}^2/\text{s} \,/\, 4.5\times10^{-3}\ \text{ft}^2/\text{s}\right](5\ \text{ft/s})$$

$$= \underline{\underline{0.134\ \text{ft/s}}}$$

Flow Around Immersed Bodies

$$\text{Dependent pi term} = \phi\left(\frac{\ell_i}{\ell}, \frac{\varepsilon}{\ell}, \frac{\rho V \ell}{\mu}\right) \quad \textbf{(7.18)}$$

- Typical problems involving flow around immersed bodies include flow around aircraft, automobiles, golf balls, and buildings.
- For flow around immersed bodies at low Mach numbers, the dependent pi term will generally be a function of the pi terms shown in Eq. 7.18.
- Frequently, the dependent variable of interest for this type of problem is the fluid force developed on the body, called the drag, $\mathfrak{D}$, and a typical corresponding dependent pi term is $\mathfrak{D}/(\frac{1}{2}\rho V^2 \ell^2)$, which is a form of *drag coefficient.*
- As shown in Eq. 7.18, geometric and Reynolds number similarity are usually required for models involving flow around immersed bodies.

EXAMPLE: The pressure, p_1, at a given point (1) on an object immersed in a stream of water moving with velocity V is to be studied with a geometrically similar model in a water tunnel. If Reynolds number similarity is to be maintained, what is the relationship between the measured pressure on the model and the corresponding predicted pressure on the prototype? The water has the same properties for both the model and the prototype, and the length scale is 1/6.
SOLUTION:

The dimensional relationship $p_1 = f(\rho, V, \ell, \mu)$ can be written in dimensionless form as

$p_1/\rho V^2 = \phi(\rho \ell V/\mu)$, where $Re = \rho \ell V/\mu = \ell V/\nu$ is the Reynolds number. Thus, for similarity $Re_m = Re$, or

$V_m \ell_m/\nu_m = V\ell/\nu$, so that with $\nu_m = \nu$,

$V_m/V = \ell/\ell_m = 6$

The prediction equation ($\Pi_{1m} = \Pi_1$) gives

$p_1/\rho V^2 = p_{1m}/\rho_m V_m^2$, so that with $\rho = \rho_m$,

$p_1 = (V/V_m)^2 p_{1m} = (1/6)^2 p_{1m}$

Hence, $\underline{\underline{p_1 = p_{1m}/36}}$

Flow with a Free Surface

$$\text{Dependent pi term} = \phi\left(\frac{\ell_i}{\ell}, \frac{\varepsilon}{\ell}, \frac{\rho V \ell}{\mu}, \frac{V}{\sqrt{g\ell}}, \frac{\rho V^2 \ell}{\sigma}\right) \qquad \textbf{(7.24)}$$

- Typical problems involving flow with a free surface include flows in canals, rivers, and spillways, as well as flow around ships.
- For flow with a free surface, the dependent pi term will generally be a function of the pi terms shown in Eq. 7.24.
- Since gravity is the driving force in these problems, *Froude number similarity* is required as indicated by Eq. 7.24. That is, $V_m/(g_m\ell_m)^{1/2} = V/(g\ell)^{1/2}$.
- As indicated by the inclusion of the Weber number, We $= \rho V^2\ell/\sigma$, and the Reynolds number, Re $= \rho V\ell/\mu$, in Eq. 7.24, surface tension and viscous effects may be important for free surface flows. For large hydraulic structures, however, these effects are often negligible and the model design is based on Froude number similarity.
- *Caution:* If both Froude number similarity and Reynolds number similarity are required, a distorted model will result if the same fluid is used in both the model and prototype.

EXAMPLE: An irrigation canal is 10 ft wide and carries water at a rate of 90 ft^3/s. A geometrically similar model that is 2 ft wide is to be used to study certain flow characteristics in the irrigation canal. What flowrate is required in the model to maintain Froude number similarity?

SOLUTION:

For Froude number similarity, $Fr_m = Fr$, or $V_m/\sqrt{g_m \ell_m} = V/\sqrt{g\ell}$.
Thus, with $g_m = g$, $V_m/V = \sqrt{\ell_m/\ell}$.
Since $Q = AV$ and $Q_m = A_m V_m$ it follows that

$Q_m/Q = (V_m A_m)/(VA) = (V_m/V)\cdot(A_m/A)$, where $A_m/A = (\ell_m/\ell)^2$

Thus, $Q_m/Q = \sqrt{\ell_m/\ell}\,(\ell_m/\ell)^2 = (\ell_m/\ell)^{5/2}$, so that

$Q_m = (2\text{ ft}/10\text{ ft})^{5/2}(90\text{ ft}^3/\text{s}) = \underline{\underline{1.61\text{ ft}^3/\text{s}}}$

7.10 Similitude Based on Governing Differential Equations

- An alternative to using dimensional analysis for developing similarity laws is based on a knowledge of the governing equations for the phenomenon of interest.
- The basic procedure consists of re-writing the governing equations (usually differential equations) in terms of dimensionless variables. It follows that the conventional similarity parameters, such as the Reynolds number and the Froude number, arise naturally from this procedure and can be used in the design of models.
- This approach has the advantage that the variables are known and the assumptions clearly identified, but it is more complicated than the use of dimensional analysis for obtaining similarity laws.

Solutions

to Review Problems

Chapter 7

7.1R

7.1R (Common Pi terms) Standard air with velocity V flows past an airfoil having a chord length, b, of 6 ft. **(a)** Determine the Reynolds number, $\rho V b/\mu$, for $V = 150$ mph. **(b)** If this airfoil were attached to an airplane flying at the same speed in a standard atmosphere at an altitude of 10,000 ft, what would be the value of the Reynolds number?

(ANS: 8.40×10^6; 6.56×10^6)

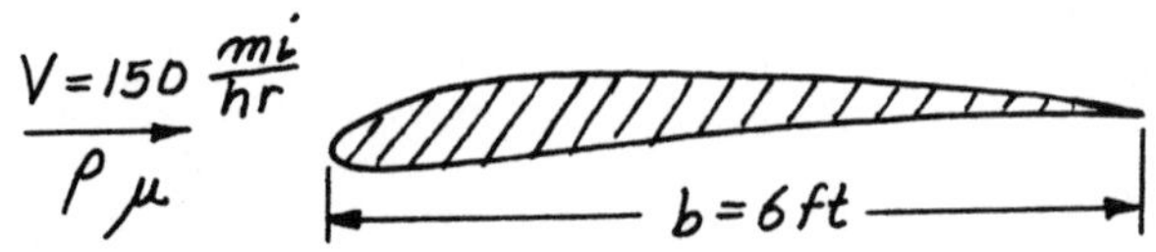

(a) $$Re = \frac{\rho V b}{\mu} = \frac{\left(2.38\times10^{-3}\ \frac{slugs}{ft^3}\right)\left(150\ \frac{mi}{hr}\times 5280\ \frac{ft}{mi}\times\frac{1}{3600\ \frac{s}{hr}}\right)(6\ ft)}{3.74\times10^{-7}\ \frac{lb\cdot s}{ft^2}}$$

or

$$Re = \underline{\underline{8.40\times10^6}}$$

(b) At 10,000 ft, $\rho = 1.756\times10^{-3}\ \frac{slugs}{ft^3}$ and $\mu = 3.534\times10^{-7}\ \frac{lb\cdot s}{ft^2}$ (Table C.1 in Appendix C) so that

$$\frac{\rho V b}{\mu} = \frac{\left(1.756\times10^{-3}\ \frac{slugs}{ft^3}\right)\left(150\ \frac{mi}{hr}\times 5280\ \frac{ft}{mi}\times\frac{1}{3600\ \frac{s}{hr}}\right)(6\ ft)}{3.534\times10^{-7}\ \frac{lb\cdot s}{ft^2}}$$

or

$$Re = \underline{\underline{6.56\times10^6}}$$

7.2R

7.2R (Dimensionless variables) Some common variables in fluid mechanics include: volume flowrate, Q, acceleration of gravity, g, viscosity, μ, density, ρ and a length, ℓ. Which of the following combinations of these variables are dimensionless? **(a)** $Q^2/g\ell^2$. **(b)** $\rho Q/\mu\ell$. **(c)** $g\ell^5/Q^2$. **(d)** $\rho Q\ell/\mu$.

(ANS: (b); (c))

$Q \doteq L^3T^{-1}$ $\quad g \doteq LT^{-2}$ $\quad \mu \doteq FL^{-2}T$ $\quad \rho \doteq FL^{-4}T^2$ $\quad \ell \doteq L$

(a) $\dfrac{Q^2}{g\ell^2} \doteq \dfrac{(L^3T^{-1})^2}{(LT^{-2})(L^2)} \doteq L^3$ **not dimensionless**

(b) $\dfrac{\rho Q}{\mu \ell} \doteq \dfrac{(FL^{-4}T^2)(L^3T^{-1})}{(FL^{-2}T)(L)} \doteq F^0L^0T^0$ **dimensionless**

(c) $\dfrac{g\ell^5}{Q^2} \doteq \dfrac{(LT^{-2})(L^5)}{(L^3T^{-1})^2} \doteq L^0T^0$ **dimensionless**

(d) $\dfrac{\rho Q\ell}{\mu} \doteq \dfrac{(FL^{-4}T^2)(L^3T^{-1})(L)}{FL^{-2}T} \doteq L^2$ **not dimensionless**

7.3R

7.3R (Determination of Pi terms) A fluid flows at a velocity V through a horizontal pipe of diameter D. An orifice plate containing a hole of diameter d is placed in the pipe. It is desired to investigate the pressure drop, Δp, across the plate. Assume that

$$\Delta p = f(D, d, \rho, V)$$

where ρ is the fluid density. Determine a suitable set of pi terms. (ANS: $\Delta p/\rho V^2 = \phi(d/D)$)

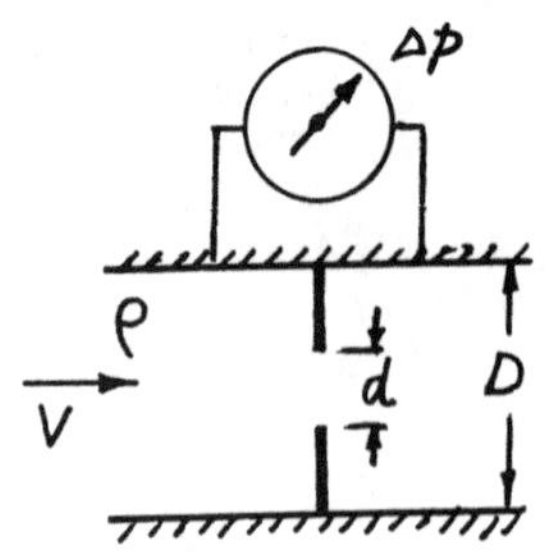

$\Delta p \doteq FL^{-2}$ $\quad D \doteq L$ $\quad d \doteq L$ $\quad \rho \doteq FL^{-4}T^{2}$ $\quad V \doteq LT^{-1}$

From the pi theorem, 5−3 = 2 pi terms required. Use D, V, and ρ as repeating variables. Thus,

$$\Pi_1 = \Delta p \, D^a V^b \rho^c$$

and

$$(FL^{-2})(L)^a(LT^{-1})^b(FL^{-4}T^2)^c \doteq F^0L^0T^0$$

so that

$$1 + c = 0 \quad \text{(for } F\text{)}$$
$$-2 + a + b - 4c = 0 \quad \text{(for } L\text{)}$$
$$-b + 2c = 0 \quad \text{(for } T\text{)}$$

It follows that $a = 0$, $b = -2$, $c = -1$, and therefore

$$\Pi_1 = \frac{\Delta p}{V^2 \rho}$$

Check dimensions using MLT system:

$$\frac{\Delta p}{V^2\rho} \doteq \frac{ML^{-1}T^{-2}}{(LT^{-1})^2(ML^{-3})} \doteq M^0L^0T^0 \quad \therefore \text{OK}$$

For Π_2:

$$\Pi_2 = d \, D^a V^b \rho^c$$

$$(L)(L)^a(LT^{-1})^b(FL^{-4}T^2)^c \doteq F^0L^0T^0$$

$$c = 0 \quad \text{(for } F\text{)}$$
$$1 + a + b - 4c = 0 \quad \text{(for } L\text{)}$$
$$-b + 2c = 0 \quad \text{(for } T\text{)}$$

It follows that $a = -1$, $b = 0$, $c = 0$, and therefore

$$\Pi_2 = \frac{d}{D}$$

which is obviously dimensionless.

Thus,

$$\frac{\Delta p}{\rho V^2} = \phi\left(\frac{d}{D}\right)$$

7.4R

7.4R (Determination of Pi terms) The flowrate, Q, in an open canal or channel can be measured by placing a plate with a V-notch across the channel as illustrated in Fig. P7.4R. This type of device is called a V-notch *weir*. The height, H, of the liquid above the crest can be used to determine Q. Assume that

$$Q = f(H, g, \theta)$$

where g is the acceleration of gravity. What are the significant dimensionless parameters for this problem?

(ANS: $Q/(gH^5)^{1/2} = \phi(\theta)$)

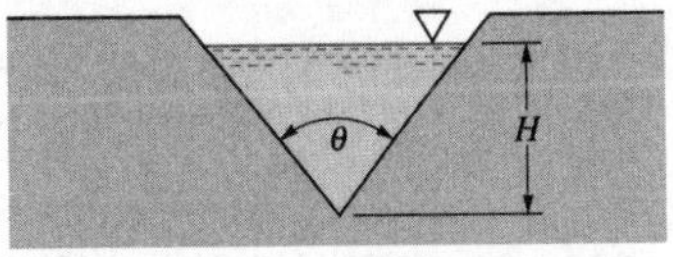

■ FIGURE P7.4R

$Q \doteq L^3T^{-1}$ $\quad H \doteq L \quad$ $g \doteq LT^{-2}$ $\quad \theta = F^0L^0T^0$

From the pi theorem, $4-2 = 2$ pi terms required: $\Pi_1 = \phi(\Pi_2)$

By inspection, for Π_1 (containing Q):

$$\Pi_1 = \frac{Q}{g^{1/2} H^{5/2}} \doteq \frac{L^3T^{-1}}{(LT^{-2})^{1/2} (L)^{5/2}} \doteq L^0T^0$$

Since the angle, θ, is dimensionless

$$\Pi_2 = \theta$$

So that

$$\frac{Q}{\sqrt{gH^5}} = \phi(\theta)$$

7.5R

7.5R (Determination of Pi terms) In a fuel injection system, small droplets are formed due to the breakup of the liquid jet. Assume the droplet diameter, d, is a function of the liquid density, ρ, viscosity, μ, and surface tension, σ, and the jet velocity, V, and diameter, D. Form an appropriate set of dimensionless parameters using μ, V, and D as repeating variables. (ANS: $d/D = \phi(\rho VD/\mu, \sigma/\mu V)$)

$$d = f(\rho, \mu, \sigma, V, D)$$

$d \doteq L \quad \rho \doteq FL^{-4}T^{2} \quad \mu \doteq FL^{-2}T \quad \sigma \doteq FL^{-1} \quad V \doteq LT^{-1} \quad D \doteq L$

From the pi theorem, 6-3 = 3 pi terms required. Use μ, V, and D as repeating variables. Thus,

$$\Pi_1 = d\,\mu^{a} V^{b} D^{c}$$

and

$$(L)(FL^{-2}T)^{a}(LT^{-1})^{b}(L)^{c} \doteq F^{0}L^{0}T^{0}$$

so that

$$a = 0 \qquad \text{(for F)}$$
$$1 - 2a + b + c = 0 \qquad \text{(for L)}$$
$$a - b = 0 \qquad \text{(for T)}$$

It follows that $a = 0$, $b = 0$, $c = -1$, and therefore

$$\Pi_1 = \frac{d}{D}$$

which is obviously dimensionless.

For Π_2:

$$\Pi_2 = \rho\,\mu^{a} V^{b} D^{c}$$

$$(FL^{-4}T^{2})(FL^{-2}T)^{a}(LT^{-1})^{b}(L)^{c} \doteq F^{0}L^{0}T^{0}$$

$$1 + a = 0 \qquad \text{(for F)}$$
$$-4 - 2a + b + c = 0 \qquad \text{(for L)}$$
$$2 + a - b = 0 \qquad \text{(for T)}$$

It follows that $a = -1$, $b = 1$, $c = 1$, and therefore

$$\Pi_2 = \frac{\rho V D}{\mu}$$

(continued)

Check dimensions using MLT system:

$$\frac{\rho VD}{\mu} \doteq \frac{(ML^{-3})(LT^{-1})(L)}{ML^{-1}T^{-1}} \doteq M^0L^0T^0 \quad \therefore \text{OK}$$

For Π_3:

$$\Pi_3 = \sigma \mu^a V^b D^c$$

$$(FL^{-1})(FL^{-2}T)^a (LT^{-1})^b (L)^c \doteq F^0L^0T^0$$

$$1 + a = 0 \qquad \text{(for } F\text{)}$$

$$-1 - 2a + b + c = 0 \qquad \text{(for } L\text{)}$$

$$a - b = 0 \qquad \text{(for } T\text{)}$$

It follows that $a = -1$, $b = -1$, $c = 0$, and therefore

$$\Pi_3 = \frac{\sigma}{\mu V}$$

Check dimensions using MLT system:

$$\frac{\sigma}{\mu V} \doteq \frac{MT^{-2}}{(ML^{-1}T^{-1})(LT^{-1})} \doteq M^0L^0T^0 \quad \therefore \text{OK}$$

Thus,

$$\frac{d}{D} = \phi\left(\frac{\rho VD}{\mu}, \frac{\sigma}{\mu V}\right)$$

7.6R

7.6R (Determination of Pi terms) The thrust, $\mathcal{T}$, developed by a propeller of a given shape depends on its diameter, D, the fluid density, ρ, and viscosity, μ, the angular speed of rotation, ω, and the advance velocity, V. Develop a suitable set of pi terms, one of which should be $\rho D^2\omega/\mu$. Form the pi terms by inspection.

(ANS: $\mathcal{T}/\rho V^2 D^2 = \phi(\rho VD/\mu,\ \rho D^2\omega/\mu)$)

$$\mathcal{T} = f(D, \rho, \mu, \omega, V)$$

$$\mathcal{T} \doteq F \quad D \doteq L \quad \rho \doteq FL^{-4}T^2 \quad \mu \doteq FL^{-2}T \quad \omega \doteq T^{-1} \quad V \doteq LT^{-1}$$

From the pi theorem, 6-3 = 3 pi terms required: $\Pi_1 = \phi(\Pi_2, \Pi_3)$.

By inspection, for Π_1 (containing $\mathcal{T}$):

$$\Pi_1 = \frac{\mathcal{T}}{\rho V^2 D^2} \doteq \frac{F}{(FL^{-4}T^2)(LT^{-1})^2(L)^2} \doteq F^0L^0T^0$$

Check using MLT:

$$\frac{\mathcal{T}}{\rho V^2 D^2} \doteq \frac{MLT^{-2}}{(ML^{-3})(LT^{-1})^2(L)^2} \doteq M^0L^0T^0 \quad \therefore \text{ok}$$

For Π_2 (containing μ):

$$\Pi_2 = \frac{\rho V D}{\mu}$$

which is the Reynolds number (known to be dimensionless).

For Π_3 (containing ω):

$$\Pi_3 = \frac{\omega \rho D^2}{\mu} \doteq \frac{(T^{-1})(FL^{-4}T^2)(L)^2}{FL^{-2}T} \doteq F^0L^0T^0$$

Check using MLT:

$$\frac{\omega \rho D^2}{\mu} \doteq \frac{(T^{-1})(ML^{-3})(L)^2}{ML^{-1}T^{-1}} \doteq M^0L^0T^0 \quad \therefore \text{ok}$$

Thus,

$$\frac{\mathcal{T}}{\rho V^2 D^2} = \phi\left(\frac{\rho V D}{\mu}, \frac{\rho D^2 \omega}{\mu}\right)$$

7.7R (Modeling/similarity) The water velocity at a certain point along a 1:10 scale model of a dam spillway is 5 m/s. What is the corresponding prototype velocity if the model and prototype operate in accordance with Froude number similarity?
(ANS: 15.8 m/s)

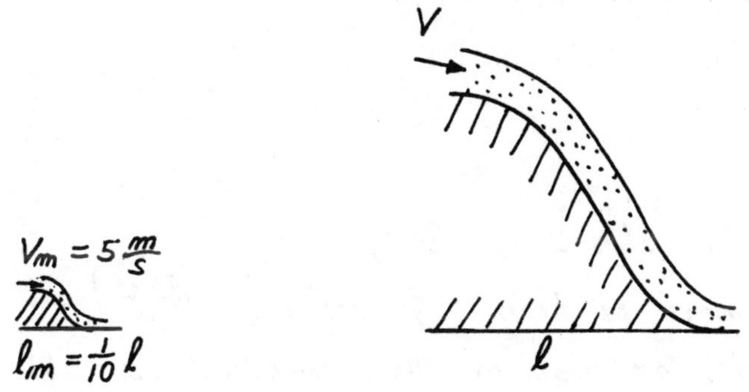

For Froude number similarity, $Fr_m = Fr$, or

$$\frac{V_m}{\sqrt{g_m l_m}} = \frac{V}{\sqrt{g l}}$$

so that

$$V = \sqrt{\left(\frac{g}{g_m}\right)\left(\frac{l}{l_m}\right)}\; V_m$$

and with $g = g_m$, $l/l_m = 10$, $V_m = 5\ m/s$, then

$$V = \sqrt{10}\left(5\frac{m}{s}\right) = \underline{\underline{15.8\ \frac{m}{s}}}$$

7.8R

7.8R (Modeling/similarity) The pressure drop per unit length in a 0.25-in.-diameter gasoline fuel line is to be determined from a laboratory test using the same tubing but with water as the fluid. The pressure drop at a gasoline velocity of 1.0 ft/s is of interest. **(a)** What water velocity is required? **(b)** At the properly scaled velocity from part (a), the pressure drop per unit length (using water) was found to be 0.45 psf/ft. What is the predicted pressure drop per unit length for the gasoline line?

(ANS: 2.45 ft/s; 0.0510 lb/ft² per ft)

For flow in a closed conduit,

$$\text{Dependent pi term} = \phi\left(\frac{\ell_i}{\ell}, \frac{\epsilon}{\ell}, \frac{\rho V \ell}{\mu}\right) \qquad \text{(Eq. 7.16)}$$

For this particular problem the dependent variable is the pressure drop per unit length, Δp_ℓ, so that

$$\text{Dependent pi term} = \frac{\Delta p_\ell\, \ell}{\rho V^2} \doteq \frac{(FL^{-2})(L)}{(FL^{-4}T^2)(L^2T^{-2})} = F^0L^0T^0$$

Also, the characteristic length for pipe flow is the pipe diameter, D. Thus,

$$\frac{\Delta p_\ell D}{\rho V^2} = \phi\left(\frac{\epsilon}{D}, \frac{\rho V D}{\mu}\right) \qquad (1)$$

(a) To maintain dynamic similarity,

$$\frac{\rho_m V_m D_m}{\mu_m} = \frac{\rho V D}{\mu}$$

and with $D_m = D$

$$V_m = \frac{\mu_m}{\mu}\frac{\rho}{\rho_m} V = \frac{\left(2.34\times 10^{-5}\ \frac{lb\cdot s}{ft^2}\right)\left(1.32\ \frac{slugs}{ft^3}\right)}{\left(6.5\times 10^{-6}\ \frac{lb\cdot s}{ft^2}\right)\left(1.94\ \frac{slugs}{ft^3}\right)}\left(1\ \frac{ft}{s}\right) = \underline{\underline{2.45\ \frac{ft}{s}}}$$

(b) With the same Reynolds number, and with $\epsilon_m / D_m = \epsilon / D$ (same tubing), then from Eq. (1)

$$\frac{\Delta p_{\ell m} D_m}{\rho_m V_m^2} = \frac{\Delta p_\ell D}{\rho V^2}$$

so that

$$\Delta p_\ell = \frac{\rho}{\rho_m}\frac{V^2}{V_m^2}\frac{D_m}{D}\Delta p_{\ell m} = \frac{\left(1.32\ \frac{slugs}{ft^3}\right)\left(1\ \frac{ft}{s}\right)^2}{\left(1.94\ \frac{slugs}{ft^3}\right)\left(2.45\ \frac{ft}{s}\right)^2}(1)\left(0.45\ \frac{psf}{ft}\right) = \underline{\underline{0.0510\ \frac{psf}{ft}}}$$

7.9R

7.9R (Modeling/similarity) A thin layer of an incompressible fluid flows steadily over a horizontal smooth plate as shown in Fig. P7.9R. The fluid surface is open to the atmosphere, and an obstruction having a square cross section is placed on the plate as shown. A model with a length scale of $\frac{1}{4}$ and a fluid density scale of 1.0 is to be designed to predict the depth of fluid, y, along the plate. Assume that inertial, gravitational, surface tension, and viscous effects are all important. What are the required viscosity and surface tension scales?

(ANS: 0.125; 0.0625)

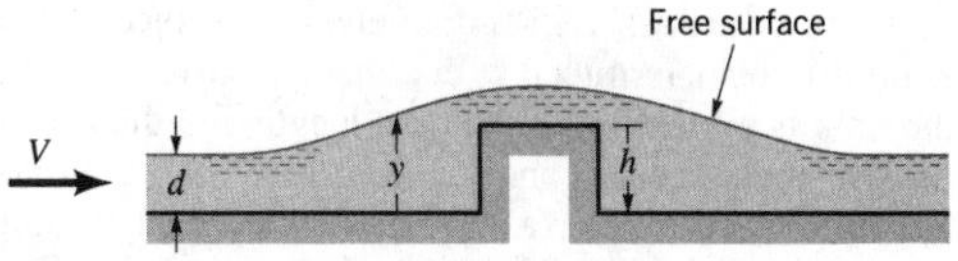

■ FIGURE P7.9R

A fluid dynamics problem for which inertial, gravitational, surface tension, and viscous effects are all important requires Froude, Reynolds, and Weber number similarity (see Table 7.1). Thus, for

$$Fr_m = Fr \quad \text{or} \quad \frac{V_m}{\sqrt{g_m d_m}} = \frac{V}{\sqrt{g d}}$$

(Froude number similarity) it follows that (with $g = g_m$)

$$\frac{V_m}{V} = \sqrt{\frac{d_m}{d}}$$

For Reynolds number similarity,

$$Re_m = Re \quad \text{or} \quad \frac{\rho_m V_m d_m}{\mu_m} = \frac{\rho V d}{\mu}$$

so that

$$\frac{\mu_m}{\mu} = \frac{\rho_m}{\rho}\frac{V_m}{V}\frac{d_m}{d} = \frac{\rho_m}{\rho}\sqrt{\frac{d_m}{d}}\,\frac{d_m}{d} = \frac{\rho_m}{\rho}\left(\frac{d_m}{d}\right)^{3/2}$$

$$= (1.0)\left(\frac{1}{4}\right)^{3/2} = \frac{1}{8} = \underline{\underline{0.125}}$$

For Weber number similarity,

$$We_m = We \quad \text{or} \quad \frac{\rho_m V_m^2 d_m}{\sigma_m} = \frac{\rho V^2 d}{\sigma}$$

Hence,

$$\frac{\sigma_m}{\sigma} = \frac{\rho_m}{\rho}\frac{V_m^2}{V^2}\frac{d_m}{d} = \frac{\rho_m}{\rho}\left(\sqrt{\frac{d_m}{d}}\right)^2\frac{d_m}{d} = \frac{\rho_m}{\rho}\left(\frac{d_m}{d}\right)^2$$

$$= (1.0)\left(\frac{1}{4}\right)^2 = \frac{1}{16} = \underline{\underline{0.0625}}$$

7.10R

7.10R (Correlation of experimental data) The drag on a 30-ft long, vertical, 1.25-ft diameter pole subjected to a 30 mph wind is to be determined with a model study. It is expected that the drag is a function of the pole length and diameter, the fluid density and viscosity, and the fluid velocity. Laboratory model tests were performed in a high-speed water tunnel using a model pole having a length of 2 ft and a diameter of 1 in. Some model drag data are shown in Fig. P7.10R. Based on these data, predict the drag on the full-sized pole.

(ANS: 52.2 lb)

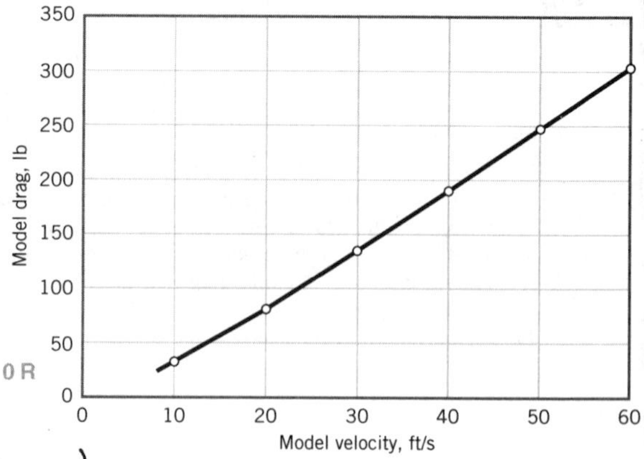

■ FIGURE P7.10R

$$\mathcal{D} = f(\ell, D, \rho, \mu, V)$$

where: $\mathcal{D}$ ~ drag $\doteq F$, ℓ ~ pole length $\doteq L$, D ~ pole diameter $\doteq L$, ρ ~ fluid density $\doteq FL^{-4}T^2$, μ ~ fluid viscosity $\doteq FL^{-2}T$, V ~ velocity $\doteq LT^{-1}$.

From the pi theorem, 6-3 = 3 pi terms required. A dimensional analysis (for example using D, V, ρ, as repeating variables yields

$$\frac{\mathcal{D}}{\rho V^2 D^2} = \phi\left(\frac{\ell}{D}, \frac{\rho V D}{\mu}\right)$$

Geometric similarity requires that $\frac{\ell_m}{D_m} = \frac{\ell}{D}$ and this condition is satisfied since

$$\frac{\ell}{D} = \frac{2\text{ ft}}{1/12\text{ ft}} = \frac{30\text{ ft}}{1.25\text{ ft}} = 24 = \frac{\ell_m}{D_m}$$

Reynolds number similarity requires that $\frac{\rho_m V_m D_m}{\mu_m} = \frac{\rho V D}{\mu}$

so that

$$V_m = \frac{\mu_m}{\mu}\frac{\rho}{\rho_m}\frac{D}{D_m}V = \left(\frac{2.34\times10^{-5}\ \frac{\text{lb}\cdot\text{s}}{\text{ft}^2}}{3.74\times10^{-7}\ \frac{\text{lb}\cdot\text{s}}{\text{ft}^2}}\right)\left(\frac{0.00238\ \frac{\text{slugs}}{\text{ft}^3}}{1.94\ \frac{\text{slugs}}{\text{ft}^3}}\right)\left(\frac{1.25\text{ ft}}{1/12\text{ ft}}\right)V$$

$$= 1.15\,V$$

and with

$$V = (30\text{ mph})\left(\frac{1\text{ hr}}{3600\text{ s}}\right)\left(\frac{5280\text{ ft}}{\text{mi}}\right) = 44.0\ \frac{\text{ft}}{\text{s}}$$

it follows that the required model velocity is $V_m = 1.15\left(44.0\ \frac{\text{ft}}{\text{s}}\right) = 50.6\ \frac{\text{ft}}{\text{s}}$

From the figure, at $V_m = 50.6\ \frac{\text{ft}}{\text{s}}$ the model drag $\mathcal{D}_m = 250$ lb. The prediction equation is

$$\frac{\mathcal{D}}{\rho V^2 D^2} = \frac{\mathcal{D}_m}{\rho_m V_m^2 D_m^2}$$

and therefore

$$\mathcal{D} = \frac{\rho}{\rho_m}\frac{V^2}{V_m^2}\frac{D^2}{D_m^2}\mathcal{D}_m = \left(\frac{0.00238\ \frac{\text{slugs}}{\text{ft}^3}}{1.94\ \frac{\text{slugs}}{\text{ft}^3}}\right)\left(\frac{44.0\ \frac{\text{ft}}{\text{s}}}{50.6\ \frac{\text{ft}}{\text{s}}}\right)^2\left(\frac{1.25\text{ ft}}{1/12\text{ ft}}\right)^2(250\text{ lb})$$

Thus, the predicted drag is

$$\mathcal{D} = \underline{\underline{52.2\text{ lb}}}$$

7.11R

7.11R (Correlation of experimental data) A liquid is contained in a U-tube as is shown in Fig. P7.11R. When the liquid is displaced from its equilibrium position and released, it oscillates with a period τ. Assume that τ is a function of the acceleration of gravity, g, and the column length, ℓ. Some laboratory measurements made by varying ℓ and measuring τ, with $g = 32.2\ \text{ft/s}^2$, are given in the following table.

τ (s)	0.548	0.783	0.939	1.174
ℓ (ft)	0.49	1.00	1.44	2.25

Based on these data, determine a general equation for the period.
(ANS: $\tau = 4.44(\ell/g)^{1/2}$)

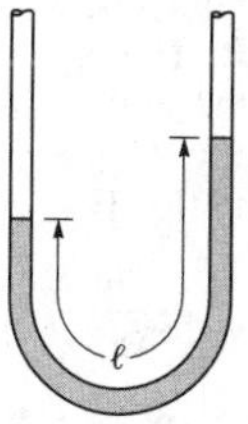

■ FIGURE P7.11R

$$\tau = f(g, \ell)$$

$$\tau \doteq T \qquad g \doteq LT^{-2} \qquad \ell \doteq L$$

From the pi theorem, 3−2 = 1 pi term required: Π_1 = constant.

By inspection:

$$\Pi_1 = \tau\sqrt{\frac{g}{\ell}} \doteq \frac{T\,(LT^{-2})^{1/2}}{(L)^{1/2}} \doteq L^0T^0$$

Since there is only 1 pi term, it follows that

$$\tau\sqrt{\frac{g}{\ell}} = C$$

where C is a constant. For the data given:

ℓ, ft	0.49	1.00	1.44	2.25
$\tau\sqrt{\frac{g}{\ell}}$	4.44	4.44	4.44	4.44

Thus, C = 4.44 and

$$\tau = 4.44\sqrt{\frac{\ell}{g}}$$

7.12 R

7.12R (Dimensionless governing equations) An incompressible fluid is contained between two large parallel plates as shown in Fig. P7.12R. The upper plate is fixed. If the fluid is initially at rest and the bottom plate suddenly starts to move with a constant velocity, U, the governing differential equation describing the fluid motion is

$$\rho \frac{\partial u}{\partial t} = \mu \frac{\partial^2 u}{\partial y^2}$$

where u is the velocity in the x direction, and ρ and μ are the fluid density and viscosity, respectively. Rewrite the equation and the initial and boundary conditions in dimensionless form using h and U as reference parameters for length and velocity, and $h^2\rho/\mu$ as a reference parameter for time.

(ANS: $\partial u^*/\partial t^* = \partial^2 u^*/\partial y^{*2}$ with $u^* = 0$ at $t^* = 0$, $u^* = 1$ at $y^* = 0$, and $u^* = 0$ at $y^* = 1$)

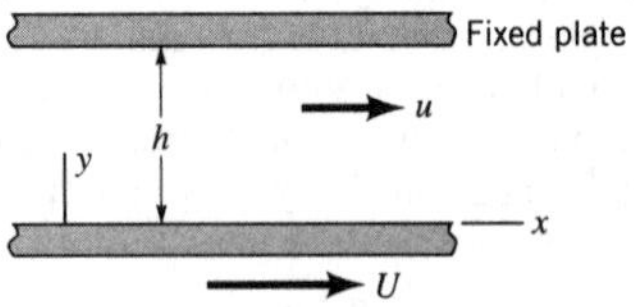

■ FIGURE P7.12R

Let $y^* = \frac{y}{h}$, $t^* = \frac{t}{\tau}$ (where $\tau = \frac{h^2\rho}{\mu}$), and $u^* = \frac{u}{U}$,

so that:

$$\frac{\partial u}{\partial t} = \frac{\partial (U u^*)}{\partial t^*}\frac{\partial t^*}{\partial t} = U\frac{\partial u^*}{\partial t^*}\left(\frac{1}{\tau}\right) = \frac{U}{\tau}\frac{\partial u^*}{\partial t^*}$$

$$\frac{\partial u}{\partial y} = \frac{\partial (U u^*)}{\partial y^*}\frac{\partial y^*}{\partial y} = U\frac{\partial u^*}{\partial y^*}\left(\frac{1}{h}\right) = \frac{U}{h}\frac{\partial u^*}{\partial y^*}$$

$$\frac{\partial^2 u}{\partial y^2} = \frac{U}{h}\frac{\partial}{\partial y^*}\left(\frac{\partial u^*}{\partial y^*}\right)\frac{\partial y^*}{\partial y} = \frac{U}{h^2}\frac{\partial^2 u^*}{\partial y^{*2}}$$

Thus, the original differential equation becomes

$$\left[\frac{\rho U}{\tau}\right]\frac{\partial u^*}{\partial t^*} = \left[\frac{\mu U}{h^2}\right]\frac{\partial^2 u^*}{\partial y^{*2}}$$

and with $\tau = h^2\rho/\mu$

$$\left[\frac{\mu U}{h^2}\right]\frac{\partial u^*}{\partial t^*} = \left[\frac{\mu U}{h^2}\right]\frac{\partial^2 u^*}{\partial y^{*2}}$$

so that

$$\frac{\partial u^*}{\partial t^*} = \frac{\partial^2 u^*}{\partial y^{*2}}$$

The initial condition $u=0$ at $t=0$ becomes $u^*=0$ at $t^*=0$, and the boundary conditions $u=U$ at $y=0$; $u=0$ at $y=h$, become $u^*=1.0$ at $y^*=0$ and $u^*=0$ at $y^*=1.0$.

7.13R

7.13R (Dimensionless governing equations) The flow between two concentric cylinders (see Fig. P7.13R) is governed by the differential equation

$$\frac{d^2 v_\theta}{dr^2} + \frac{d}{dr}\left(\frac{v_\theta}{r}\right) = 0$$

where v_θ is the tangential velocity at any radial location, r. The inner cylinder is fixed and the outer cylinder rotates with an angular velocity ω. Express the equation in dimensionless form using R_o and ω as reference parameters.

(ANS: $d^2v_\theta^*/dr^{*2} + d(v_\theta^*/r^*)dr^* = 0$)

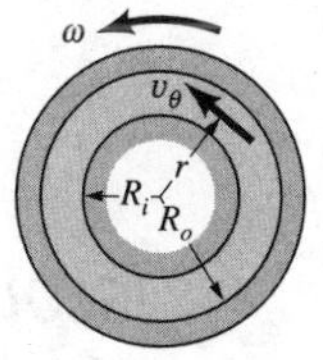

■ FIGURE P7.13R

Let $r^* = \frac{r}{R_o}$ and $v_\theta^* = \frac{v_\theta}{R_o\omega}$ so that

$$\frac{dv_\theta}{dr} = \frac{d(R_o\omega\, v_\theta^*)}{dr^*}\frac{dr^*}{dr} = R_o\omega\frac{dv_\theta^*}{dr^*}\left(\frac{1}{R_o}\right) = \omega\frac{dv_\theta^*}{dr^*}$$

and

$$\frac{d^2 v_\theta}{dr^2} = \omega\frac{d}{dr^*}\left(\frac{dv_\theta^*}{dr^*}\right)\frac{dr^*}{dr} = \frac{\omega}{R_o}\frac{d^2 v_\theta^*}{dr^{*2}}$$

Similarly,

$$\frac{d}{dr}\left(\frac{v_\theta}{r}\right) = \frac{d}{dr^*}\left(\frac{R_o\omega\, v_\theta^*}{R_o r^*}\right)\frac{dr^*}{dr} = \frac{\omega}{R_o}\frac{d}{dr^*}\left(\frac{v_\theta^*}{r^*}\right)$$

Thus, the original differential equation becomes

$$\left[\frac{\omega}{R_o}\right]\frac{d^2 v_\theta^*}{dr^{*2}} + \left[\frac{\omega}{R_o}\right]\frac{d}{dr^*}\left(\frac{v_\theta^*}{r^*}\right) = 0$$

or

$$\frac{d^2 v_\theta^*}{dr^{*2}} + \frac{d}{dr^*}\left(\frac{v_\theta^*}{r^*}\right) = 0$$

8
Viscous Flow in Pipes

8.1 General Characteristics of Pipe Flow

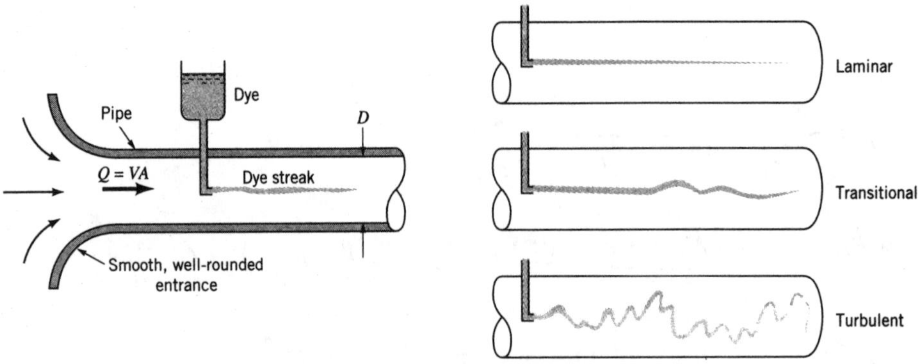

(a) (b)

■ FIGURE 8.3 **(a) Experiment to illustrate type of flow. (b) Typical dye streaks.**

$$\frac{\ell_e}{D} = 0.06 \text{ Re for laminar flow} \qquad \textbf{(8.1)}$$

$$\frac{\ell_e}{D} = 4.4\,(\text{Re})^{1/6} \text{ for turbulent flow} \qquad \textbf{(8.2)}$$

- Pipe flow may be *laminar*, *transitional*, or *turbulent* as illustrated in Fig. 8.3.
- The flow in a round pipe is laminar if the Reynolds number, $\text{Re} = \rho VD/\mu$, is less than approximately 2100. If the Reynolds number is greater than approximately 4000 the flow will usually be turbulent.
- The region of flow near where a fluid enters the pipe is termed the *entrance region*. In this region the velocity profile changes along the pipe.
- Beyond the entrance region the flow is said to be *fully developed* and the velocity profile remains constant along the pipe.
- As shown by Eqs. 8.1 and 8.2, the entrance length is a function of the Reynolds number.
- For horizontal pipe flow, it is the pressure difference, $\Delta p = p_1 - p_2$, between one section of pipe and the other which forces the fluid to flow.
- *Caution:* When calculating a Reynolds number be sure to use a consistent system of units.

EXAMPLE: Ethyl alcohol is pumped through a 4-in.-diameter pipe at the rate of 0.90 ft^3/s. Is the flow laminar, transitional, or turbulent?

SOLUTION:

$Q = AV$ so that $V = Q/A = (0.90\,ft^3/s)/[(\pi/4)(4/12\,ft)^2] = 10.3\,ft/s$.

Thus, $Re = \rho VD/\mu = (1.53\,slugs/ft^3)(10.3\,ft/s)(4/12\,ft)/(2.49\times10^{-5}\,lb\cdot s/ft^2)$

or $Re = 2.11\times10^5 \gg 4000$. Hence, the flow is turbulent.

EXAMPLE: Glycerin flows from a large tank into a pipe having a diameter of 0.25 m. The mean velocity in the pipe is 2.5 m/s. Approximately how long is the entrance length in this pipe?

SOLUTION:

Since $Re = \rho VD/\mu = (1260\,kg/m^3)(2.5\,m/s)(0.25\,m)/(1.5\,N\cdot s/m^2)$, or

$Re = 525 < 2100$, the flow is laminar.

Thus, from Eq. 8.1, $\ell_e/D = 0.06\,Re = 0.06(525) = 31.5$, or

$\ell_e = 31.5\,(0.25\,m) = 7.88\,m$

8.2 Fully Developed Laminar Flow

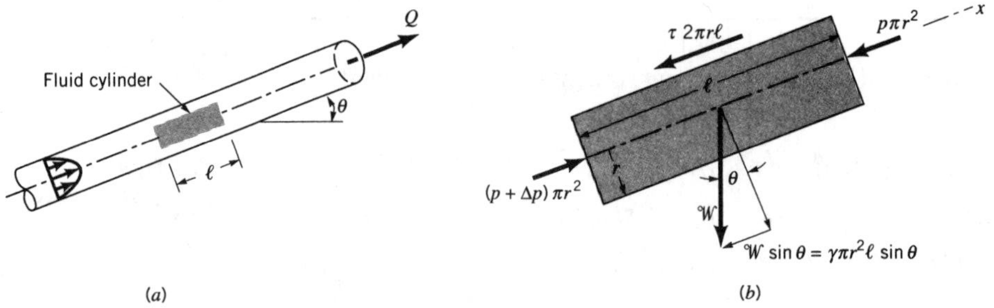

■ **FIGURE 8.10 Free-body diagram of a fluid cylinder for flow in a nonhorizontal pipe.**

$$\Delta p = \frac{4\ell\tau_w}{D} \qquad \textbf{(8.5)}$$

$$u(r) = \left(\frac{\Delta p D^2}{16\mu\ell}\right)\left[1 - \left(\frac{2r}{D}\right)^2\right] = V_c\left[1 - \left(\frac{2r}{D}\right)^2\right] \qquad \textbf{(8.7)}$$

$$Q = \frac{\pi D^4\,\Delta p}{128\mu\ell} \qquad \textbf{(8.9)}$$

$$Q = \frac{\pi(\Delta p - \gamma\ell\sin\theta)D^4}{128\mu\ell} \qquad \textbf{(8.12)}$$

- For steady, fully developed, laminar flow in a horizontal pipe, the pressure drop, Δp, is balanced by the wall shear stress, τ_w, as shown by Eq. 8.5.
- As shown by Eq. 8.7, for steady, fully developed, laminar pipe flow the velocity profile is parabolic. If in addition the pipe is horizontal, the flowrate and pressure drop are linearly related as shown by Eq. 8.9. This relationship is termed Poiseuille's law.
- For steady, fully developed, laminar flow in an inclined pipe (Fig. 8.10), the relationship between the flowrate and pressure drop is given by Eq. 8.12.
- *Caution:* When using the equations that are valid for laminar flow be sure to check the Reynolds number to make sure the flow is laminar (i. e., Re < 2100).

EXAMPLE: A viscous liquid (ρ = 2.0 slugs/ft^3 and μ = 0.039 lb·s/ft^2) flows through a horizontal tube having a diameter D. The pressure drop over a 3-ft length of the tube is 0.80 psi when the flowrate is 0.038 ft^3/s. Determine the tube diameter. Assume laminar flow.

SOLUTION:

From Eq. 8.9, $Q = \pi D^4 \Delta p / 128\mu\ell$ provided Re < 2100. Thus,
$D^4 = (128)(0.039\ \text{lb·s/ft}^2)(3\ \text{ft})(0.038\ \text{ft}^3/\text{s}) / [\pi(0.80\ \text{lb/in.}^2)(144\ \text{in.}^2/\text{ft}^2)]$
or $D = \underline{0.199\ \text{ft}}$.

Note that with $V = Q/A = (0.038\ \text{ft}^3/\text{s})/[(\pi/4)(0.199\ \text{ft})^2] = 1.22\ \text{ft/s}$
it follows that $Re = \rho VD/\mu = (2.0\ \text{slugs/ft}^3)(1.22\ \text{ft/s})(0.199\ \text{ft})/(0.039\ \text{lb·s/ft}^2)$
or Re = 12.5 < 2100, so the flow is laminar.

8.3 Fully Developed Turbulent Flow

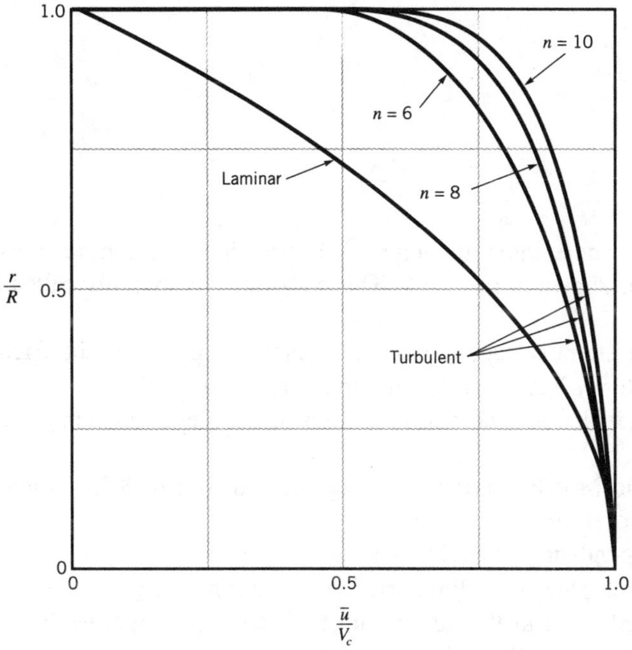

■ FIGURE 8.18 **Typical laminar flow and turbulent flow velocity profiles.**

$$\tau = \mu \frac{d\bar{u}}{dy} - \rho\overline{u'v'} = \tau_{\text{lam}} + \tau_{\text{turb}} \tag{8.26}$$

$$\frac{\bar{u}}{V_c} = \left(1 - \frac{r}{R}\right)^{1/n} \tag{8.31}$$

- Turbulent flow involves randomness on a *macroscopic* scale, whereas laminar flow involves randomness on a *microscopic* scale.
- Turbulent flow characteristics, such as velocity and pressure, can be described in terms of mean and fluctuating components.
- As shown by Eq. 8.26, for turbulent flow, the shear stress is the sum of a laminar portion and a turbulent portion, where the turbulent portion (turbulent shear stress) is related to the fluctuating components of the velocity.
- As shown in Fig. 8.18, turbulent velocity profiles are significantly different than laminar velocity profiles. A power-law velocity profile (Eq. 8.31) is often used as an approximation for a turbulent velocity profile. The parameter n depends on the Reynolds number, Re = $\rho VD/\mu$.

8.4 Dimensional Analysis of Pipe Flow

Moody Chart

$$h_L = f\frac{\ell}{D}\frac{V^2}{2g} \qquad \textbf{(8.34)}$$

$$\frac{1}{\sqrt{f}} = -2.0\log\left(\frac{\varepsilon/D}{3.7}+\frac{2.51}{\mathrm{Re}\sqrt{f}}\right) \qquad \textbf{(8.35)}$$

- The head loss, h_L, for steady, fully developed, incompressible flow in a round pipe is given in terms of the *friction factor*, *f*, as shown in Eq. 8.34. This equation is commonly called the *Darcy-Weisbach equation.*
- The Moody chart, Fig. 8.20 (see the next page), shows the relationship among the friction factor, f, the Reynolds number, $\mathrm{Re} = \rho VD/\mu$, and the *relative roughness*, ε/D.
- In general, for turbulent flow, the friction factor is a function of the Reynolds number and the relative rougnhess.
- The turbulent flow portion of the Moody chart can be represented by Eq. 8.35, which is commonly called the *Colebrook formula.*
- For laminar flow, $f = 64/\mathrm{Re}$, independent of the relative roughness.
- *Caution:* Note that even for smooth pipes ($\varepsilon = 0$) the friction factor is not zero.
- *Caution:* The characteristic length used in the definition of the Reynolds number for pipe flow is the pipe diameter, *not* the pipe length.

EXAMPLE: Water flows through a horizontal 0.5-ft-diameter pipe with a mean velocity of 8 ft/s. The surface roughness of the pipe is $\varepsilon = 0.0005$ ft. Determine the head loss in a 100-ft length of the pipe.

SOLUTION:

From Eq. 8.34, $h_L = f(\ell/D)(V^2/2g)$, where from Fig. 8.20 (Moody chart) with $\varepsilon/D = 0.0005\text{ ft}/0.5\text{ ft} = 0.001$ and $\mathrm{Re} = \rho VD/\mu$ $= (1.94\text{ slugs/ft}^3)(8\text{ ft/s})(0.5\text{ ft})/(2.34\times10^{-5}\text{ lb}\cdot\text{s/ft}^2) = 3.32\times10^5$ it follows that $f = 0.021$.

Thus, $h_L = (0.021)(100\text{ ft}/0.5\text{ ft})(8\text{ ft/s})^2/[2(32.2\text{ ft/s}^2)] = 4.17\text{ ft}$

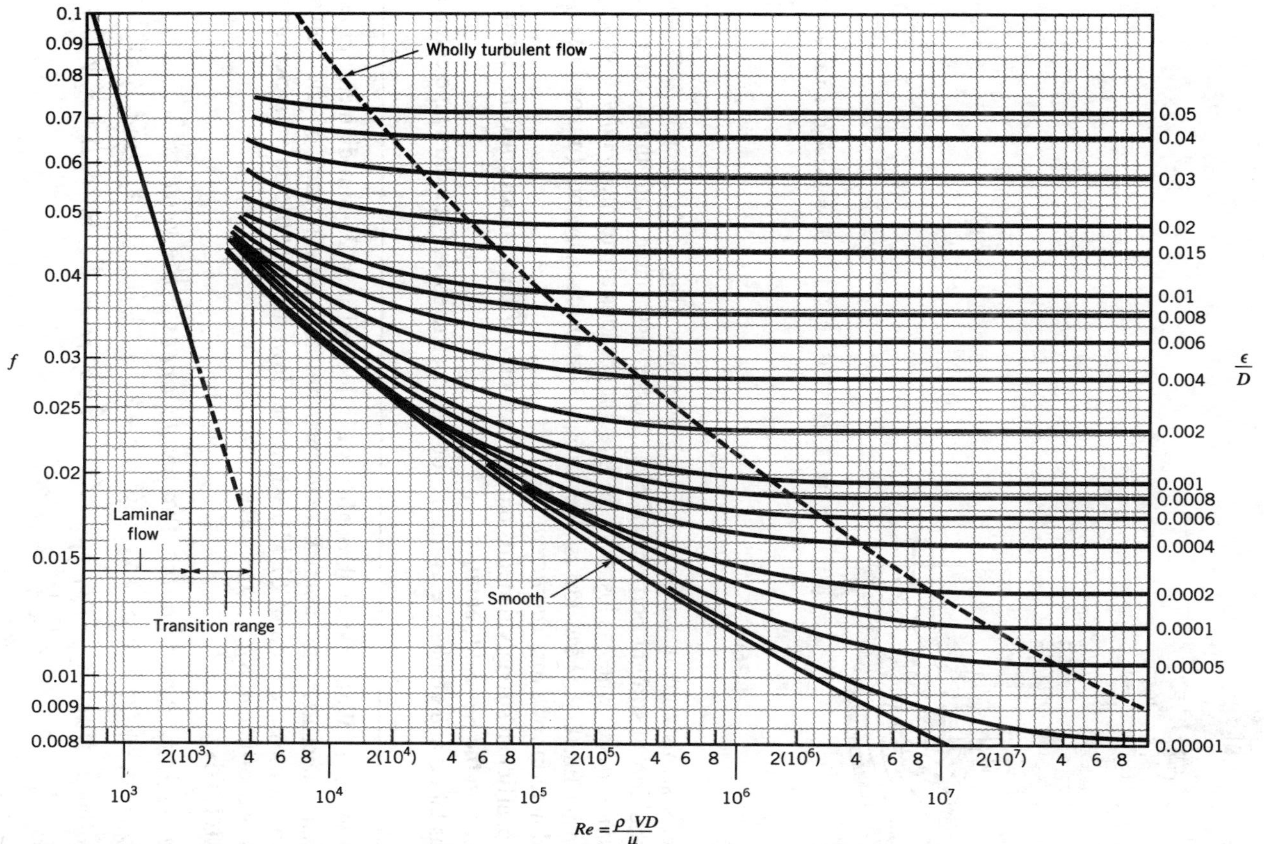

■ **FIGURE 8.20** **Friction factor as a function of Reynolds number and relative roughness for round pipes—the Moody chart. (Data from Ref. 7 with permission.)**

Minor Losses

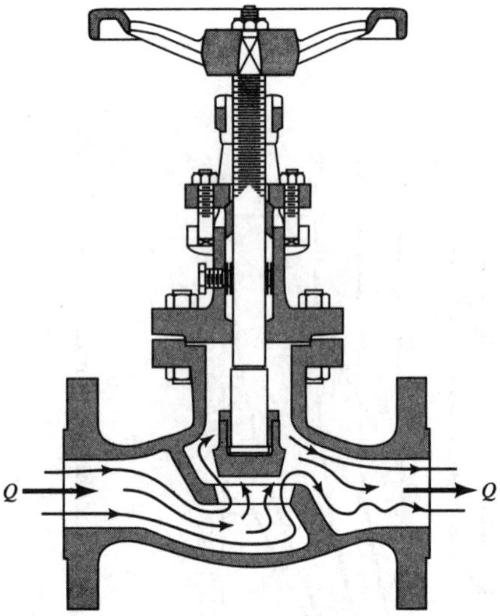

■ FIGURE 8.21 Flow through a valve.

$$h_L = K_L \frac{V^2}{2g} \qquad \textbf{(8.36)}$$

- Most pipe systems contain components such as valves (Fig. 8.21), bends, tees, entrances, exits in addition to straight portions of pipe. Losses associated with these components are termed *minor losses*, whereas losses in the straight portions are termed *major losses*.
- As shown by Eq. 8.36, head losses due to pipe system components (minor losses) can be expressed in terms of a *loss coefficient*, K_L.
- The loss coefficient strongly depends on the geometry of the component considered. It may also depend on the Reynolds number.
- Numerous tables and graphs are available for loss coefficients of standard pipe components.

EXAMPLE: Water flows through a valve located in a 25-mm-diameter pipe. Pressure measurements immediately upstream and downstream of the valve indicate a pressure drop of 67 kPa when the flowrate is 0.0015 m^3/s. Determine the loss coefficient, K_L, for the valve.

SOLUTION:

The energy equation for two points across the valve is
$p_1/\gamma + z_1 + V_1^2/2g = p_2 + z_2 + V_2^2/2g + h_L$. With $z_1 = z_2$ and $V_1 = V_2$
it follows that $h_L = (p_1 - p_2)/\gamma$. Also, from Eq. 8.36, $h_L = K_L V^2/2g$.
Thus, $(p_1 - p_2)/\gamma = K_L V^2/2g$, or
$K_L = 2g(p_1 - p_2)/\gamma V^2$
Also, $V = Q/A = (0.0015\,m^3/s)/[(\pi/4)(25\times10^{-3}m)^2] = 3.06\,m/s$, so that
$K_L = (2)(9.81\,m/s^2)(67\times10^3 N/m^2)/[(9.8\times10^3 N/m^3)(3.06\,m/s)^2] = \underline{\underline{14.3}}$

Noncircular Ducts

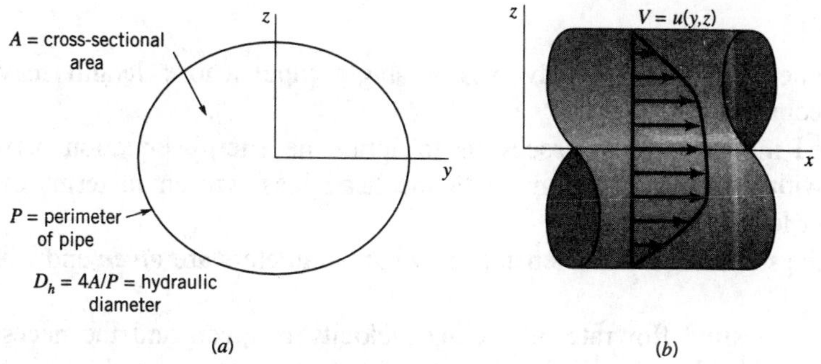

■ **FIGURE 8.34** **Noncircular duct.**

- As illustrated in Fig. 8.34, many ducts used for conveying fluids are not circular in cross section.
- To describe the geometry of a noncircular duct, the *hydraulic diameter*, D_h, is used. The hydraulic diameter, defined as $D_h = 4A/P$, is four times the ratio of the cross-sectional flow area, A, divided by the wetted perimeter, P, of the duct.
- For noncircular ducts the Reynolds number, $\text{Re}_h = \rho V D_h/\mu$, is based on the hydraulic diameter.
- For laminar flow in noncircular ducts, the friction factor can be expressed as $f = C/\text{Re}_h$ where the constant C depends on the particular shape of the duct and is known from theory or experiment for various shapes.
- For fully developed turbulent flow in noncircular ducts, an approximate value for the friction factor can be obtained from the Moody chart for round pipes, with the diameter replaced by the hydraulic diameter and the Reynolds number based on the hydraulic diameter.

EXAMPLE: Air at standard temperature and pressure flows through a 1-m by 2-m rectangular, smooth-walled duct with an average velocity of 9 m/s. Determine the head loss per unit length, h_L/ℓ, in the duct.

SOLUTION:

$h_L = f(\ell/D_h)(V^2/2g)$, where f is a function of $Re_h = \rho V D_h/\mu$ and ε/D_h, and the hydraulic diameter is

$D_h = 4A/P = 4(1m)(2m)/[2(1m+2m)] = 1.33m$. Since the duct is smooth, $\varepsilon/D_h = 0$. Also,

$Re_h = (1.23\ kg/m^3)(9\ m/s)(1.33m)/(1.79\times10^{-5}\ N\cdot s/m^2) = 8.23\times10^5$

Thus, from the Moody chart (with a smooth pipe) $f = 0.012$, so that

$h_L/\ell = (0.012)(9m/s)^2/[(1.33m)(2)(9.81m/s^2)] = \underline{\underline{0.0372}}$

8.5 Pipe Flow Examples

Single Pipes

- The most common pipe flow problem involves a single pipe whose length may be interrupted by various components.
- The main idea involved in the solution process is to apply the energy equation between appropriate locations within the flow system, with the head loss written in terms of the friction factor and minor loss coefficients.
- The specific nature of the solution process depends on what parameters are *given* and what is to be *calculated.*
- In a Type I problem the desired flowrate or average velocity is given and the necessary pressure difference, head added by a pump, head removed by a turbine, or head loss is to be calculated.
- In a Type II problem the applied driving pressure or head loss is given and the resulting flowrate or average velocity is to be calculated. Type II problems often require trial-and-error solution techniques since the friction factor is a function of the unknown velocity (and, therefore, the Reynolds number).
- In a Type III problem the pressure drop and flowrate are given and the required pipe diameter is to be calculated. Type III problems often require trial-and-error solution techniques since the friction factor is a function of the unknown diameter (and, therefore, the Reynolds number).

EXAMPLE (TYPE I): Water is pumped through a horizontal, 0.5-ft-diameter pipe ($\varepsilon/D = 0.01$) and discharges into the atmosphere with a mean velocity of 12 ft/s. At a distance of 500 ft upstream from the end of the pipe a pressure gage reads 0 psi. A pump and a valve ($K_L = 10$) are located in the section of pipe between the pressure gage and the end of the pipe. Determine the head added by the pump.

SOLUTION:

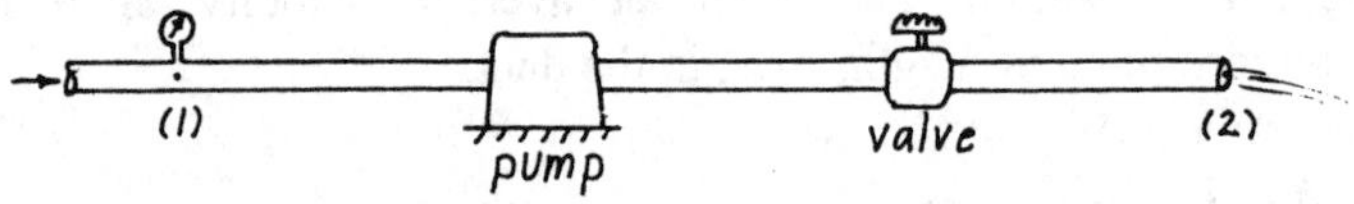

From the energy equation,

$$p_1/\gamma + z_1 + V_1^2/2g + h_p = p_2/\gamma + z_2 + V_2^2/2g + f(\ell/D)(V^2/2g) + K_L(V^2/2g),$$

where $p_1 = p_2 = 0$, $z_1 = z_2$, and $V_1 = V_2$.

Thus, $h_p = [f(\ell/D) + K_L](V^2/2g)$ (1)

Also, $Re = \rho VD/\mu = (1.94\ \text{slugs/ft}^3)(12\ \text{ft/s})(0.5\ \text{ft})/(2.34\times10^{-5}\ \text{lb}\cdot\text{s/ft}^2) = 4.97\times10^5$,

so that with $\varepsilon/D = 0.01$, from the Moody chart $f = 0.038$

Hence, from Eq.(1), $h_p = [(0.038)(500\text{ft})/(0.5\text{ft}) + 10](12\ \text{ft/s})^2/[2(32.2\ \text{ft/s}^2)]$

$= \underline{107\ \text{ft}}$

EXAMPLE (TYPE II): Water flows by gravity from a large open tank through 80 m of a 0.05-m-diameter pipe (ε/D = 0.0008) and discharges into the atmosphere. The difference in elevation between the open surface of the tank and the discharge end of the pipe is 2 m. Determine the average velocity in the pipe. Neglect minor losses.

SOLUTION:

From the energy equation,

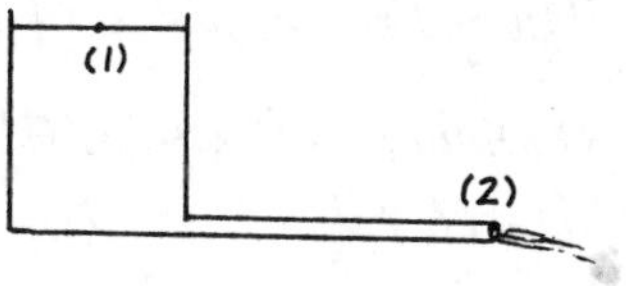

$p_1/\gamma + z_1 + V_1^2/2g = p_2/\gamma + z_2 + V_2^2/2g + f(\ell/D)V^2/2g,$

where $p_1 = p_2 = 0$, $V_1 = 0$, $z_1 = 2m$, $z_2 = 0$ and $V_2 = V$.

Thus, $z_1 = [1 + f(\ell/D)]V^2/2g$, or

$2m = [1 + f(80m)/(0.05m)]\,V^2/[2(9.81\,m/s^2)]$, so that

$V = [39.2/(1+1600f)]^{1/2}$ (1)

Also, $Re = \rho VD/\mu = (999\,kg/m^3)(V)(0.05m)/(1.12\times10^{-3}\,N\cdot s/m^2)$, so that

$Re = 4.46\times10^4\,V$ (2)

Trial and error solution:

Assume $f = 0.018$: from Eq. (1) $V = 1.15\,m/s$; from Eq. (2) $Re = 5.13\times10^4$; and from the Moody chart (with $\varepsilon/D = 0.0008$), $f = 0.024 \neq 0.018$. Try again.

Assume $f = 0.024$: from Eq. (1) $V = 0.997\,m/s$; from Eq. (2) $Re = 4.45\times10^4$; and from the Moody chart $f = 0.024$, which is the same as the assumed value.

Thus, the solution to Eq. (1), (2), and the Moody chart is $\underline{\underline{V = 0.997\,m/s}}$.

EXAMPLE (TYPE III): Gasoline flows through a horizontal, smooth-walled pipe with an average velocity of 4 ft/s. The pressure drop per unit length along the pipe is 0.65 (lb/ft^2)/ft. Determine the diameter, *D*, of the pipe.

SOLUTION:

From the energy equation,

$p_1/\gamma + z_1 + V_1^2/2g = p_2/\gamma + z_2 + V_2^2/2g + f(\ell/D)V^2/2g$, where $z_1 = z_2$, $V_1 = V_2$.

Thus, $(p_1 - p_2)/\ell = \gamma f V^2/(2gD)$, so that with $(p_1 - p_2)/\ell = 0.65\ lb/ft^3$,

$0.65\ lb/ft^3 = (42.5\ lb/ft^3)(f)(4\ ft/s)^2/[2(32.2\ ft/s^2)D]$. Hence,

$D = 16.2\ f$ (1)

Also, $Re = \rho VD/\mu = (1.32\ slugs/ft^3)(D)(4\ ft/s)/(6.5 \times 10^{-6}\ lb \cdot s/ft^2)$, or

$Re = 8.12 \times 10^5\ D$ (2)

Trial and error solution:

Assume $f = 0.02$: from Eq. (1) $D = 0.324\ ft$; from Eq. (2) $Re = 2.63 \times 10^5$; and from the Moody chart (with $\varepsilon/D = 0$) $f = 0.015 \neq 0.02$. Try again.

Assume $f = 0.015$: from Eq. (1) $D = 0.243\ ft$; from Eq. (2) $Re = 1.97 \times 10^5$; and from the Moody chart $f = 0.016 \neq 0.015$ (but close).

With $f = 0.016$ Eqs. (1), (2), and the Moody chart give $D = 0.259\ ft$, $Re = 2.10 \times 10^5$, and $f = 0.016$ (as assumed). Thus, $D = \underline{0.259\ ft}$.

Multiple Pipe Systems

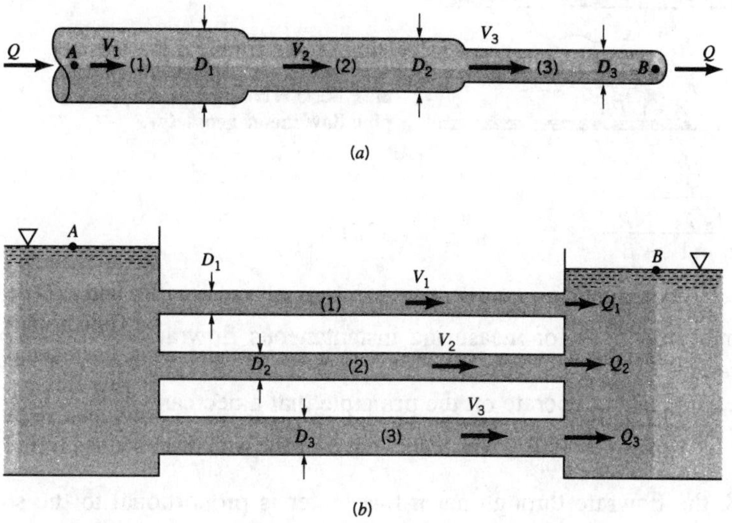

■ **FIGURE 8.35 Series (*a*) and parallel (*b*) pipe systems.**

- Many pipe systems contain more than a single pipe and may be a complex combination of numerous interconnected pipes of different sizes, lengths, etc.
- The simplest *multiple pipe system* is one in which the pipes are in *series* as shown in Fig. 8.35*a*. In this case the flowrate is the same through each pipe ($Q_1 = Q_2 = Q_3$), and the total head loss is the sum of the head losses in each one of the pipes ($h_{L(A\text{-}B)} = h_{L1} + h_{L2} + h_{L3}$).
- Another common multiple pipe system is one in which the pipes are in *parallel* as shown in Fig. 8.35*b*. In this case the total flowrate is equal to the sum of the flowrates in each of the pipes ($Q = Q_1 + Q_2 + Q_3$), and the head loss is the same through each pipe ($h_{L1} = h_{L2} = h_{L3}$).
- Other multiple pipe systems can vary from a relatively simple branching system with one branch to a complex network of pipes that could be found in a city water distribution system.

EXAMPLE: A liquid having a kinematic viscosity of $\nu = 1.5 \times 10^{-5}$ ft²/s flows through a series of three pipes as shown in Fig. 8.35*a*. When the flowrate through section (1) is 0.6 ft³/s, the total head loss between points A and B is 75 ft, with $h_{L1} = 3$ ft and $h_{L2} = 21$ ft. If the diameter of the pipe in section (3) is 3 in., determine its length. Assume the friction factor is 0.03 for all three sections of pipe. Neglect minor losses.

SOLUTION:

The total head loss is $h_{L(A\text{-}B)} = h_{L1} + h_{L2} + h_{L3}$, or
$75 \text{ ft} = 3 \text{ ft} + 21 \text{ ft} + h_{L3}$ so that $h_{L3} = 51 \text{ ft}$. Also, $h_{L3} = f(\ell/D)V^2/2g$,
where $V = Q/A$ and $Q = Q_1 = Q_2 = Q_3 = 0.6 \text{ ft}^3/\text{s}$.
Thus, $V = 0.6 \text{ ft}^3/\text{s}/[(\pi/4)(3/12 \text{ ft})^2] = 12.2 \text{ ft/s}$, so that
$h_{L3} = 51 \text{ ft} = 0.03[\ell/(3/12 \text{ ft})](12.2 \text{ ft/s})^2/[2(32.2 \text{ ft/s}^2)]$
Hence, $\ell = \underline{\underline{184 \text{ ft}}}$.

8.6 Pipe Flowrate Measurement

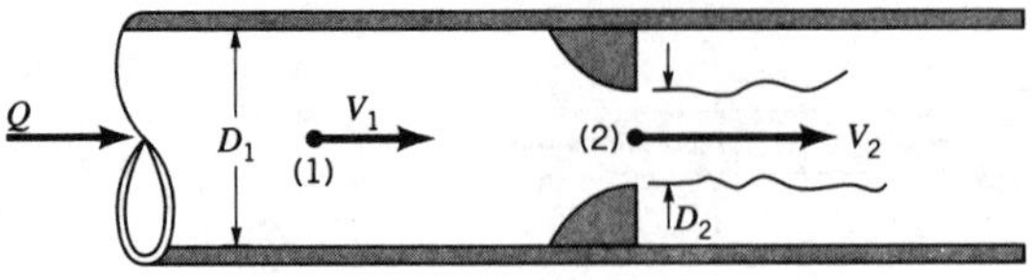

■ FIGURE 8.39 Typical pipe flow meter geometry.

$$Q = C_o Q_{\text{ideal}} = C_o A_o \sqrt{\frac{2(p_1 - p_2)}{\rho(1 - \beta^4)}} \qquad \textbf{(8.38)}$$

- Three of the most common devices for measuring instantaneous flowrate in pipes are the *orifice meter*, the *nozzle meter*, and the *Venturi meter*.
- Orifice, nozzle, and Venturi meters operate on the principle that a decrease in flow area in a pipe, as illustrated in Fig. 8.39, causes an increase in velocity that is accompanied by a decrease in pressure.
- As shown by Eq. 8.38, the flowrate through an orifice meter is proportional to the square root of the pressure difference, $p_1 - p_2$. To account for real fluid effects an *orifice discharge coefficient,* C_o, is used. In the notation of Fig. 8.38, the diameter ratio is $\beta = D_2/D_1$.
- For the nozzle meter and the Venturi meter, Eq. 8.38 can also be used with the orifice discharge coefficient, C_o, replaced with the appropriate *nozzle discharge coefficient,* C_n, or the *Venturi discharge coefficient,* C_v.
- Flow meter discharge coefficients typically depend on the specific geometry of the device as well as the Reynolds number.
- Other common types of flow meters include the rotameter and the turbine meter, as well as devices to measure the amount (volume rather than volume flowrate) of fluid that has passed through a pipe in a given time interval.
- *Caution:* To use Eq. 8.38 for the determination of a flowrate, it may be necessary to use a trial and error technique since the discharge coefficient depends on the unknown flowrate (and, therefore, the Reynolds number). For some problems, however, an approximate constant value for the discharge coefficient can be used to provide, with sufficient accuracy, a direct (no trial and error) determination of a flowrate.

EXAMPLE: Standard air flows through a nozzle meter similar to that shown in Fig. 8.39. The measured pressure drop is $p_1 - p_2 = 9.6$ kPa. The nozzle meter is located in a pipe with a diameter of 50 mm and the diameter of the opening in the nozzle meter is 25 mm. Assume a nozzle discharge coefficient of 0.96 and that the air is incompressible. Determine the flowrate through the pipe.

SOLUTION:

From Eq. 8.38, $Q = C_n A_n \sqrt{2(p_1 - p_2)/\rho(1-\beta^4)}$, where $A_n = A_2$, $C_n = 0.96$, and $\beta = D_2/D_1 = 25\,\text{mm}/50\,\text{mm} = 0.5$.

Thus, $Q = (0.96)(\pi/4)(25\times10^{-3}\,\text{m})^2 \sqrt{2(9.6\times10^3\,\text{N/m}^2)/[(1.23\,\text{kg/m}^3)(1-0.5^4)]}$

$= \underline{0.0608\ \text{m}^3/\text{s}}$

Solutions

to Review Problems

Chapter 8

8.1R (Laminar flow) Asphalt at 120 °F, considered to be a Newtonian fluid with a viscosity 80,000 times that of water and a specific gravity of 1.09, flows through a pipe of diameter 2.0 in. If the pressure gradient is 1.6 psi/ft determine the flowrate assuming the pipe is **(a)** horizontal; **(b)** vertical with flow up.

(ANS: $4.69 \times 10^{-3}\ ft^3/s$; $3.30 \times 10^{-3}\ ft^3/s$)

If the flow is laminar, then $Q = \dfrac{\pi(\Delta p - \gamma \ell \sin\theta)D^4}{128 \mu \ell}$ (1)

where $\gamma = SG\, \gamma_{H_2O} = 1.09\left(62.4 \frac{lb}{ft^3}\right) = 68.0 \frac{lb}{ft^3}$ and from Table B.1 for μ_{H_2O} at 120°F,

$$\mu = 80{,}000\, \mu_{H_2O}\Big|_{120°F} = 8 \times 10^4 \left(1.164 \times 10^{-5} \frac{lb \cdot s}{ft^2}\right) = 0.931 \frac{lb \cdot s}{ft^2}$$

a) For horizontal flow, $\theta = 0$

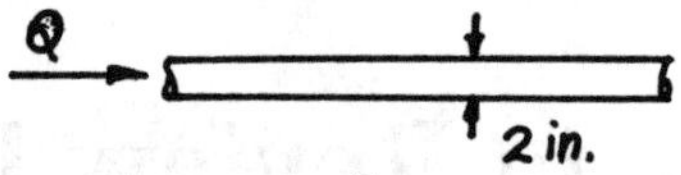

Thus, from Eq.(1)

$$Q = \frac{\pi\left(1.6 \times 144 \frac{lb}{ft^2}\right)\left(\frac{2}{12} ft\right)^4}{128\left(0.931 \frac{lb \cdot s}{ft^2}\right)(1 ft)} = \underline{\underline{4.69 \times 10^{-3} \frac{ft^3}{s}}}$$

b) For vertical flow up, $\theta = 90$

Thus, from Eq.(1)

$$Q = \frac{\pi\left(1.6 \times 144 \frac{lb}{ft^2} - 68 \frac{lb}{ft^3}(1 ft)\right)\left(\frac{2}{12} ft\right)^4}{128\left(0.931 \frac{lb \cdot s}{ft^2}\right)(1 ft)} = \underline{\underline{3.30 \times 10^{-3} \frac{ft^3}{s}}}$$

Note: We must check to see if our assumption of laminar flow is correct.

Since $V = \frac{Q}{A} = \dfrac{4.69 \times 10^{-3} \frac{ft^3}{s}}{\frac{\pi}{4}\left(\frac{2}{12}\right)^2} = 0.215 \frac{ft}{s}$ it follows that

$$Re = \frac{\rho V D}{\mu} = \frac{1.09\left(1.94 \frac{slug}{ft^3}\right)(0.215)\left(\frac{2}{12} ft\right)}{0.931 \frac{lb \cdot s}{ft^2}} = 0.0814 < 2100$$

The flow is laminar when $\theta = 0$. It is laminar for $\theta = 90°$ also.

8.2R

8.2R (Laminar flow) A fluid flows through two horizontal pipes of equal length which are connected together to form a pipe of length 2ℓ. The flow is laminar and fully developed. The pressure drop for the first pipe is 1.44 times greater than it is for the second pipe. If the diameter of the first pipe is D, determine the diameter of the second pipe.

(ANS: 1.095 D)

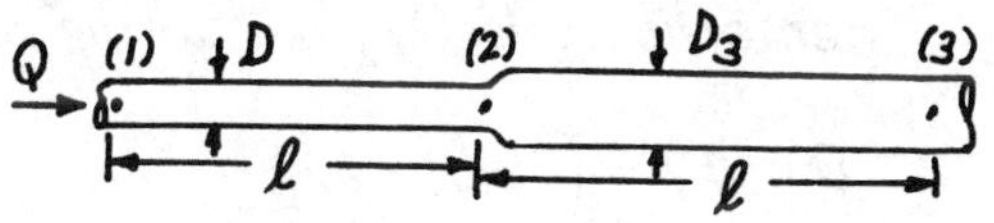

For laminar flow, $Q = \frac{\pi D^4 \Delta p}{128 \mu \ell}$, where $Q_1 = Q_3$ and $\Delta p_{1-2} = 1.44 \Delta p_{2-3}$

Thus,

$$Q_1 = \frac{\pi D^4 \Delta p_{1-2}}{128 \mu \ell} = Q_3 = \frac{\pi D_3^4 \Delta p_{2-3}}{128 \mu \ell}$$

or

$$D_3 = D\left(\frac{\Delta p_{1-2}}{\Delta p_{2-3}}\right)^{1/4} = D(1.44)^{1/4} = \underline{\underline{1.095\,D}}$$

8.3R (Velocity profile) A fluid flows through a pipe of radius R with a Reynolds number of 100,000. At what location, r/R, does the fluid velocity equal the average velocity? Repeat if the Reynolds number is 1000.

(ANS: 0.758; 0.707)

For $Re = 10^5$, $\frac{\bar{u}}{V_c} = (1-\frac{r}{R})^{\frac{1}{n}}$, where $n = 7.1$ (see Fig. 8.17)

The relationship between the average velocity, V, and the centerline velocity, V_c, can be obtained from

$$Q = AV = \int \bar{u}\, dA \quad \text{or}$$

$$\pi R^2 V = \int \bar{u}\,(2\pi r\, dr) = 2\pi V_c \int_{r=0}^{r=R} (1-\frac{r}{R})^{\frac{1}{n}} r\, dr \qquad (1)$$

But, with

$x \equiv 1-\frac{r}{R}$ so that $dx = -\frac{1}{R}dr$ and $r = R(1-x)$, we have

$$\int_{r=0}^{r=R} (1-\frac{r}{R})^{\frac{1}{n}} r\, dr = -R^2 \int_{x=1}^{x=0} x^{\frac{1}{n}}(1-x)\,dx = R^2 \int_0^1 [x^{\frac{1}{n}} - x^{\frac{n+1}{n}}]\,dx$$

$$= R^2 \left[\frac{n}{(n+1)} x^{\frac{n+1}{n}} - \frac{n}{(2n+1)} x^{\frac{2n+1}{n}}\right]_0^1 = R^2 \frac{n^2}{(n+1)(2n+1)} \qquad (2)$$

Thus, from Eqs. (1) and (2):

$$\pi R^2 V = 2\pi V_c R^2 \frac{n^2}{(n+1)(2n+1)} \quad \text{or} \quad V = \frac{2n^2}{(n+1)(2n+1)} V_c$$

Hence, $\bar{u} = V$ when

$$\bar{u} = V_c (1-\frac{r}{R})^{\frac{1}{n}} \quad \text{or} \quad V = \frac{(n+1)(2n+1)}{2n^2} V (1-\frac{r}{R})^{\frac{1}{n}}$$

or

$$\frac{r}{R} = 1 - \left[\frac{2n^2}{(n+1)(2n+1)}\right]^n = 1 - \left[\frac{2(7.1)^2}{(7.1+1)(2\times 7.1+1)}\right]^{7.1} = \underline{\underline{0.758}}$$

For $Re = 10^3$ the flow is laminar with $u = V_c[1-(\frac{r}{R})^2]$ and

$$V = \frac{1}{2} V_c$$

Thus, $u = V$ when

$$\frac{1}{2}V_c = V_c[1-(\frac{r}{R})^2], \quad \text{or} \quad \frac{r}{R} = \sqrt{\frac{1}{2}} = \underline{\underline{0.707}}$$

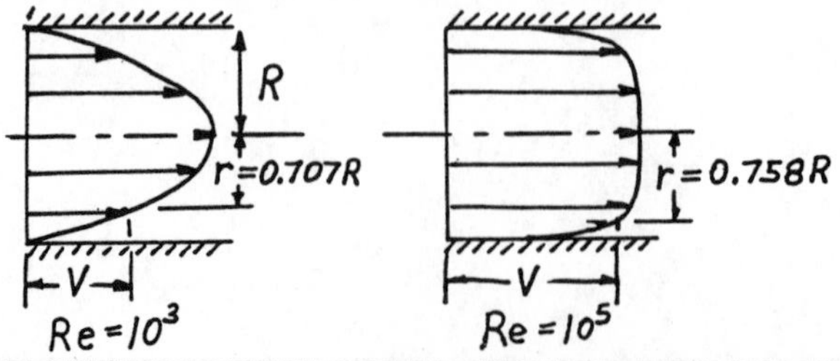

8.4R (Turbulent velocity profile) Water at 80 °C flows through a 120-mm-diameter pipe with an average velocity of 2 m/s. If the pipe wall roughness is small enough so that it does not protrude through the laminar sublayer, the pipe can be considered as smooth. Approximately what is the largest roughness allowed to classify this pipe as smooth?

(ANS: 2.31×10^{-5} m)

Let h = roughness. Thus, $h = \delta_s$, where $\delta_s = \frac{5\nu}{u^*}$ with $u^* = \left(\frac{\tau_w}{\rho}\right)^{1/2}$ and $\tau_w = \frac{D\Delta p}{4\ell}$. Since $\Delta p = f\frac{\ell}{D}\frac{1}{2}\rho V^2$ we obtain

$$\tau_w = \frac{\rho f V^2}{8} \quad \text{or} \quad u^* = \left(\frac{fV^2}{8}\right)^{1/2}$$

With $Re = \frac{VD}{\nu} = \frac{(2\,\frac{m}{s})(0.12\,m)}{3.65\times10^{-7}\,\frac{m^2}{s}} = 6.58\times10^5$ and a smooth pipe we obtain $f = 0.0125$ (see Fig. 8.20).

Hence, $u^* = \left(\frac{0.0125\,(2\,\frac{m}{s})^2}{8}\right)^{1/2} = 0.0791\,\frac{m}{s}$

and

$$\delta_s = \frac{5\,(3.65\times10^{-7}\,\frac{m^2}{s})}{0.0791\,\frac{m}{s}} = \underline{\underline{2.31\times10^{-5}\,m}}$$

$\nu\big|_{80°C} = 3.65\times10^{-7}\,\frac{m^2}{s}$ (see Table B.2)

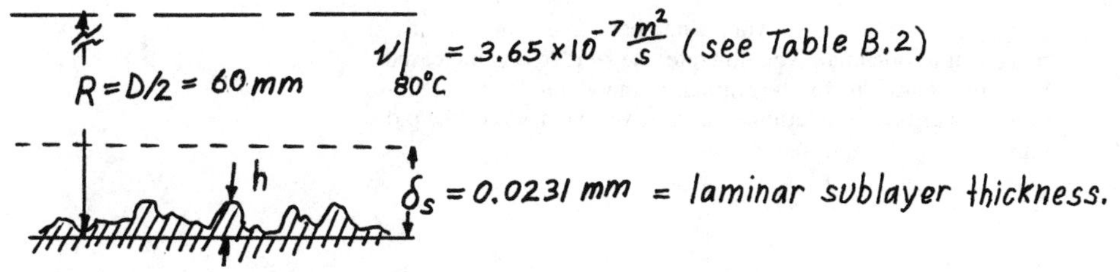

If the roughness element is smaller than 0.0231 mm it lies within the laminar sublayer.

8.5R

8.5R (Moody chart) Water flows in a smooth plastic pipe of 200-mm diameter at a rate of 0.10 m^3/s. Determine the friction factor for this flow.

(ANS: 0.0128)

For a $D = 0.200\,m$ plastic pipe, $\frac{\varepsilon}{D} = 0$ (see Table 8.1)

Also,

$$Re = \frac{VD}{\nu}, \text{ where } V = \frac{Q}{A} = \frac{0.10\,\frac{m^3}{s}}{\frac{\pi}{4}(0.2m)^2} = 3.18\,\frac{m}{s}$$

Hence,

$$Re = \frac{(3.18\,\frac{m}{s})(0.2m)}{1.12\times10^{-6}\,\frac{m^2}{s}} = 5.68\times10^5,$$ so from Fig. 8.20 we obtain

$f = \underline{\underline{0.0128}}$

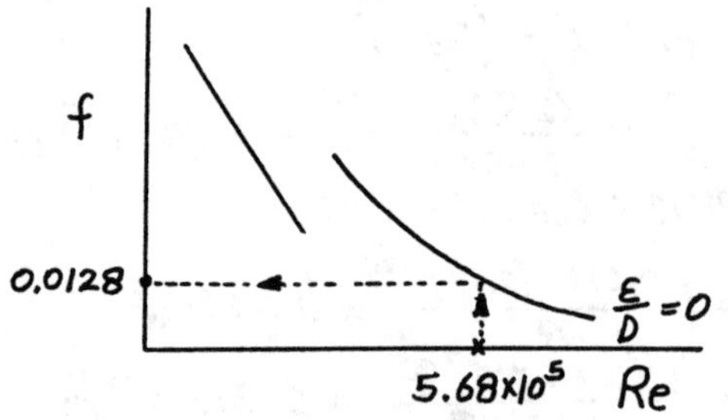

8.6R

8.6R (Moody chart) After a number of years of use, it is noted that to obtain a given flowrate, the head loss is increased to 1.6 times its value for the originally smooth pipe. If the Reynolds number is 10^6, determine the relative roughness of the old pipe.

(ANS: 0.00070)

Let $(\ )_o$ denote the old pipe and $(\ )_s$ the smooth pipe. Thus,

$\left(\frac{\varepsilon}{D}\right)_s = 0$, $Q_o = Q_s$, $V_o = V_s$ and $h_{L_o} = 1.6\,h_{L_s}$. Hence, with $D_o = D_s$

$f_o \frac{\ell_o}{D_o}\frac{V_o^2}{2g} = f_s \frac{\ell_s}{D_s}\frac{V_s^2}{2g}$, or $f_o = 1.6\,f_s$, where from Fig. 8.20 with $Re = 10^6$

$f_s = 0.0115$ Thus, $f_o = 1.6\,(0.0115) = 0.0184$, which with $Re = 10^6$ implies

$\left(\frac{\varepsilon}{D}\right)_o = \underline{\underline{0.00070}}$

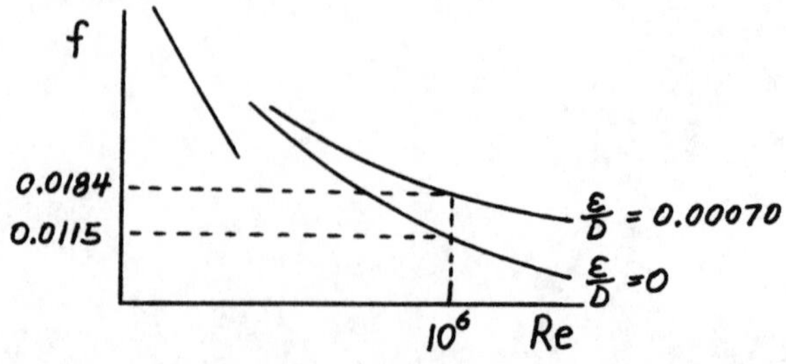

8.7R

8.7R (Minor losses) Air flows through the fine mesh gauze shown in Fig. P8.7R with an average velocity of 1.50 m/s in the pipe. Determine the loss coefficient for the gauze.

(ANS: 56.7)

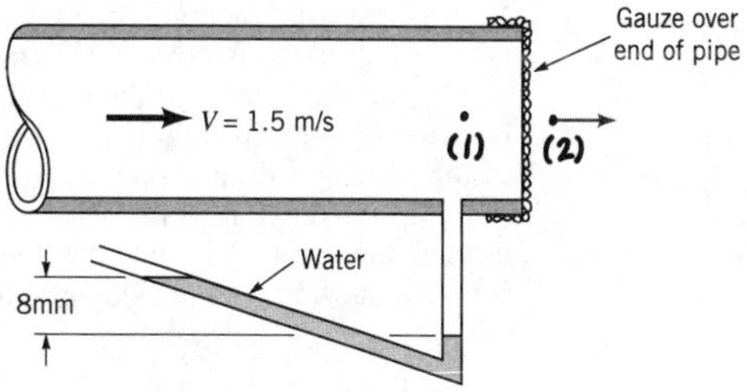

■ FIGURE P8.7R

$$\frac{p_1}{\gamma}+\frac{V_1^2}{2g}+z_1=\frac{p_2}{\gamma}+\frac{V_2^2}{2g}+z_2+K_L\frac{V^2}{2g}, \text{ where } z_1=z_2,\ V_1=V_2=V=1.5\,\frac{m}{s}$$

Thus, $K_L=\dfrac{2(p_1-p_2)}{\rho V^2}$

where $p_2=0$ and $p_1=8\text{mm water}$

or $p_1=(8\times10^{-3}\,m)(9.80\times10^3\,\frac{N}{m^3})=78.4\,\frac{N}{m^2}$

Hence,

$$K_L=\frac{2\,(78.4\,\frac{N}{m^2})}{(1.23\,\frac{kg}{m^3})(1.5\,\frac{m}{s})^2}=\underline{\underline{56.7}}$$

8.8R

8.8R (Noncircular conduits) A manufacturer makes two types of drinking straws: one with a square cross-sectional shape, and the other type the typical round shape. The amount of material in each straw is to be the same. That is, the length of the perimeter of the cross section of each shape is the same. For a given pressure drop, what is the ratio of the flowrates through the straws? Assume the drink is viscous enough to ensure laminar flow and neglect gravity.

(ANS: $Q_{round} = 1.83\, Q_{square}$)

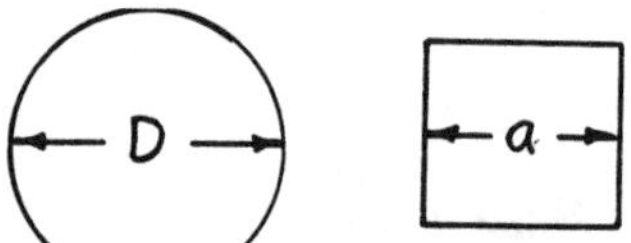

$\Delta p_1 = \Delta p_2$ where $\Delta p = \gamma h_L$

Since $h_{L_1} = h_{L_2}$,

$f_1 \frac{\ell_1}{D_{h_1}} \frac{V_1^2}{2g} = f_2 \frac{\ell_2}{D_{h_2}} \frac{V_2^2}{2g}$, where $\ell_1 = \ell_2$, $D_{h_1} = D$, and $D_{h_2} = \frac{4A_2}{P_2} = \frac{4a^2}{4a} = a$

Thus,

$$\frac{f_1 V_1^2}{D} = \frac{f_2 V_2^2}{a} \qquad (1)$$

Since the perimeters are equal, $P_1 = P_2$, or $\pi D = 4a$. Hence, $a = \frac{\pi}{4} D$

For laminar flow $f = \frac{C}{Re_h}$, where $C_1 = 64$ and $C_2 = 56.9$ (Table 8.3)

and $Re_{h_1} = \frac{V_1 D_{h_1}}{\nu} = \frac{V_1 D}{\nu}$, $Re_{h_2} = \frac{V_2 D_{h_2}}{\nu} = \frac{V_2 a}{\nu} = \frac{\pi V_2 D}{4\nu}$

Thus, from Eq.(1)

$$\frac{\frac{64}{\left(\frac{V_1 D}{\nu}\right)} V_1^2}{D} = \frac{\frac{56.9}{\left(\frac{\pi V_2 D}{4\nu}\right)} V_2^2}{\frac{\pi D}{4}} \quad \text{or} \quad V_1 = 1.441\, V_2$$

Also, $Q_1 = A_1 V_1 = \frac{\pi}{4} D^2 V_1$ and $Q_2 = A_2 V_2 = a^2 V_2 = \frac{\pi^2}{16} D^2 V_2$

so that

$$\frac{Q_1}{Q_2} = \frac{\frac{\pi}{4} D^2 V_1}{\frac{\pi^2}{16} D^2 V_2} = \frac{4}{\pi} \frac{V_1}{V_2} = \frac{4}{\pi}(1.441) = 1.83$$

$$\underline{\underline{Q_{round} = 1.83\, Q_{square}}}$$

8.9R

8.9R (Single pipe—determine pressure drop) Determine the pressure drop per 300-m length of new 0.20-m-diameter horizontal cast iron water pipe when the average velocity is 1.7 m/s.

(ANS: 47.6 kN/m²)

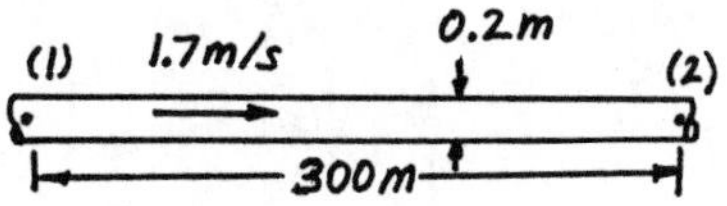

$$\frac{p_1}{\gamma} + \frac{V_1^2}{2g} + z_1 = \frac{p_2}{\gamma} + \frac{V_2^2}{2g} + z_2 + f\frac{\ell}{D}\frac{V^2}{2g}$$

or with $V_1 = V_2 = V = 1.7\,\frac{m}{s}$, $\ell = 300m$, $D = 0.2m$ this gives

$$\Delta p = p_1 - p_2 = \gamma(z_2 - z_1) + f\frac{\ell}{D}\frac{1}{2}\rho V^2 = f\frac{\ell}{D}\frac{1}{2}\rho V^2$$ if the pipe is horizontal.

But from Table 8.1, $\frac{\varepsilon}{D} = \frac{0.26\times10^{-3}\,m}{0.2\,m} = 1.3\times10^{-3}$

Also, $Re = \frac{VD}{\nu} = \frac{(1.7\,\frac{m}{s})(0.2\,m)}{1.12\times10^{-6}\,\frac{m^2}{s}} = 3.04\times10^5$

so that from Fig. 8.20, $f = 0.022$

Thus,

$$\Delta p = 0.022\left(\frac{300\,m}{0.2\,m}\right)\frac{1}{2}\left(999\,\frac{kg}{m^3}\right)\left(1.7\,\frac{m}{s}\right)^2 = 4.76\times10^4\,\frac{N}{m^2} = 47.6\,\frac{kN}{m^2}$$

8.10R

8.10R (Single pipe—determine pressure drop) A fire protection association code requires a minimum pressure of 65 psi at the outlet end of a 250-ft-long, 4-in.-diameter hose when the flowrate is 500 gal/min. What is the minimum pressure allowed at the pumper truck that supplies water to the hose? Assume a roughness of $\varepsilon = 0.03$ in.

(ANS: 94.0 psi)

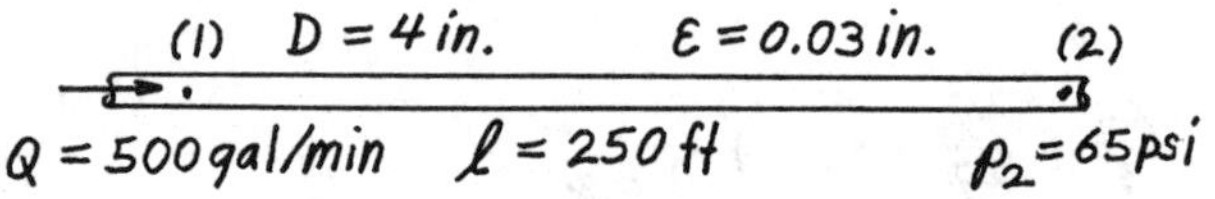

$$\frac{p_1}{\gamma} + \frac{V_1^2}{2g} + z_1 = \frac{p_2}{\gamma} + \frac{V_2^2}{2g} + z_2 + f\frac{\ell}{D}\frac{V^2}{2g}, \text{ where } z_1 = z_2,\ V_1 = V_2 = V$$

and $Q = 500\,\frac{gal}{min}\left(231\,\frac{in.^3}{gal}\right)\left(\frac{1\,ft^3}{1728\,in.^3}\right)\left(\frac{1\,min}{60\,s}\right) = 1.114\,\frac{ft^3}{s}$

Thus,

$$p_1 = p_2 + f\frac{\ell}{D}\frac{1}{2}\rho V^2 \text{ where } V = \frac{Q}{A} = \frac{1.114\,\frac{ft^3}{s}}{\frac{\pi}{4}\left(\frac{4}{12}\,ft\right)^2} = 12.8\,\frac{ft}{s} \qquad (1)$$

From Fig. 8.20 with

$$Re = \frac{VD}{\nu} = \frac{\left(12.8\,\frac{ft}{s}\right)\left(\frac{4}{12}\,ft\right)}{1.21\times 10^{-5}\,\frac{ft^2}{s}} = 3.53\times 10^5 \text{ and } \frac{\varepsilon}{D} = \frac{0.03\,in.}{4\,in.} = 0.0075$$

we obtain

$f = 0.035$ Hence, from Eq. (1)

$$p_1 = 65\,\frac{lb}{in.^2} + (0.035)\left(\frac{250\,ft}{\frac{4}{12}\,ft}\right)\frac{\left(12.8\,\frac{ft}{s}\right)^2}{2}\left(1.94\,\frac{slugs}{ft^3}\right)\left(\frac{1\,ft^2}{144\,in.^2}\right)$$

or

$$p_1 = 65\,psi + 29.0\,psi = \underline{\underline{94.0\,psi}}$$

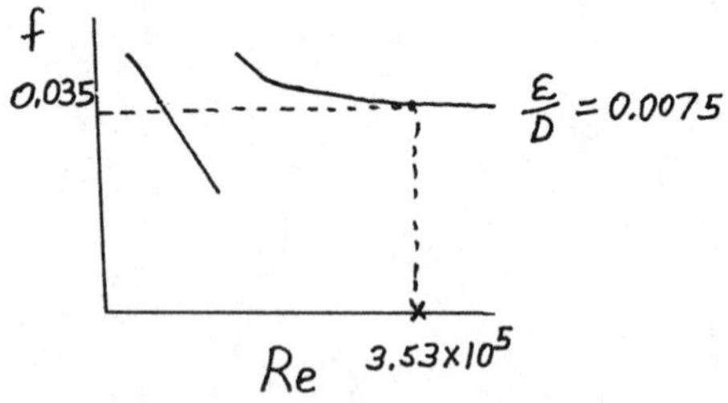

8.11R

8.11R (Single pipe—determine flowrate) An above ground swimming pool of 30 ft diameter and 5 ft depth is to be filled from a garden hose (smooth interior) of length 100 ft and diameter 5/8 in. If the pressure at the faucet to which the hose is attached remains at 55 psi, how long will it take to fill the pool? The water exits the hose as a free jet 6 ft above the faucet.

(ANS: 32.0 hr)

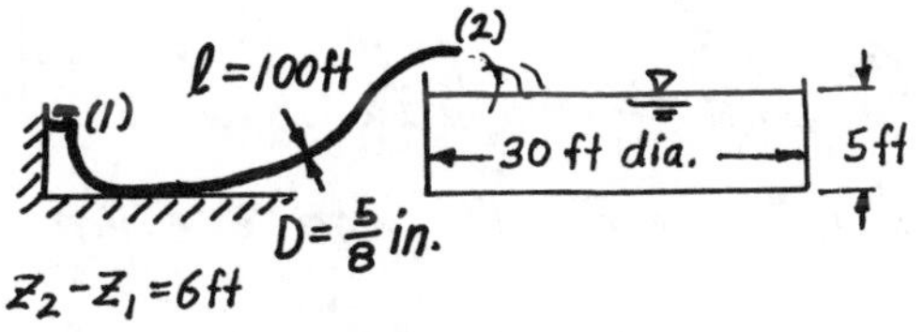

$z_2 - z_1 = 6\text{ft}$

$\frac{p_1}{\gamma} + \frac{V_1^2}{2g} + z_1 = \frac{p_2}{\gamma} + \frac{V_2^2}{2g} + z_2 + f\frac{\ell}{D}\frac{V^2}{2g}$, where $p_1 = 55\,\frac{\text{lb}}{\text{in.}^2}$, $p_2 = 0$, $z_1 = 0$,

$z_2 = 6\,\text{ft}$, and $V_1 = V_2 = V$

Thus,

$$\frac{p_1}{\gamma} = z_2 + f\frac{\ell}{D}\frac{V^2}{2g} \quad \text{or} \quad \frac{(55\frac{\text{lb}}{\text{in.}^2})(144\frac{\text{in.}^2}{\text{ft}^2})}{62.4\frac{\text{lb}}{\text{ft}^3}} = 6\,\text{ft} + f\left(\frac{100\,\text{ft}}{\frac{5}{8(12)}\,\text{ft}}\right)\frac{V^2}{2(32.2\frac{\text{ft}}{\text{s}^2})}$$

or

$fV^2 = 4.06$, where $V \sim \text{ft}$ (1)

Also, $Re = \frac{VD}{\nu} = \frac{(\frac{5}{8(12)}\,\text{ft})V}{1.21\times10^{-5}\frac{\text{ft}^2}{\text{s}}}$, or $Re = 4300V$, where $V \sim \text{ft}$ (2)

and from Fig. 8.20 with $\frac{\varepsilon}{D} = 0$

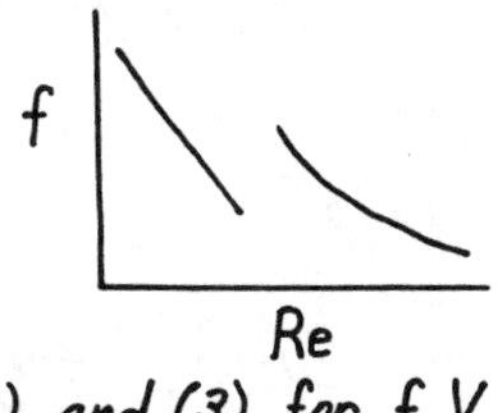

(3)

Trial and error solution of Eqs. (1), (2), and (3) for f, V, and Re:

Assume $f = 0.02$; from Eq. (1) $V = 14.2\,\frac{\text{ft}}{\text{s}}$; from Eq. (2) $Re = 6.11\times10^4$, so from Eq. (3) $f = 0.0196 \neq 0.02$ (but close!)

Assume $f = 0.0196$; from Eq. (1) $V = 14.4\,\frac{\text{ft}}{\text{s}}$; from Eq. (2) $Re = 6.19\times10^4$, so from Eq. (3) $f = 0.0196$, which agrees with the assumed value.

Thus, $V = 14.4\,\frac{\text{ft}}{\text{s}}$, or $Q = AV = \frac{\pi}{4}\left(\frac{5}{8(12)}\,\text{ft}\right)^2\left(14.4\,\frac{\text{ft}}{\text{s}}\right) = 0.0307\,\frac{\text{ft}^3}{\text{s}}$

Also, $\mathcal{V} = Qt$, where t = filling time and $\mathcal{V}$ = pool volume,

or

$$t = \frac{\mathcal{V}}{Q} = \frac{\frac{\pi}{4}(30\,\text{ft})^2(5\,\text{ft})}{0.0307\frac{\text{ft}^3}{\text{s}}} = 1.151\times10^5\,\text{s}\left(\frac{1\,\text{hr}}{3600\,\text{s}}\right) = \underline{\underline{32.0\,\text{hr}}}$$

8.12R

8.12R (Single pipe—determine pipe diameter) Water is to flow at a rate of 1.0 m^3/s through a rough concrete pipe (ε = 3 mm) that connects two ponds. Determine the pipe diameter if the elevation difference between the two ponds is 10 m and the pipe length is 1000 m. Neglect minor losses.

(ANS: 0.748 m)

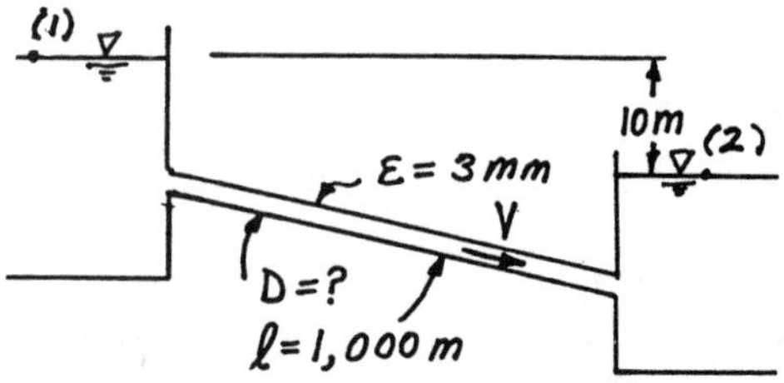

$$\frac{p_1}{\gamma}+\frac{V_1^2}{2g}+z_1=\frac{p_2}{\gamma}+\frac{V_2^2}{2g}+z_2+f\frac{\ell}{D}\frac{V^2}{2g}$$

where $p_1=p_2=V_1=V_2=z_2=0$, and $z_1=10\,m$

Thus, $f\frac{\ell}{D}\frac{V^2}{2g}=10\,m$ with (1)

$V=\frac{Q}{A}=\frac{1\,m^3/s}{\frac{\pi}{4}D^2}=\frac{1.27}{D^2}\,\frac{m}{s}$ with $D \sim m$

Hence, from Eq. (1)

$$f\cdot\frac{(1{,}000\,m)}{D}\,\frac{(1.27/D^2\,\frac{m}{s})^2}{2\,(9.81\,\frac{m}{s^2})}=10\,m \quad \text{or} \quad \frac{f}{D^5}=0.121$$

or

$D=1.53\,f^{1/5}$, where $D \sim m$ (2)

Also,

$$Re=\frac{VD}{\nu}=\frac{(\frac{1.27}{D^2})D}{\nu}=\frac{1.27\,\frac{m^2}{s}}{(1.12\times10^{-6}\,\frac{m^2}{s})D}$$

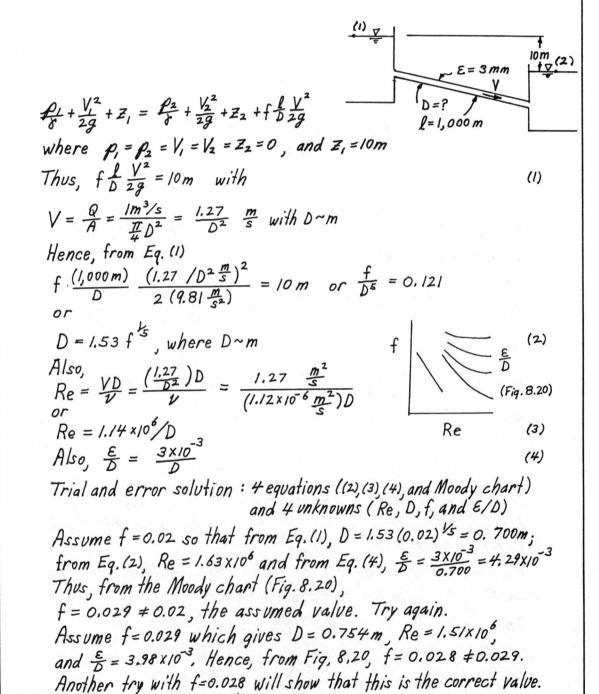

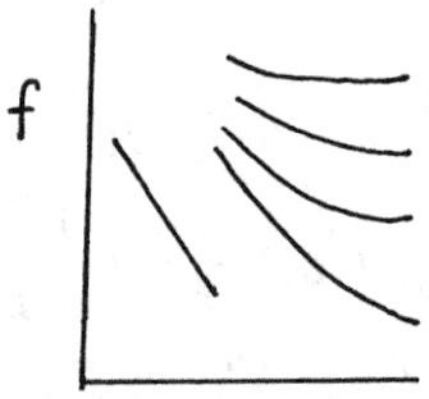

(Fig. 8.20)

or

$Re=1.14\times10^6/D$ (3)

Also, $\frac{\varepsilon}{D}=\frac{3\times10^{-3}}{D}$ (4)

Trial and error solution: 4 equations ((2), (3), (4), and Moody chart) and 4 unknowns (Re, D, f, and ε/D)

Assume $f=0.02$ so that from Eq. (1), $D=1.53(0.02)^{1/5}=0.700\,m$; from Eq. (2), $Re=1.63\times10^6$ and from Eq. (4), $\frac{\varepsilon}{D}=\frac{3\times10^{-3}}{0.700}=4.29\times10^{-3}$

Thus, from the Moody chart (Fig. 8.20),

$f=0.029\neq0.02$, the assumed value. Try again.

Assume $f=0.029$ which gives $D=0.754\,m$, $Re=1.51\times10^6$, and $\frac{\varepsilon}{D}=3.98\times10^{-3}$. Hence, from Fig. 8.20, $f=0.028\neq0.029$.

Another try with $f=0.028$ will show that this is the correct value.

Thus, $D=1.53\,(0.028)^{1/5}=\underline{\underline{0.748\,m}}$

8.13R

8.13R (Single pipe with pump) Without the pump shown in Fig. P8.13R it is determined that the flowrate is too small. Determine the horsepower added to the fluid if the pump causes the flowrate to be doubled. Assume the friction factor remains at 0.020 in either case.

(ANS: 1.51 hp)

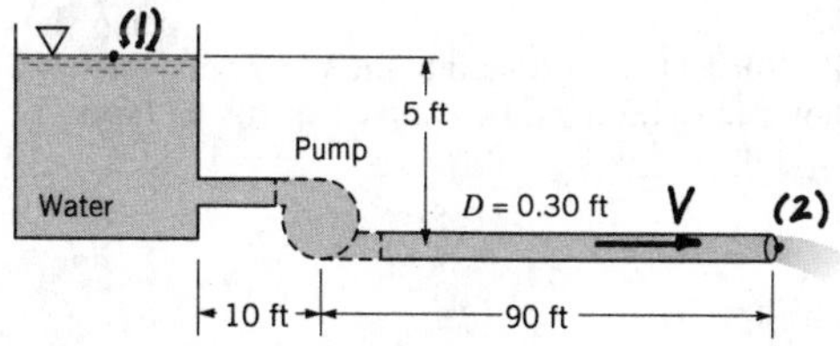

■ FIGURE P8.13R

$$h_p + \frac{p_1}{\gamma} + \frac{V_1^2}{2g} + z_1 = \frac{p_2}{\gamma} + \frac{V_2^2}{2g} + z_2 + \left(f\frac{\ell}{D} + K_L\right)\frac{V^2}{2g}$$, where $p_1 = p_2 = 0$, $V_1 = 0$, $z_1 = 5\,ft$, $z_2 = 0$, and $V_2 = V$. Assume a sharp edged entrance: $K_L = 0.5$ (Fig. 8.22)

Without the pump, $h_p = 0$ Thus,

$$z_1 = \left(1 + f\frac{\ell}{D} + 0.5\right)\frac{V^2}{2g} \quad \text{or} \quad 5\,ft = \left(1 + 0.02\left(\frac{100\,ft}{0.3\,ft}\right) + 0.5\right)\frac{V^2}{2(32.2\frac{ft}{s^2})}$$

Thus, $V = 6.28\,\frac{ft}{s}$

With the pump $V = 2(6.28\frac{ft}{s}) = 12.56\frac{ft}{s}$

Thus, $h_p + z_1 = \left(1 + f\frac{\ell}{D} + 0.5\right)\frac{V^2}{2g}$

or

$$h_p = -5\,ft + \left(1 + 0.02\left(\frac{100\,ft}{0.3\,ft}\right) + 0.5\right)\frac{(12.56\frac{ft}{s})^2}{2(32.2\frac{ft}{s^2})} = -5\,ft + 20\,ft = 15\,ft$$

Hence, from Eq. 5.85

$$\dot{W}_p = \gamma Q h_p = (62.4\frac{lb}{ft^3})(\frac{\pi}{4})(0.3\,ft)^2(12.56\frac{ft}{s})(15\,ft) = 831\,\frac{ft\cdot lb}{s}\left(\frac{1\,hp}{550\,\frac{ft\cdot lb}{s}}\right)$$

or

$\dot{W}_p = 1.51\,hp$

8.14R

8.14R (Single pipe with pump) The pump shown in Fig. P8.14R adds a 15-ft head to the water being pumped when the flowrate is 1.5 ft^3/s. Determine the friction factor for the pipe.
(ANS: 0.0306)

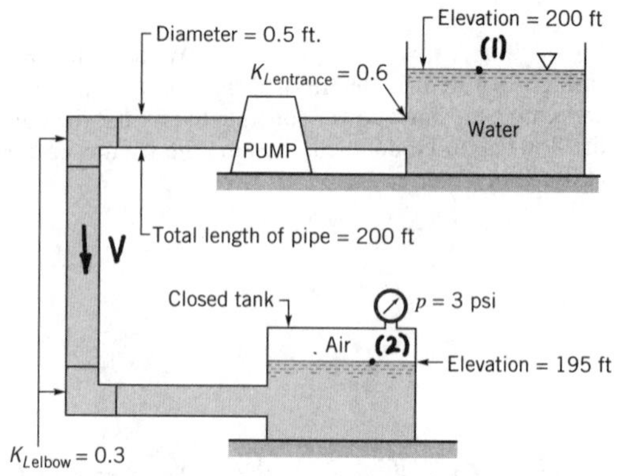

■ FIGURE P8.14R

For flow from the upper tank to the lower tank:

$$\frac{p_1}{\gamma} + \frac{V_1^2}{2g} + z_1 + h_p = \frac{p_2}{\gamma} + \frac{V_2^2}{2g} + z_2 + \left(f\frac{\ell}{D} + \Sigma K_L\right)\frac{V^2}{2g} \qquad (1)$$

where $p_1 = 0$, $V_1 = 0$, $z_1 = 200\ ft$, $h_p = 15\ ft$, $z_2 = 195\ ft$, $V_2 = 0\ \frac{ft}{s}$, and

$\Sigma K_L = K_{L_{ent}} + 2K_{L_{elbow}} + K_{L_{exit}} = 0.6 + 2(0.3) + 1 = 2.2$ (see Fig. 8.25)

Thus, Eq. (1) becomes

$$200\ ft + 15\ ft = \frac{(3\frac{lb}{in.^2})(144\frac{in.^2}{ft^2})}{62.4\frac{lb}{ft^3}} + 195\ ft + \left(f\frac{(200 ft)}{(0.5 ft)} + 2.2\right)\frac{V^2}{2(32.2\frac{ft}{s^2})} \qquad (2)$$

but with

$$V = \frac{Q}{A} = \frac{1.5\frac{ft^3}{s}}{\frac{\pi}{4}(0.5 ft)^2} = 7.64\frac{ft}{s}, \text{ Eq. (2) gives } f = \underline{\underline{0.0306}}$$

Note: If the flow was from the lower tank (2) to the upper tank (1), then

$$\frac{p_2}{\gamma} + \frac{V_2^2}{2g} + z_2 + h_p = \frac{p_1}{\gamma} + \frac{V_1^2}{2g} + z_1 + \left[f\frac{\ell}{D} + \Sigma K_L\right]\frac{V^2}{2g}$$

or

$$\frac{(3\frac{lb}{in.^2})(144\frac{in.^2}{ft^2})}{62.4\ lb/ft^3} + 195\ ft + 15\ ft = 200\ ft + \left[f\frac{(200 ft)}{(0.5 ft)} + 2.2\right]\frac{(7.64\frac{ft}{s})^2}{2(32.2\ ft/s^2)}$$

which gives $f = \underline{0.0412}$

Both the assumed downflow ($f = 0.0306$) and the upflow ($f = 0.0412$) give reasonable friction factor values.

8.15R

8.15R (Single pipe with turbine) Water drains from a pressurized tank through a pipe system as shown in Fig. P8.15R. The head of the turbine is equal to 116 m. If entrance effects are negligible, determine the flow rate.

(ANS: 3.71×10^{-2} m³/s)

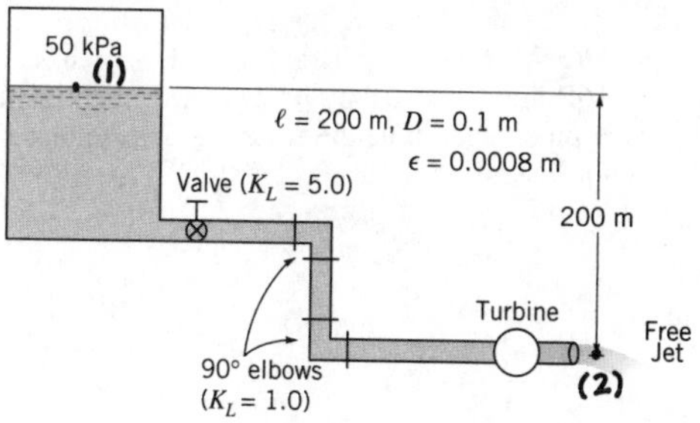

■ FIGURE P8.15R

$$\frac{p_1}{\gamma} + \frac{V_1^2}{2g} + z_1 + h_s = \frac{p_2}{\gamma} + \frac{V_2^2}{2g} + z_2 + \left[f\frac{\ell}{D} + \Sigma K_L\right]\frac{V^2}{2g}$$

where $z_1 = 200\,m$,

$p_2 = V_1 = z_2 = 0$, and $h_s = -116\,m$ ($h_s < 0$ since it is a turbine)

Thus,

$$\frac{50\times10^3\,\frac{N}{m^2}}{9.8\times10^3\,\frac{N}{m^3}} + 200\,m - 116\,m = \frac{V^2\,\frac{m^2}{s^2}}{2(9.81\frac{m}{s^2})}\left[1 + f\left(\frac{200\,m}{0.1\,m}\right) + 5 + 2(1)\right]$$

or

$$V^2 = \frac{1748}{(8 + 2000f)}, \text{ where } V \sim \frac{m}{s} \qquad (1)$$

Also,

$$Re = \frac{\rho V D}{\mu} = \frac{(999\,\frac{kg}{m^3})V\frac{m}{s}(0.1m)}{1.12\times10^{-3}\,\frac{N\cdot s}{m^2}}, \text{ or } Re = 8.92\times10^4\,V \qquad (2)$$

The final equation is the Moody chart (Eq. 8.20) with

$$\frac{\varepsilon}{D} = \frac{0.0008\,m}{0.1\,m} = 0.008 \qquad (3)$$

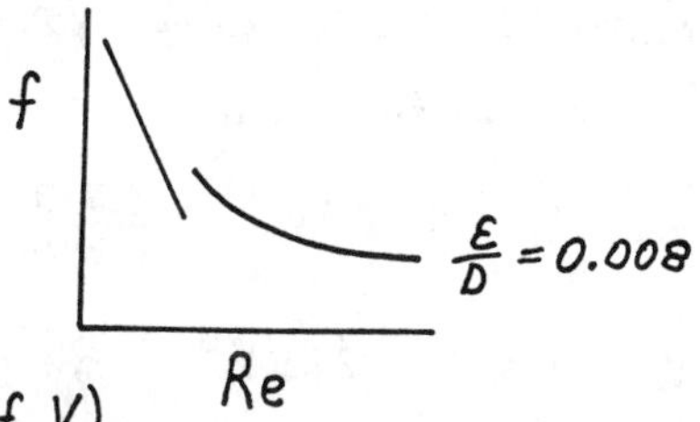

Trial and error solution:
(3 equations; 3 unknowns: Re, f, V)

Assume $f = 0.038$ so that from Eq. (1), $V = 4.56\,\frac{m}{s}$; from Eq. (2), $Re = 4.07\times10^5$; and from the Moody chart, $f = 0.035$ which does not equal the assumed value. Try again.

Assume $f = 0.035$ which gives $V = 4.73\,\frac{m}{s}$ and $Re = 4.22\times10^5$. Hence, from the Moody chart $f = 0.035$ which agrees with the assumed value. Thus, $V = 4.73\,\frac{m}{s}$ and

$$Q = AV = \frac{\pi}{4}(0.1m)^2\left(4.73\frac{m}{s}\right) = 3.71\times10^{-2}\,\frac{m^3}{s}$$

8.16R

8.16R (Multiple pipes) The three tanks shown in Fig. P8.16R are connected by pipes with friction factors of 0.03 for each pipe. Determine the water velocity in each pipe. Neglect minor losses.

(ANS: (A) 4.73 ft/s, (B) 8.35 ft/s, (C) 10.3 ft/s)

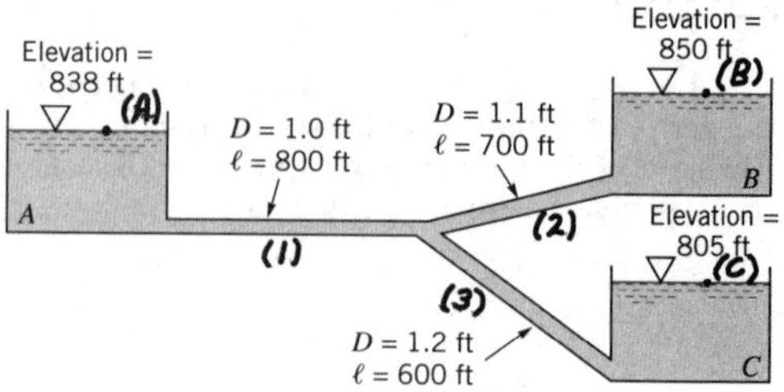

■ FIGURE P8.16R

Assume the flow from both tanks A and B is into tank C, or $Q_3 = Q_1 + Q_2$

Thus, $\frac{\pi}{4}D_3^2 V_3 = \frac{\pi}{4}D_1^2 V_1 + \frac{\pi}{4}D_2^2 V_2$, or $1.2^2 V_3 = 1.0^2 V_1 + 1.1^2 V_2$

Hence, $V_3 = 0.694\,V_1 + 0.840\,V_2$ (1)

For the flow from A to C, with $p_A = p_C = 0$, $V_A = V_C = 0$, we obtain

$$z_A = z_C + f_1 \frac{\ell_1}{D_1}\frac{V_1^2}{2g} + f_3 \frac{\ell_3}{D_3}\frac{V_3^2}{2g}, \text{ or } 838\text{ ft} = 805\text{ ft} + \frac{0.03}{2(32.2\frac{\text{ft}}{\text{s}^2})}\left[\frac{800\text{ft}}{1\text{ft}}V_1^2 + \frac{600\text{ft}}{1.2\text{ft}}V_3^2\right]$$

or

$$33 = 0.373\,V_1^2 + 0.233\,V_3^2 \quad (2)$$

Similarly for the flow from B to C, with $p_B = p_C = 0$, $V_B = V_C = 0$, we obtain

$$z_B = z_C + f_2 \frac{\ell_2}{D_2}\frac{V_2^2}{2g} + f_3 \frac{\ell_3}{D_3}\frac{V_3^2}{2g}, \text{ or } 850\text{ft} = 805\text{ ft} + \frac{0.03}{2(32.2\frac{\text{ft}}{\text{s}^2})}\left[\frac{700\text{ft}}{1.1\text{ft}}V_2^2 + \frac{600\text{ft}}{1.2\text{ft}}V_3^2\right]$$

or

$$45 = 0.296\,V_2^2 + 0.233\,V_3^2 \quad (3)$$

Thus, 3 equations ((1),(2), and (3)) for V_1, V_2, and V_3. Solve as follows:

Subtract (2) from (3) to obtain

$$12 = 0.296\,V_2^2 - 0.373\,V_1^2 \quad (4)$$

From (2): $V_3 = \sqrt{141.6 - 1.6V_1^2}$, or when combined with (1):

$$\sqrt{141.6 - 1.6V_1^2} = 0.694V_1 + 0.840V_2, \text{ or } V_2 = \sqrt{200 - 2.27V_1^2} - 0.826\,V_1 \quad (5)$$

Combine Eqs. (4) and (5) to obtain:

$$\frac{12}{0.296} = \left[\sqrt{200 - 2.27V_1^2} - 0.826\,V_1\right]^2 - \frac{0.373}{0.296}, \text{ which can be simplified to}$$

$$V_1\sqrt{200 - 2.27V_1^2} = 96.5 - 1.725V_1^2 \quad (6)$$ By squaring this equation we obtain (after simplification):

$$V_1^4 - 101.5V_1^2 + 1774 = 0 \quad \text{Hence: } V_1^2 = \frac{101.5 \pm \sqrt{101.5^2 - 4(1774)}}{2} = 79.1 \text{ or } 22.4$$

Thus, $V_1 = 8.89\,\frac{\text{ft}}{\text{s}}$ or $V_1 = 4.73\,\frac{\text{ft}}{\text{s}}$

Note: The $V_1 = 8.89$ solution is an extra root introduced by squaring Eq. (6). It is not a solution of the original Eqs. (1), (2), (3). For this value, Eq. (6) becomes $8.89\sqrt{200 - 2.27(8.89^2)} \overset{?}{=} 96.5 - 1.725(8.89)^2$, or "$40 = -40$"

Thus $V_1 = \underline{4.73\,\frac{\text{ft}}{\text{s}}}$, from Eq. (2) $V_3 = \left[\frac{33 - 0.373(4.73)^2}{0.233}\right]^{1/2} = \underline{10.3\,\frac{\text{ft}}{\text{s}}}$,

and from Eq. (1) $V_2 = \frac{10.3 - 0.694(4.73)}{0.840} = \underline{8.35\,\frac{\text{ft}}{\text{s}}}$

8.17R

8.17R (Flow meters) Water flows in a 0.10-m-diameter pipe at a rate of 0.02 m^3/s. If the pressure difference across the orifice meter in the pipe is to be 28 kPa, what diameter orifice is needed?

(ANS: 0.070 m)

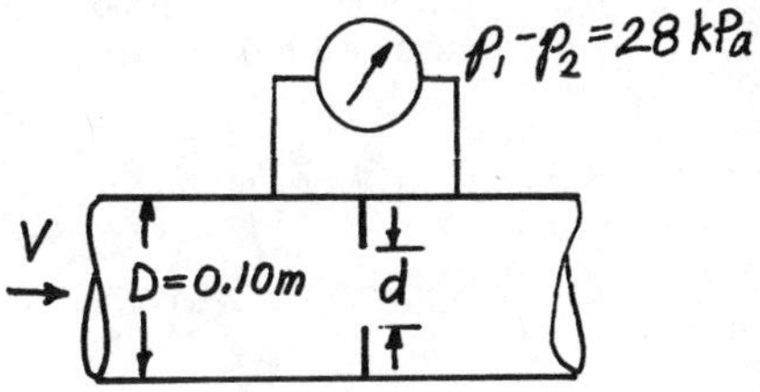

$$Q = C_o A_o \sqrt{\frac{2(p_1 - p_2)}{\rho(1-\beta^4)}}, \text{ where } Q = 0.02\,\frac{m^3}{s},\ p_1 - p_2 = 28 \text{ kPa},\ \beta = \frac{d}{D},$$
and $\rho = 999\,\frac{kg}{m^3}$.

Thus,

$$0.02\,\frac{m^3}{s} = C_o \frac{\pi}{4} d^2 \sqrt{\frac{2(28\times10^3\,\frac{N}{m^2})}{999\,\frac{kg}{m^3}\left(1-\left(\frac{d}{0.1}\right)^4\right)}}, \text{ with } d \sim m$$

Hence,

$$3.40\times10^{-3} = \frac{C_o d^2}{\sqrt{1-(10d)^4}}, \text{ where } C_o = C_o\left(Re, \frac{d}{D}\right) \text{ from Fig. 8.41.} \quad (1)$$

Also, $Re = \frac{VD}{\nu}$, where $V = \frac{Q}{\frac{\pi}{4}D^2} = \frac{0.02\,\frac{m^3}{s}}{\frac{\pi}{4}(0.1m)^2} = 2.55\,\frac{m}{s}$

Thus, $Re = \frac{(2.25\,\frac{m}{s})(0.1m)}{1.12\times10^{-6}\,\frac{m^2}{s}} = 2.28\times10^5$

Trial and error solution: Assume a value of d; calculate $\beta = \frac{d}{D}$; look up C_o in Fig. 8.41 and calculate $\frac{C_o d^2}{\sqrt{1-(10d)^4}}$; compare with 3.40×10^{-3} (see Eq. (1)).

d, m	β	C_o	$\frac{C_o d^2}{\sqrt{1-(10d)^4}}$	
0.070	0.70	0.61	3.43×10^{-3}	← checks
0.069	0.69	0.61	3.30×10^{-3}	
0.071	0.71	0.61	3.56×10^{-3}	

Thus, $d = \underline{\underline{0.070\,m}}$

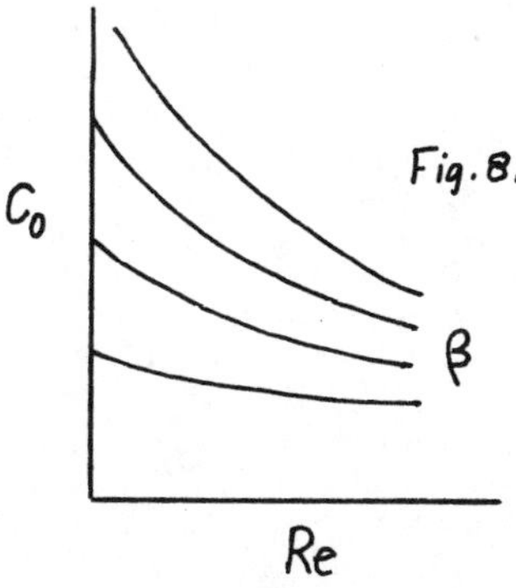

Fig. 8.41

8.18R

8.18R (Flow meters) A 2.5-in.-diameter flow nozzle is installed in a 3.8-in.-diameter pipe that carries water at 160 °F. If the flowrate is 0.78 cfs, determine the reading on the inverted air-water U-tube manometer used to measure the pressure difference across the meter.

(ANS: 6.75 ft)

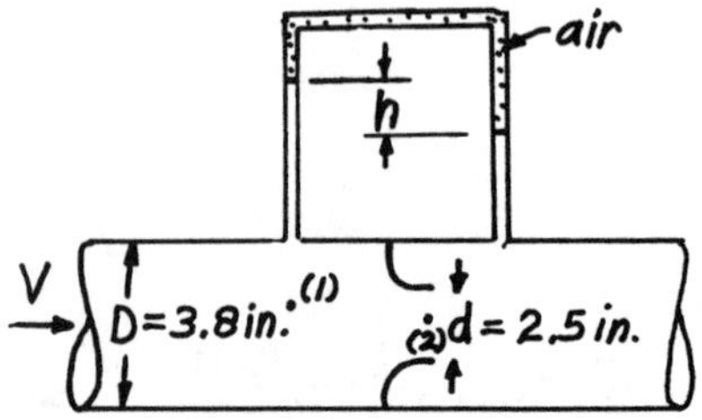

$$Q = C_n A_n \sqrt{\frac{2(p_1 - p_2)}{\rho(1-\beta^4)}}, \text{ where } \beta = \frac{d}{D} = \frac{2.5 \text{ in.}}{3.8 \text{ in.}} = 0.658 \qquad (1)$$

$$V = \frac{Q}{\frac{\pi}{4}D^2} = \frac{0.78 \frac{ft^3}{s}}{\frac{\pi}{4}\left(\frac{3.8}{12} ft\right)^2} = 9.90 \frac{ft}{s} \text{ and from Table B.1: } \rho = 1.896 \frac{slugs}{ft^3}$$

$$\mu = 8.32 \times 10^{-6} \frac{lb \cdot s}{ft^2} \text{ at } T = 160°F$$

Thus, with

$$Re = \frac{\rho V D}{\mu} = \frac{(1.896 \frac{slugs}{ft^3})(9.90 \frac{ft}{s})(\frac{3.8}{12} ft)}{8.32 \times 10^{-6} \frac{lb \cdot s}{ft^2}} = 7.14 \times 10^5 \text{ we obtain from Fig. 8.43:}$$

$C_n = 0.989$ Thus, from Eq. (1)

$$0.78 \frac{ft^3}{s} = (0.989)\frac{\pi}{4}\left(\frac{2.5}{12} ft\right)^2 \sqrt{\frac{2(p_1 - p_2)}{(1.896 \frac{slugs}{ft^3})(1 - 0.658^4)}}$$

or

$$p_1 - p_2 = 412 \frac{lb}{ft^2}$$

However, $p_1 - p_2 = \gamma_{H_2O} h$ so that $h = \frac{412 \frac{lb}{ft^2}}{(32.2 \frac{ft}{s^2})(1.896 \frac{slugs}{ft^3})} = \underline{\underline{6.75 \text{ ft}}}$

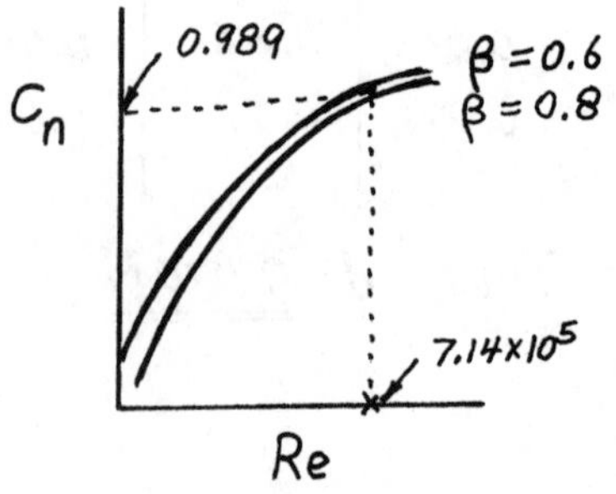

9 Flow Over Immersed Bodies

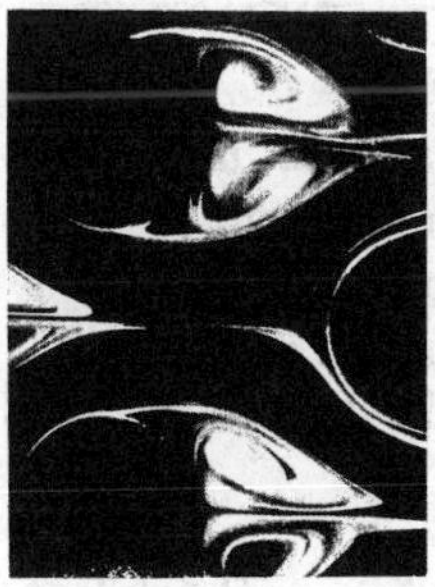

9.1 General External Flow Characteristics

$$\mathscr{D} = \int dF_x = \int p \cos\theta \, dA + \int \tau_w \sin\theta \, dA \qquad \textbf{(9.1)} \qquad C_D = \frac{\mathscr{D}}{\frac{1}{2}\rho U^2 A}$$

$$\mathscr{L} = \int dF_y = -\int p \sin\theta \, dA + \int \tau_w \cos\theta \, dA \qquad \textbf{(9.2)} \qquad C_L = \frac{\mathscr{L}}{\frac{1}{2}\rho U^2 A}$$

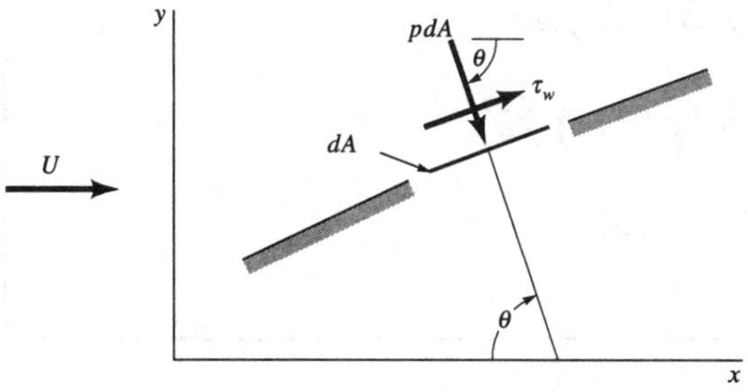

■ **FIGURE 9.4 Pressure and shear forces on a small element of the surface of a body.**

- A fluid flowing past an object interacts with the object through the pressure and shear stress on the surface of the object. The resultant fluid forces normal to and parallel to the upstream velocity are termed the *lift*, $\mathscr{L}$, and *drag*, $\mathscr{D}$, respectively.
- As indicated in Eqs. 9.1 and 9.2 and Fig. 9.4, the lift and drag can be obtained by integrating the pressure and viscous forces over the surface of the object.
- The lift and drag are often obtained by use of the *lift coefficient*, C_L, and the drag coefficient, C_D, as shown by the above equations.
- The character of flow past an object is often dependent on the value of the Reynolds number of the flow.
- *Caution:* The pressure on the front portion of most objects is positive; the pressure on the rear portion is often negative (less than the upstream pressure).

EXAMPLE: A fluid flows past a thin circular disk of diameter D that is oriented normal to the upstream velocity whose speed is U as shown in the figure. The average pressure on the front side of the disk is 0.8 times the stagnation pressure and the average pressure on the back side of the disk is a negative pressure (a vacuum) with a magnitude 0.5 times the stagnation pressure. Determine the drag coefficient for this disk.

SOLUTION:

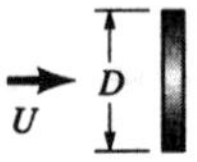

From Eq. 9.1, $\mathcal{D} = \int p \cos\theta \, dA + \int \tau_w \sin\theta \, dA$

But on the front of the disk $\theta = 0$ so that $\cos\theta = 1$ and $\sin\theta = 0$. On the back of the disk $\theta = 180°$ (see Fig. 9.4) so that $\cos\theta = -1$ and $\sin\theta = 0$. Hence, from the above equation $\int \tau_w \sin\theta \, dA = 0$ and

$$\mathcal{D} = \int p \cos\theta \, dA = \int_{front} p \, dA + \int_{back} (-p) \, dA$$

$$= \int_{front} 0.8(\rho U^2/2) \, dA + \int_{back} -(-0.5)(\rho U^2/2) \, dA$$

$$= (0.8 + 0.5)(\rho U^2/2)\frac{\pi}{4}D^2 = 1.3 \, (\rho U^2/2)\frac{\pi}{4}D^2$$

Therefore, $C_D = \mathcal{D} / \left[(\rho U^2/2)\frac{\pi}{4}D^2\right] = \underline{\underline{1.3}}$

Characteristics of Flow Past an Object

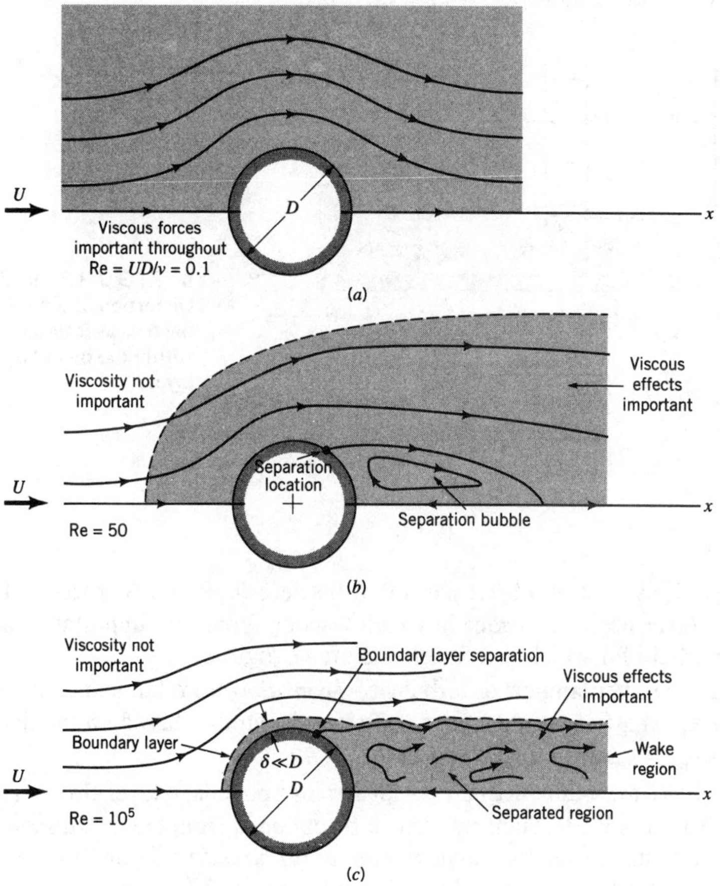

■ **FIGURE 9.6 Character of the steady, viscous flow past a circular cylinder: (*a*) low Reynolds number flow, (*b*) moderate Reynolds number flow, (*c*) large Reynolds number flow.**

- The characteristics of external flow past an object depend of several factors, including the shape of the object and the numerical values of various dimensionless parameters such as the Reynolds number, Re, the Mach number, Ma, and the Froude number, Fr.
- Several typical external flow characteristics are shown in Fig. 9.6 for incompressible, steady flow past a circular cylinder.
- For low Reynolds number flows, Fig. 9.6*a*, the presence of the cylinder and the accompanying viscous effects are felt throughout the entire flow field.
- For moderate Reynolds number flows, Fig. 9.6*b*, the region in which viscous effects are important becomes smaller and the flow separates from the body.
- For large Reynolds number flows, Fig. 9.6*c*, viscous effects are important only in the thin boundary layer next to the object and in the wake region downstream of the object. The flow in these regions may be laminar or turbulent, depending on various factors.
- *Caution:* Most "every day type flows" have a large Reynolds number and are of the type shown in Fig. 9.6*c*.

9.2 Boundary Layer Characteristics

Boundary Layer Structure and Thickness on a Flat Plate

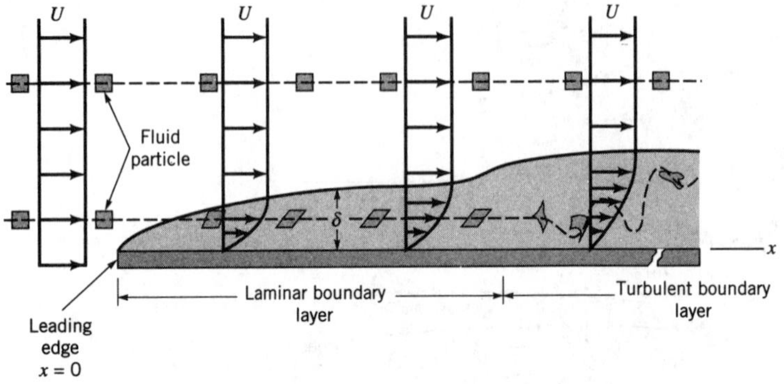

■ **FIGURE 9.7** **Distortion of a fluid particle as it flows within the boundary layer.**

$$\delta = 5\sqrt{\frac{\nu x}{U}} \qquad \textbf{(9.15)}$$

$$\tau_w = 0.332U^{3/2}\sqrt{\frac{\rho\mu}{x}} \qquad \textbf{(9.18)}$$

- As indicated in Fig. 9.7, flow past an object can often be described as two parts: (1) the relatively thin *boundary layer* near the surface in which viscous forces are important, and (2) the remainder of the flow field for which viscous effects are negligible.
- Boundary layer flow can be either laminar or turbulent. *Transition* from laminar to turbulent flow in the boundary layer on a flat plate occurs at a Reynolds number based on the distance from the leading edge, $Re_{xcr} = x_{cr}U/\nu$, of approximately 5 x 10^5.
- The *boundary layer thickness* and *wall shear stress* for laminar boundary layer flow on a flat plate are given by Eqs. 9.15 and 9.18. Such results can be obtained from the *Blasius solution* to the boundary layer equations or from the *momentum-integral boundary layer equation.*
- *Caution:* Equations 9.15 and 9.18 are valid only for laminar flow on a flat plate.

EXAMPLE: Air flows past a flat plate that is orientated parallel to the flow with a steady upstream velocity of 20 ft/s. **(a)** At approximately what distance from the leading edge of the plate does the boundary layer flow become turbulent? **(b)** Determine the boundary layer thickness at a distance of 3 ft from the leading edge.

SOLUTION:

(a) Transition occurs at approximately $Re_{x_{cr}} = x_{cr}U/\nu = 5 \times 10^5$.

Thus, $x_{cr} = 5\times10^5 \nu/U$

$= 5\times10^5 (1.57\times10^{-4} \text{ft}^2/\text{s})/20 \text{ ft/s} = \underline{\underline{3.93 \text{ ft}}}$

(b) From Eq. 9.15, at $x = 3 \text{ ft} < x_{cr} = 3.93 \text{ ft}$ the laminar boundary layer thickness is

$$\delta = 5\sqrt{\frac{\nu x}{U}} = 5\left[1.57\times10^{-4} \text{ft}^2/\text{s}\,(3 \text{ ft})/(20 \text{ ft/s})\right]^{1/2} = \underline{\underline{0.0243 \text{ ft}}}$$

Turbulent Boundary Layers

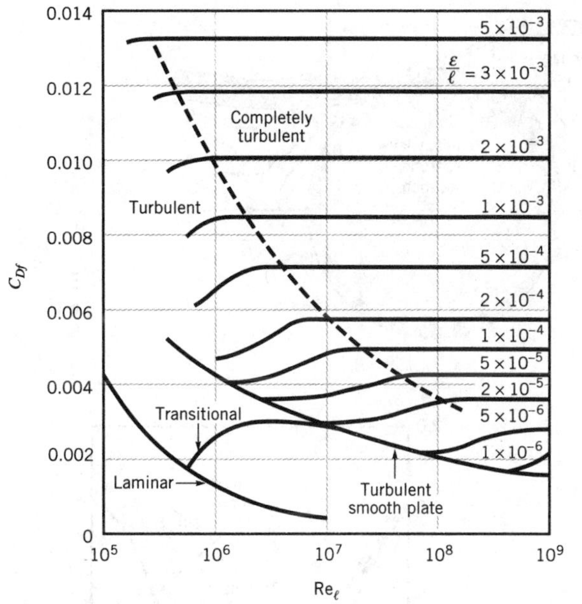

■ **FIGURE 9.15**
Friction drag coefficient for a flat plate parallel to the upstream flow (Ref. 18, with permission).

- The structure of turbulent boundary layer flow is very complex. It involves the random, irregular convection of fluid properties such as mass, momentum, and energy in both the downstream direction and across the boundary layer.
- The random transport of finite-sized fluid particles associated with the turbulent eddies causes considerable mixing in a turbulent flow. Thus, the drag associated with a turbulent boundary layer is considerably greater than that for a laminar boundary layer.
- As shown in Fig. 9.15, the friction drag coefficient for turbulent boundary layer flow on a flat plate is a function of the Reynolds number based on the plate length, $Re_\ell = U\ell/\nu$, and relative roughness, ε/ℓ.
- *Caution:* More often than not, a boundary layer flow will be turbulent. Do not use the laminar boundary layer flow equations if the flow is turbulent.

EXAMPLE: Water with a velocity of 2 m/s flows past a rough flat plate that is 3 m long and 1 m wide. If the roughness of the plate is $\varepsilon = 3$ mm, determine the drag on one side of the plate.

SOLUTION:

$\mathscr{D} = \frac{1}{2}\rho U^2 A C_D$ where $U = 2\,m/s$ and $A = 3m(1m) = 3m^2$.

Also, from Fig. 9.15, C_D is a function of Re_ℓ and ε/ℓ, where

$Re_\ell = \ell U/\nu = 3m\,(2m/s)/1.12\times10^{-6}\,m^2/s = 5.36\times10^6$ and

$\varepsilon/\ell = 3\times10^{-3}m/3\,m = 10^{-3}$.

Hence, from Fig. 9.15, $C_D = 0.0085$ so that

$\mathscr{D} = \frac{1}{2}(999\,kg/m^3)(2m/s)^2(3m^2)(0.0085) = \underline{\underline{50.9N}}$

Effects of Pressure Gradient

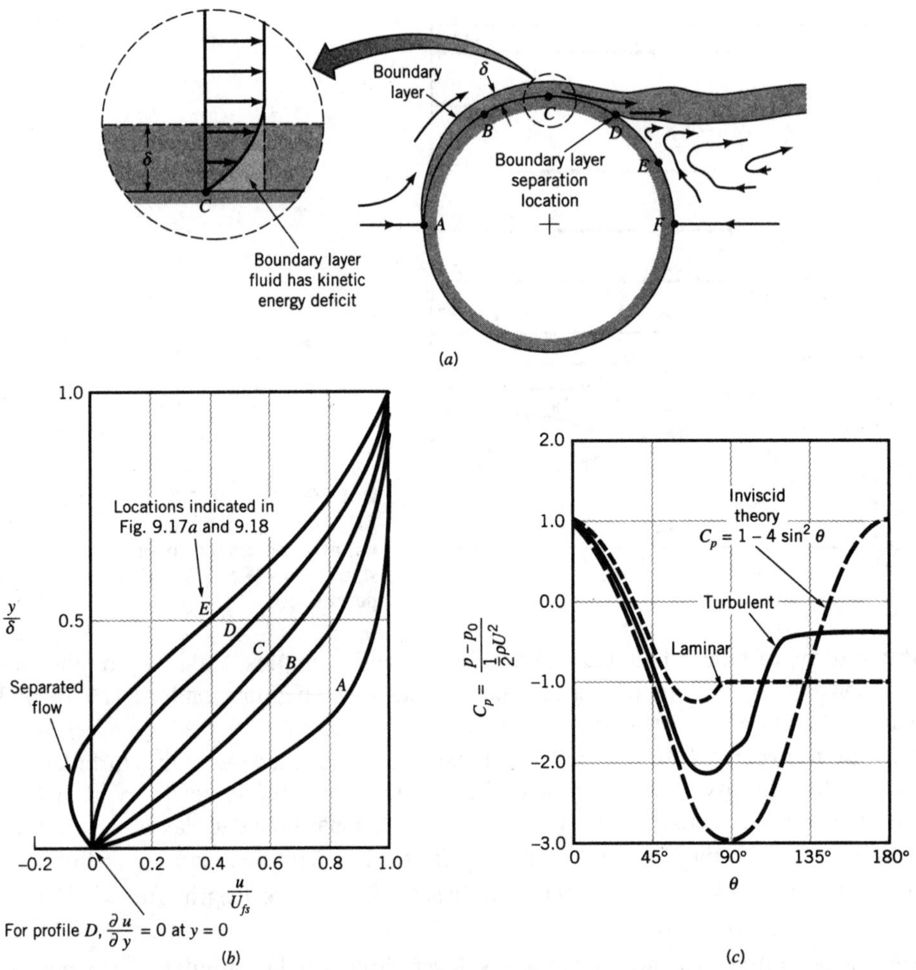

■ **FIGURE 9.17** **Boundary layer characteristics on a circular cylinder: (*a*) boundary layer separation location, (*b*) typical boundary layer velocity profiles at various locations on the cylinder, (*c*) surface pressure distributions for inviscid flow and boundary layer flow.**

- In general, for objects other than flat plates the pressure is not constant along the surface. The resulting pressure gradient can cause *boundary layer separation*, which significantly alters the structure of the flow both within and outside of the boundary layer (see Fig. 9.17).
- If the *adverse pressure gradient* on the rear portion of a body is too large, the boundary layer fluid will not contain enough momentum to overcome the increasing pressure in the flow direction, and the boundary layer will *separate* from the object.
- The low pressure in the separated wake region behind the object causes a large pressure drag.
- The location of the separation point and the width of the wake region may depend on whether the boundary layer is laminar or turbulent.
- Streamlined bodies are those designed to eliminate (or reduce) the effects of separation.

9.3 Drag

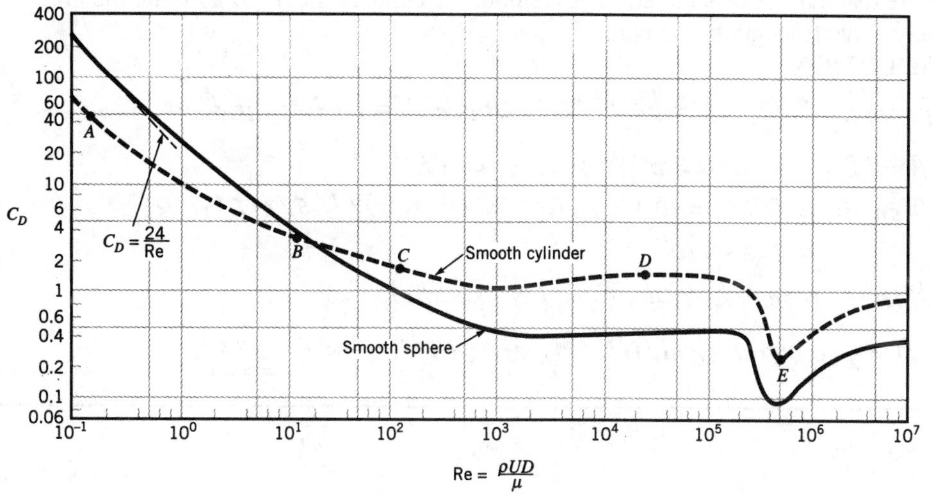

■ FIGURE 9.21 **Drag coefficient as a function of Reynolds number for a smooth circular cylinder and a smooth sphere.**

$$C_D = \frac{\mathscr{D}}{\frac{1}{2}\rho U^2 A} \qquad \textbf{(9.36)}$$

- The drag on an object, which is the sum of the friction (viscous) drag and the pressure (form) drag, can be determined by use of the *drag coefficient*, C_D, as shown in Eq. 9.36.
- The drag coefficient is often a strong function of the shape of the object, with streamlined objects having smaller drag coefficients than blunt ones.
- The drag coefficient is often a function of various other dimensionless parameters including the Reynolds number (see Fig. 9.21), the Mach number, the Froude number, and the relative roughness of the surface.
- The drag of a complex body can often be well approximated as the sum of the drag of the various components making up the entire body (sometimes called the composite body).
- *Caution:* The characteristic area, A, used in the drag coefficient is *usually* the frontal area, although for some objects the planform area is used.

EXAMPLE: A 60 mph (88 ft/s) wind blows past (normal to) a 0.5 in.-diameter wire that may be considered to be a smooth circular cylinder. Determine the drag on a 1000 ft length of the wire.

SOLUTION:

From Eq. 9.36, $\mathcal{D} = \frac{1}{2}\rho U^2 A C_D$ where $U = 88\ ft/s$ and

$A = \ell D = 1000ft(0.5\ in.)(1\ ft/12\ in.) = 41.7\ ft^2$

Also, $Re = DU/\nu = (0.5in.)(1\ ft/12in.)(88\ ft/s)/(1.57\times10^{-4}ft^2/s)$

$= 2.34\times10^4$

Hence, from Fig. 9.21, $C_D = 1.5$ so that

$\mathcal{D} = \frac{1}{2}(0.00238\ slugs/ft^3)(88ft/s)^2(41.7ft^2)(1.5) = \underline{\underline{576\ lb}}$

EXAMPLE: A 10 m/s wind blows past a 2-m by 4-m flag that is attached to a flag pole that is 15 m tall and 0.2 m in diameter. The drag coefficient for the flag (based on the planform area of the flag) is 0.12; the drag coefficient for the cylinder (based on the frontal area) is 1.1. Determine the total drag on the flag and flag pole.

SOLUTION:

$$\mathcal{D} = \mathcal{D}_{flag} + \mathcal{D}_{pole} = \frac{1}{2}\rho U^2\left[A_{pole}C_{D_{pole}} + A_{flag}C_{D_{flag}}\right]$$

$$= \frac{1}{2}(1.23\,kg/m^3)(10m/s)^2\left[15m(0.2m)(1.1) + 2m(4m)(0.12)\right]$$

or

$\mathcal{D} = 203\ N + 59\ N = \underline{\underline{262\ N}}$

9.4 Lift

$$C_L = \frac{\mathcal{L}}{\frac{1}{2}\rho U^2 A} \qquad \textbf{(9.39)}$$

- *Lift* is the result of a larger average pressure on the lower part of the object compared to the smaller average pressure on the upper surface.
- As indicated in Eq. 9.39, the lift can be determined if the *lift coefficient*, C_L, the dynamic pressure, $\rho U^2/2$, and area, A, are known.
- The lift coefficient is a strong function of the shape of the object. For an airfoil, this includes the basic cross-sectional shape, its angle of attack, α, and whether leading edge or trailing edge flaps are extended.
- If the angle of attack becomes too large, the boundary layer flow separates from the upper surface. This condition, called *stall*, results in a sudden decrease in the lift coefficient and an increase in the drag coefficient.
- *Caution:* The characteristic area, A, used in the lift and drag coefficients for airfoils is the planform area.

EXAMPLE: A rectangular planform wing with an aspect ratio (span, b, divided by chord length, c) of 8 is to generate 5000 lb lift when flying at 200 mph (293 ft/s). Determine the wing span if the lift coefficient is 0.60. Assume standard sea level air.

SOLUTION:

From Eq. 9.39, $\mathcal{L} = \frac{1}{2}\rho U^2 A C_L$ where $A = bc = b(b/8) = b^2/8$

Thus, $5000\ \text{lb} = \frac{1}{2}(0.00238\ \text{slugs/ft}^3)(293\ \text{ft/s})^2(b^2/8)\text{ft}^2\,(0.60)$

or $b = \underline{\underline{25.5\ \text{ft}}}$

Circulation

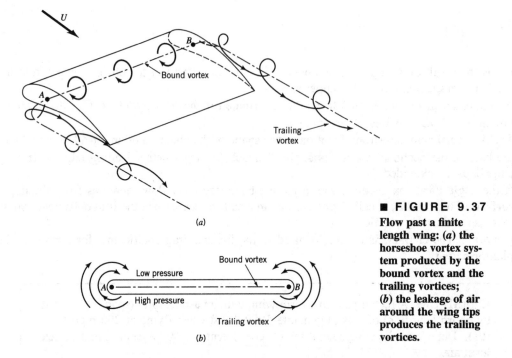

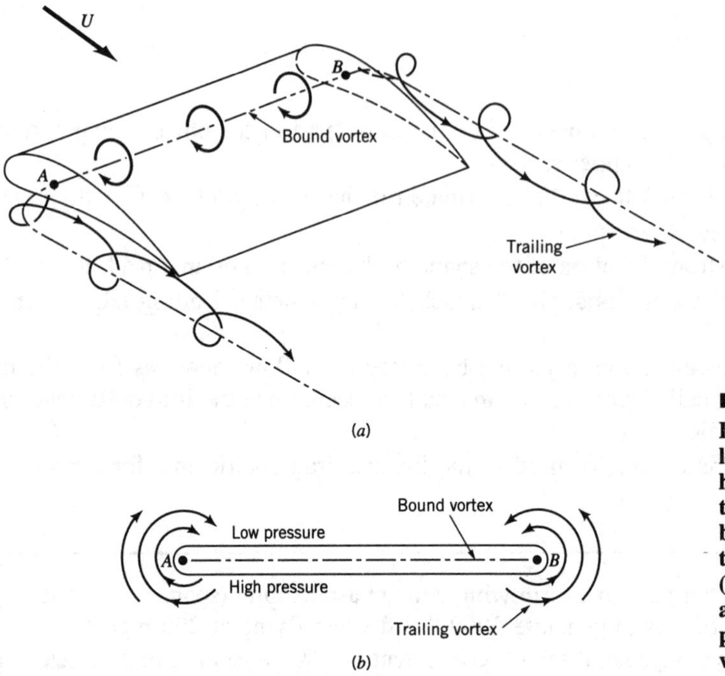

■ **FIGURE 9.37**
Flow past a finite length wing: (*a*) the horseshoe vortex system produced by the bound vortex and the trailing vortices; (*b*) the leakage of air around the wing tips produces the trailing vortices.

- When a wing generates lift, the pressure difference between the high-pressure lower surface of the wing and the low-pressure upper surface causes some of the fluid to migrate from the lower to the upper surface as shown in Fig. 9.37.
- The swirling fluid from the wing tips is then swept downstream, producing a U-shaped *horeshoe vortex* that consists of the trailing vortex and the bound vortex. The strength of these vorticies is termed the *circulation.*

Solutions

to Review Problems

Chapter 9

9.1R

9.1R (Lift/drag calculation) Determine the lift and drag coefficients (based on frontal area) for the triangular two-dimensional object shown in Fig. P9.1R. Neglect shear forces.

(ANS: 0; 1.70)

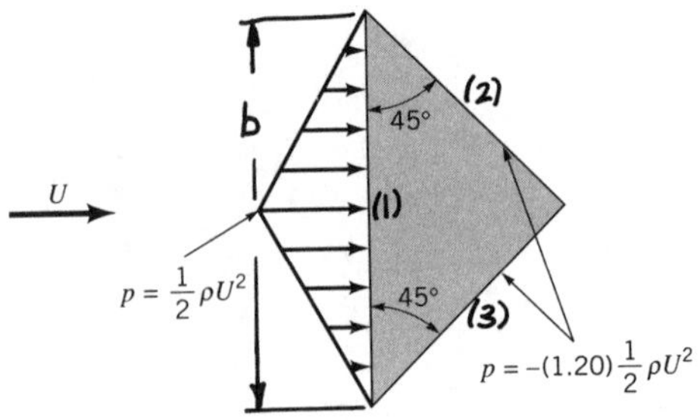

■ FIGURE P9.1R

$$\mathcal{D} = \int p \cos\theta \, dA + \int \tau_w \sin\theta \, dA, \text{ where } \tau_w = 0$$

Thus,

$$\mathcal{D} = \int_1 p \cos\theta \, dA + \int_2 p \cos\theta \, dA + \int_3 p \cos\theta \, dA$$

$$= \int_1 p \, dA - 2\int_2 p \cos 45° \, dA = \tfrac{1}{2}\left(\tfrac{1}{2}\rho U^2\right) \ell b - 2\left(-1.20\left(\tfrac{1}{2}\rho U^2\right)\cos 45° A_2\right)$$

where

$$A_2 = \ell \frac{b/2}{\cos 45°}$$

ℓ = length of object

Hence,

$$\mathcal{D} = 1.7\left(\tfrac{1}{2}\rho U^2\right)\ell b$$

or

$$C_D = \frac{\mathcal{D}}{\frac{1}{2}\rho U^2 A} = \frac{1.7\left(\frac{1}{2}\rho U^2\right)\ell b}{\frac{1}{2}\rho U^2 \ell b} = \underline{\underline{1.70}}$$

Because of symmetry of the object, $\mathcal{L} = 0$

or

$$C_L = \frac{\mathcal{L}}{\frac{1}{2}\rho U^2 A} = \underline{\underline{0}}$$

Note:

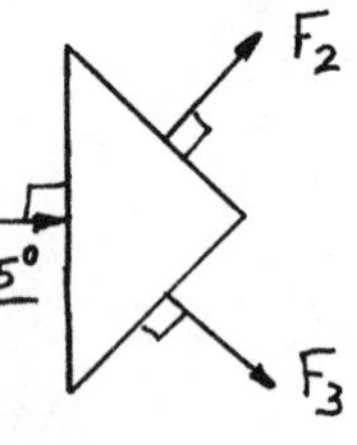

$$\mathcal{D} = F_1 + (F_2 + F_3)\cos 45°$$

$$= \tfrac{1}{2}\left(\tfrac{1}{2}\rho U^2\right)\ell b + 2(1.2)\left(\tfrac{1}{2}\rho U^2\right)\frac{\ell b \cos 45°}{2\cos 45°}$$

$$= 1.7\left(\tfrac{1}{2}\rho U^2\right)\ell b$$

which checks with the above answer.

9.2R

9.2R (External flow character) A 0.23-m-diameter soccer ball moves through the air with a speed of 10 m/s. Would the flow around the ball be classified as low, moderate, or large Reynolds number flow? Explain.

(ANS: Large Reynolds number flow)

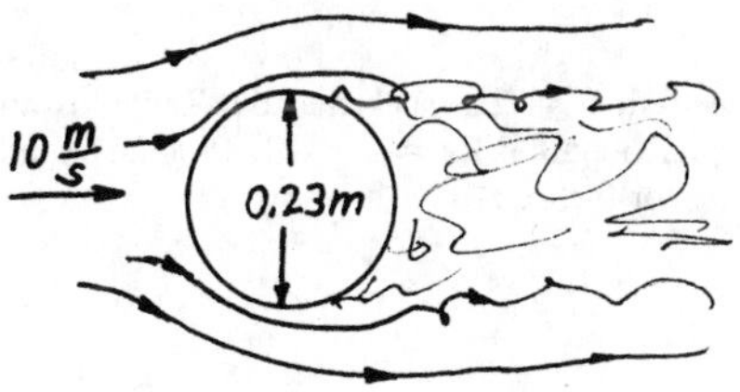

$Re = \frac{UD}{\nu}$, where $D = 0.23\,m$, $U = 10\,\frac{m}{s}$, and $\nu = 1.46\times10^{-5}\,\frac{m^2}{s}$ for standard air.

Thus,

$$Re = \frac{(10\frac{m}{s})(0.23m)}{1.46\times10^{-5}\,\frac{m^2}{s}} = 1.58\times10^5$$

This is a large Reynolds number flow. (See Fig. 9.6c)

9.3R

9.3R (External flow character) A small 15-mm-long fish swims with a speed of 20 mm/s. Would a boundary layer type flow be developed along the sides of the fish? Explain.

(ANS: No)

$Re = \frac{U\ell}{\nu}$, or with $\ell = 15\times10^{-3}\,m$, $U = 20\times10^{-3}\,\frac{m}{s}$ and $\nu = 1.12\times10^{-6}\,\frac{m^2}{s}$ (i.e., 15.5 °C water)

$$Re = \frac{(20\times10^{-3}\frac{m}{s})(15\times10^{-3}m)}{1.12\times10^{-6}\,\frac{m^2}{s}} = 268$$

This Reynolds number is not large enough to have true boundary layer type flow. ($Re \approx 1000$ is often assumed to be the lower limit.)

9.4R

9.4R (Boundary layer flow) Air flows over a flat plate of length $\ell = 2$ ft such that the Reynolds number based on the plate length is Re $= 2 \times 10^5$. Plot the boundary layer thickness, δ, for $0 \leq x \leq \ell$.

$Re_\ell = \frac{U\ell}{\nu} = 2 \times 10^5$, where $\ell = 2\,ft$ and $\nu = 1.57 \times 10^{-4} \frac{ft^2}{s}$

Thus, $U = \frac{(2 \times 10^5)(1.57 \times 10^{-4} \frac{ft^2}{s})}{2\,ft} = 15.7 \frac{ft}{s}$

For $Re_\ell \leq 5 \times 10^5$ the boundary layer flow is laminar and

$\delta = 5\sqrt{\frac{\nu x}{U}} = 5\sqrt{\frac{(1.57 \times 10^{-4})x}{15.7}}$, or $\delta = 0.0158\sqrt{x}$, where $x \sim ft$, $\delta \sim ft$

This result is plotted below for $0 \leq x \leq 2\,ft$. Note the x-axis and δ-axis are different scale (the boundary layer is "thin").

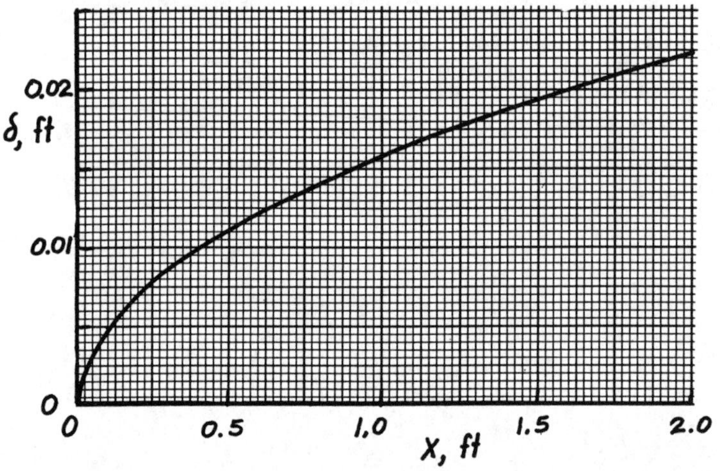

The boundary layer drawn to scale for $0 \leq x \leq 0.5\,ft$ is shown below. Note: It is "thin" (i.e., $\frac{\delta}{x} \ll 1$).

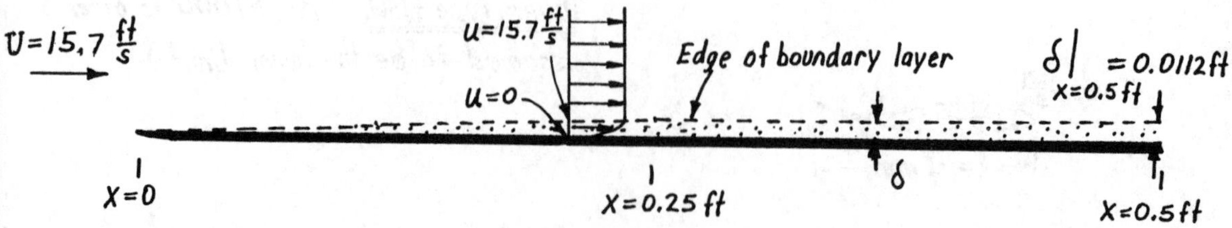

9.5R

9.5R (Boundary layer flow) At a given location along a flat plate the boundary layer thickness is $\delta = 45$ mm. At this location, what would be the boundary layer thickness if it were defined as the distance from the plate where the velocity is 97% of the upstream velocity rather than the standard 99%? Assume laminar flow.

(ANS: 38.5 mm)

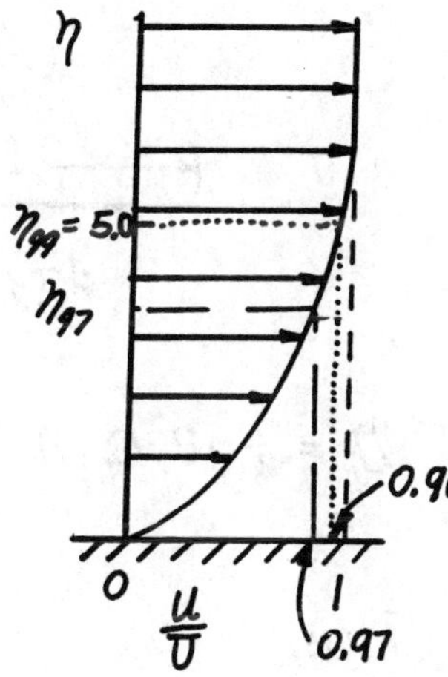

The standard definition of δ is the value of y where $u = 0.99\,U$. This occurs at a value of $\eta = \sqrt{\frac{U}{\nu x}}\,y = 5.0$ That is,

$$\delta_{99} = 45\text{ mm} \quad \text{when } \eta_{99} = 5.0$$

For given U, ν, and x

$$\frac{\eta_{97}}{\eta_{99}} = \frac{\delta_{97}(U/\nu x)^{1/2}}{\delta_{99}(U/\nu x)^{1/2}}$$

or

$$\delta_{97} = \delta_{99}\left(\frac{\eta_{97}}{\eta_{99}}\right) \qquad (1)$$

From Table 9.1 we interpolate to find η_{97}, the value of η when $\frac{u}{U} = 0.97$
Thus,

$$\eta_{97} = 4.0 + 0.4\left(\frac{0.9700 - 0.9555}{0.9759 - 0.9555}\right) = 4.28$$

so that Eq.(1) gives:

$$\delta_{97} = 45\text{ mm}\left(\frac{4.28}{5.00}\right) = \underline{\underline{38.5\text{ mm}}}$$

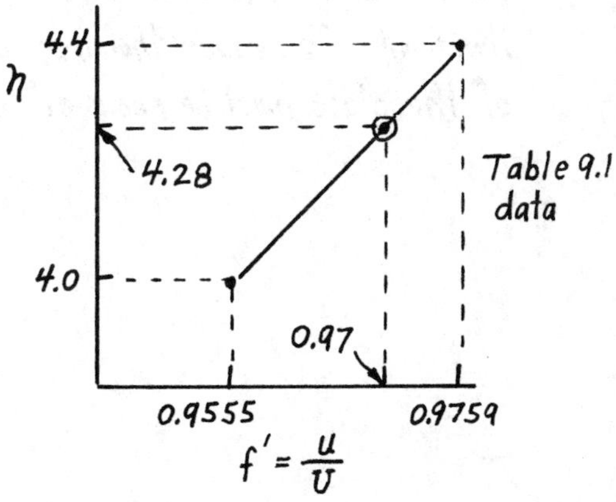

9.6R

9.6R (Friction drag) A laminar boundary layer formed on one side of a plate of length ℓ produces a drag $\mathcal{D}$. How much must the plate be shortened if the drag on the new plate is to be $\mathcal{D}/4$? Assume the upstream velocity remains the same. Explain your answer physically.

(ANS: $\ell_{new} = \ell/16$)

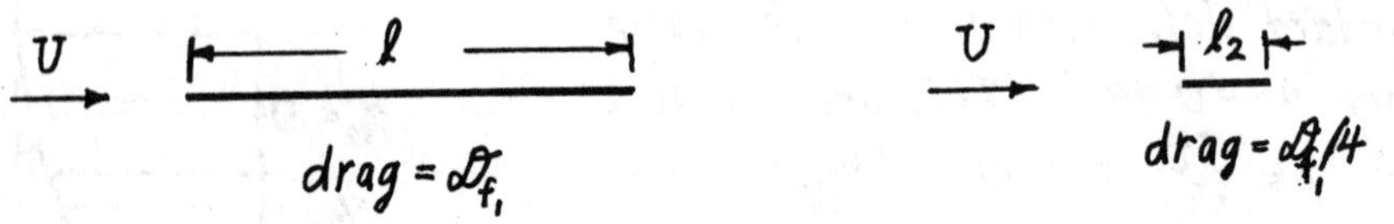

$\mathcal{D}_f = \frac{1}{2}\rho U^2 C_{Df} A$ where $C_{Df} = \frac{1.328}{\sqrt{Re_\ell}} = \frac{1.328}{\sqrt{\frac{U\ell}{\nu}}}$ and

$A = b\ell$ where b = plate width.

Thus,

$$\mathcal{D}_f = \frac{1}{2}\rho U^2 \frac{1.328}{\sqrt{\frac{U\ell}{\nu}}} b\ell = 0.664\,\rho U^{\frac{3}{2}} b\sqrt{\nu}\sqrt{\ell} \qquad (1)$$

Consider two flows with $\rho_1 = \rho_2$, $U_1 = U_2$, $b_1 = b_2$, $\nu_1 = \nu_2$, and $\ell = \ell_1$ or ℓ_2 so from Eq. (1)

$$\frac{\mathcal{D}_{f1}}{\mathcal{D}_{f2}} = \sqrt{\frac{\ell_1}{\ell_2}} \text{ so that with } \mathcal{D}_{f_2} = \frac{1}{4}\mathcal{D}_{f_1},\ \sqrt{\frac{\ell}{\ell_2}} = 4 \text{ or } \underline{\underline{\ell_2 = \frac{\ell}{16}}}$$

Since the boundary layer is thinner on the front portion of the plate than on the rear portion, the wall shear stress is greater near the front. To reduce the drag by a factor of four, more than three-fourths of the plate must be removed (i.e., from length ℓ to length $\frac{\ell}{16}$).

9.7R

9.7R (Momentum integral equation) As is indicated in Table 9.2, the laminar boundary layer results obtained from the momentum integral equation are relatively insensitive to the shape of the assumed velocity profile. Consider the profile given by $u = U$ for $y > \delta$, and $u = U\{1 - [(y - \delta)/\delta]^2\}^{1/2}$ for $y \leq \delta$ as shown in Fig. P9.7R. Note that this satisfies the conditions $u = 0$ at $y = 0$ and $u = U$ at $y = \delta$. However, show that such a profile produces meaningless results when used with the momentum integral equation. Explain.

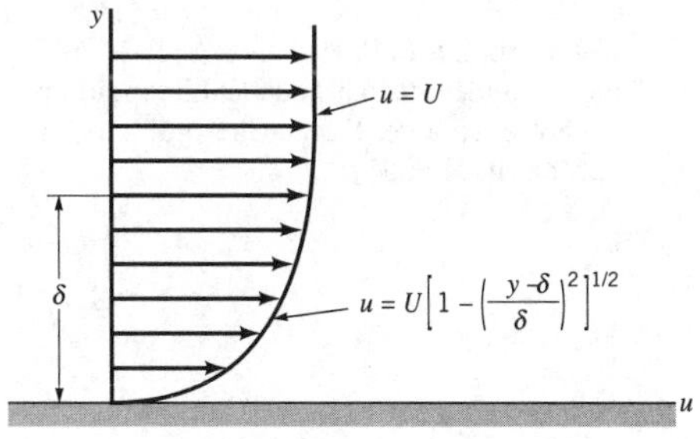

FIGURE P9.7R

From the momentum integral equation,

$$\delta = \sqrt{\frac{2C_2 \nu x}{U C_1}}\ , \quad \text{where } \frac{u}{U} = g(Y) = \left[1 - (Y-1)^2\right]^{1/2} \text{ with } Y = \frac{y}{\delta}. \qquad (1)$$

Note: $\frac{u}{U} = 0$ at $Y = 0$ and $\frac{u}{U} = 1$ and $Y = 1$, as required.

Also, $C_1 = \int_0^1 g(1-g)\,dY$ which can be evaluated for the given $g(Y)$.

However,

$$C_2 = \left.\frac{dg}{dY}\right|_{Y=0}, \quad \text{where } \frac{dg}{dY} = \frac{1}{2}\left[1-(Y-1)^2\right]^{-\frac{1}{2}}(-2)(Y-1) = \frac{(1-Y)}{\left[1-(Y-1)^2\right]^{1/2}}$$

Thus, as $Y \to 0$, $\frac{dg}{dY} \to \infty$ so that

$C_2 = \infty$, which from Eq.(1) gives $\delta = \infty$

This profile cannot be used since it gives $\delta = \infty$ due to the physically unrealistic $\frac{\partial u}{\partial y} = \infty$ at the surface ($y = 0$). See the figure below.

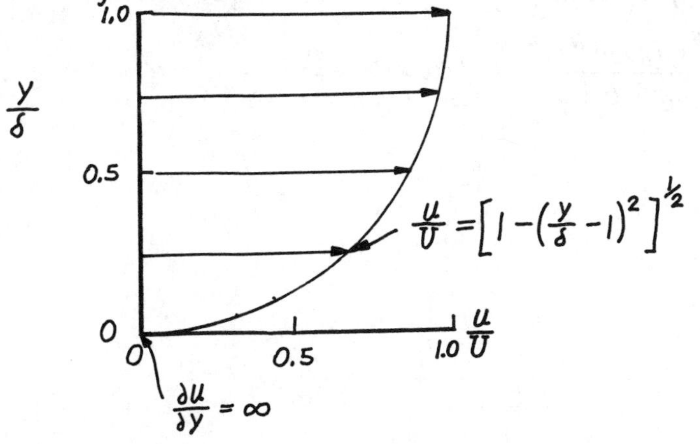

9.8R

9.8R (Drag—low Reynolds number) How fast do small water droplets of 0.06-μm (6×10^{-8} m) diameter fall through the air under standard sea-level conditions? Assume the drops do not evaporate. Repeat the problem for standard conditions at 5000-m altitude.

(ANS: 1.10×10^{-7} m/s; 1.20×10^{-7} m/s)

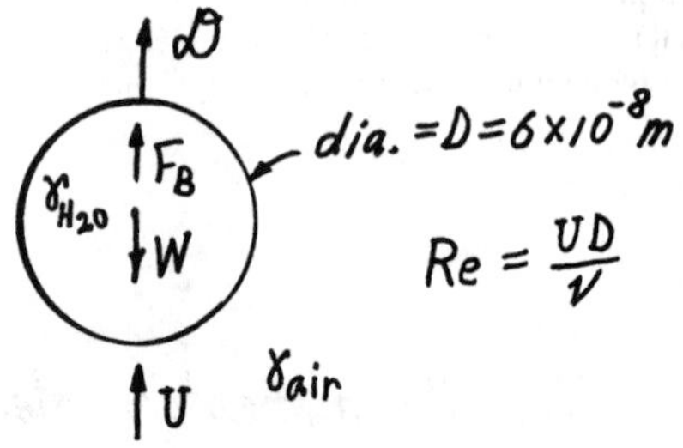

For steady conditions, $\mathcal{D} + F_B = W$,
where, if $Re = \frac{UD}{\nu} < 1$,

$\mathcal{D}$ = drag = $3\pi D U \mu$ Also, $W = \gamma_{H_2O} \forall = \gamma_{H_2O} \frac{4}{3}\pi\left(\frac{D}{2}\right)^3$ = weight

and $F_B = \gamma_{air} \forall = \gamma_{air} \frac{4}{3}\pi\left(\frac{D}{2}\right)^3$ = buoyant force

Since $\gamma_{air} \ll \gamma_{H_2O}$, we can neglect the buoyant force.
That is, $\mathcal{D} = W$, or

$$3\pi D U \mu = \gamma_{H_2O} \frac{4\pi}{3}\left(\frac{D}{2}\right)^3, \text{ or } U = \frac{\gamma_{H_2O} D^2}{18\mu} \qquad (1)$$

At sea level $\mu = 1.789 \times 10^{-5} \frac{N\cdot s}{m^2}$ (see Table C.2) so that

$$U = \frac{(9.80 \times 10^3 \frac{N}{m^3})(6 \times 10^{-8} m)^2}{18(1.789 \times 10^{-5} \frac{N\cdot s}{m^2})} = \underline{\underline{1.10 \times 10^{-7} \frac{m}{s}}}$$

Note that $Re = \frac{(1.10 \times 10^{-7} \frac{m}{s})(6 \times 10^{-8} m)}{1.46 \times 10^{-5} \frac{m^2}{s}} = 4.52 \times 10^{-10} \ll 1$ so the use of the low Re drag equation is valid.

At an altitude of 5000 m, $\mu = 1.628 \times 10^{-5} \frac{N\cdot s}{m^2}$ and from Eq. (1)

$$U = \frac{(9.80 \times 10^3 \frac{N}{m^3})(6 \times 10^{-8} m)^2}{18(1.628 \times 10^{-5} \frac{N\cdot s}{m^2})} = \underline{\underline{1.20 \times 10^{-7} \frac{m}{s}}}$$

9.9R

9.9R (Drag) A 12-mm-diameter cable is strung between a series of poles that are 40 m apart. Determine the horizontal force this cable puts on each pole if the wind velocity is 30 m/s.
(ANS: 372 N)

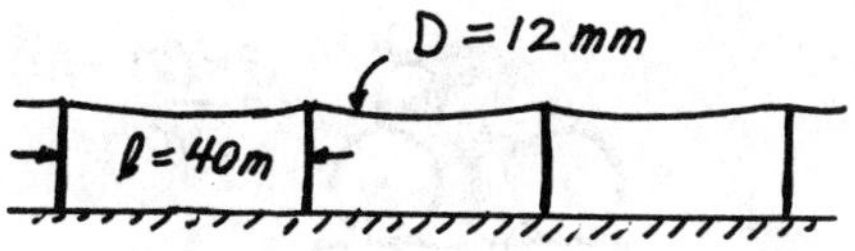

F_p = force on one pole = $\mathscr{D}$
where $\mathscr{D} = C_D \frac{1}{2}\rho U^2 A$
Since $Re = \frac{UD}{\nu} = \frac{(30\frac{m}{s})(0.012m)}{1.46\times10^{-5}\frac{m^2}{s}} = 2.47\times10^4$ it follows from Fig. 9.21 that
$C_D = 1.4$. Hence, $F_p = 1.4(\frac{1}{2})(1.23\frac{kg}{m^3})(30\frac{m}{s})^2(40\,m)(0.012m) = \underline{\underline{372\,N}}$

9.10R

9.10R (Drag) How much less power is required to pedal a racing-style bicycle at 20 mph with a 10-mph tail wind than at the same speed with a 10-mph head wind? (See Fig. 9.30.)

(ANS: 0.375 hp)

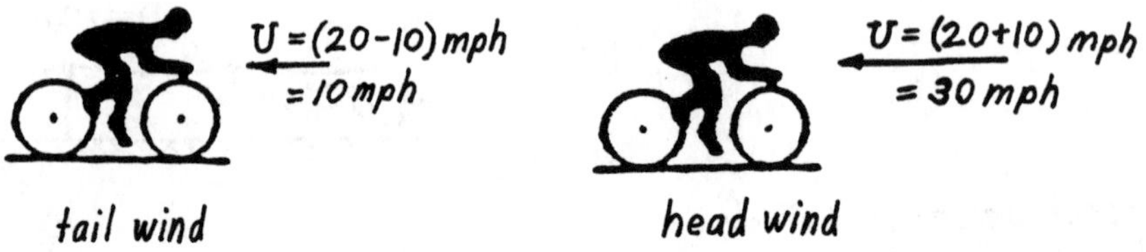

$\mathcal{P}$ = power to overcome aerodynamic drag $= U_B \mathcal{D}$, where U_B = bike speed and $\mathcal{D} = \frac{1}{2}\rho U^2 C_D A$, with U = wind speed relative to bike.

From Fig. 9.30, for a racing bike $C_D = 0.88$ and $A = 3.9\ ft^2$

With a tail wind $U_B = (20\ mph)\left(\frac{88\ \frac{ft}{s}}{60\ mph}\right) = 29.3\ \frac{ft}{s}$

and $U = (20-10)\ mph = 14.7\ \frac{ft}{s}$

Thus, $\mathcal{P}_{tail} = (29.3\ \frac{ft}{s})\frac{1}{2}(0.00238\ \frac{slugs}{ft^3})(14.7\ \frac{ft}{s})^2(3.9\ ft^2)(0.88)$

$= (25.9\ \frac{ft \cdot lb}{s})\left(\frac{1\ hp}{550\ \frac{ft \cdot lb}{s}}\right) = 0.0471\ hp$

With a head wind $U = (20+10)\ mph = 44\ \frac{ft}{s}$

Thus, $\mathcal{P}_{head} = (29.3\ \frac{ft}{s})\frac{1}{2}(0.00238\ \frac{slugs}{ft^3})(44\ \frac{ft}{s})^2(3.9\ ft^2)(0.88)$

$= (232\ \frac{ft \cdot lb}{s})\left(\frac{1\ hp}{550\ \frac{ft \cdot lb}{s}}\right) = 0.422\ hp$

Hence,

$\mathcal{P}_{head} - \mathcal{P}_{tail} = (0.422 - 0.0471)\ hp = \underline{\underline{0.375\ hp}}$

9.11R

9.11R (Drag) A rectangular car-top carrier of 1.6-ft height, 5.0-ft length (front to back), and a 4.2-ft width is attached to the top of a car. Estimate the additional power required to drive the car with the carrier at 60 mph through still air compared with the power required to drive only the car at 60 mph.
(ANS: 12.9 hp)

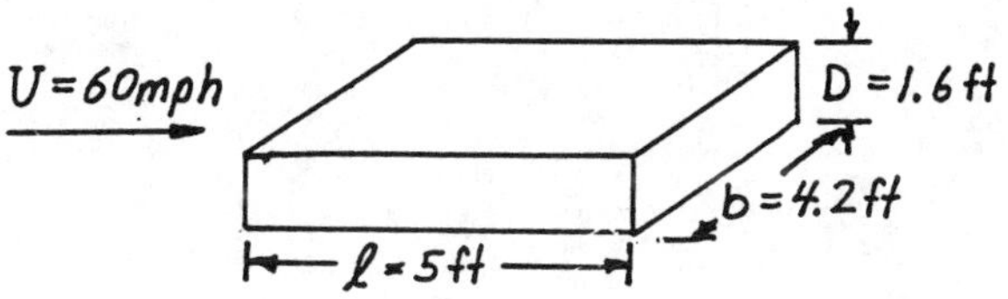

$\mathcal{P}$ = power = $U\mathcal{D}$ where $\mathcal{D}$ = drag = $C_D \frac{1}{2}\rho U^2 A$ (1)

with $U = \left(60\frac{mi}{hr}\right)\left(\frac{1\ hr}{3600\,s}\right)\left(\frac{5280\,ft}{mi}\right) = 88\frac{ft}{s}$

From Fig. 9.28 with $\frac{\ell}{D} = \frac{5}{1.6} = 3.13$ we obtain $C_D = 1.3$

(Note: $Re = \frac{UD}{\nu} = \frac{(88\frac{ft}{s})(1.6\,ft)}{1.57\times10^{-4}\frac{ft^2}{s}} = 8.97\times10^5$, whereas the C_D value from Fig. 9.28 is that for $Re = 10^5$. However, for a blunt object like this box, the value of C_D is essentially independent of Re.)

Thus, from Eq. (1)

$\mathcal{D} = 1.3\left(\frac{1}{2}\right)\left(0.00238\frac{slugs}{ft^3}\right)(1.6\,ft)(4.2\,ft)\left(88\frac{ft}{s}\right)^2 = 80.5\,lb$

so that

$\mathcal{P} = U\mathcal{D} = \left(88\frac{ft}{s}\right)(80.5\,lb)\frac{1\ hp}{550\frac{ft\cdot lb}{s}} = \underline{\underline{12.9\ hp}}$

9.12R

9.12R (Drag) Estimate the wind velocity necessary to blow over the 250-kN boxcar shown in Fig. P9.12R.

(ANS: approximately 32.6 m/s to 35.1 m/s)

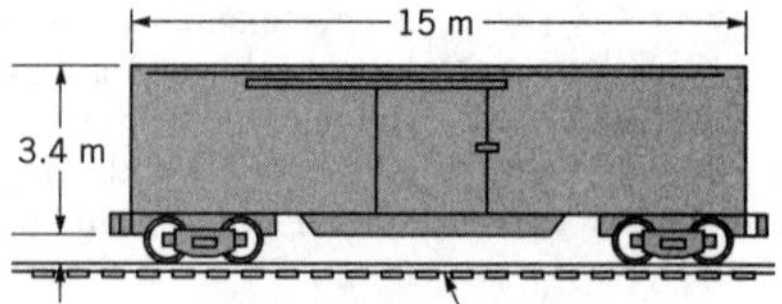

■ FIGURE P9.12R

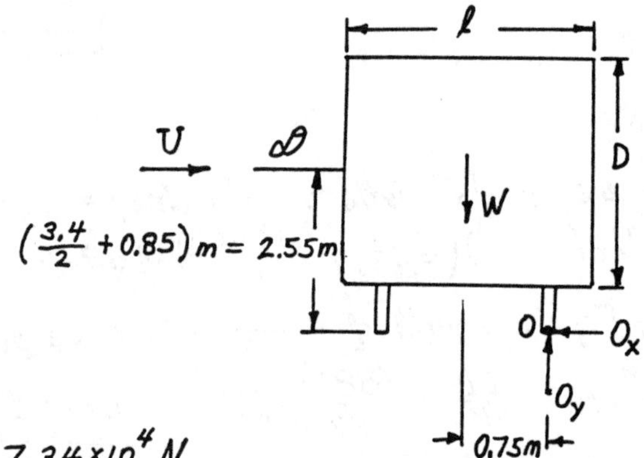

If the boxcar is about to tip around point O, then $\Sigma M_0 = 0$, or

$$2.55\,\mathcal{D} = 0.75\,W$$

Hence,

$$\mathcal{D} = \frac{0.75(250\times10^3 N)}{2.55} = 7.34\times10^4 N,$$

where

$$\mathcal{D} = C_D \tfrac{1}{2}\rho U^2 A \quad \text{or} \quad U = \sqrt{\frac{2\mathcal{D}}{C_D \rho A}}$$

If we consider the boxcar as a flat plate, then from Fig. 9.19 $C_D = 1.9$ so that

$$U = \left[\frac{2\,(7.34\times10^4 N)}{1.9\,(1.23\,\frac{kg}{m^3})(3.4m)(15m)}\right]^{1/2} = \underline{\underline{35.1\ \frac{m}{s}}}$$

Note : The C_D value used above is for a thin plate normal to the flow (i.e. $\ell = 0$ in the above figure). From Fig. 9.28 with $\ell/D = 1$ (i.e. assume the width equals the height) we have $C_D = 2.2$ which gives $U = 32.6\,\frac{m}{s}$. The two values are not too different.

9.13R

9.13R (Drag) A 200-N rock (roughly spherical in shape) of specific gravity $SG = 1.93$ falls at a constant speed U. Determine U if the rock falls through **(a)** air; **(b)** water.

(ANS: 176 m/s; 5.28 m/s)

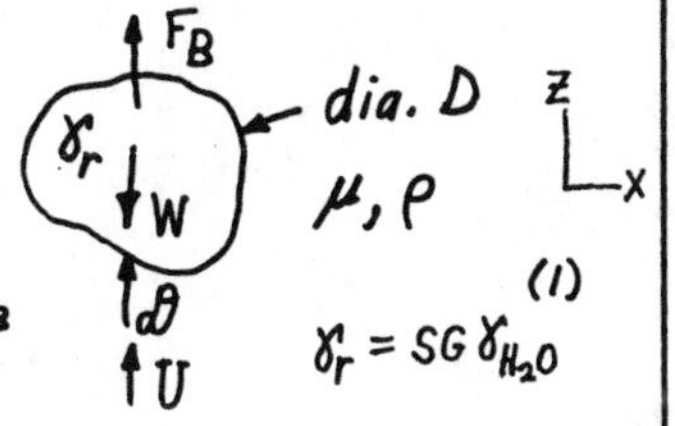

For steady fall $\sum F_z = 0$ or $F_B + \mathcal{D} = W$ (1)

where $\mathcal{D} = C_D \frac{1}{2}\rho U^2 \frac{\pi}{4} D^2$ and $W = \gamma_r \forall = \gamma_r \frac{4\pi}{3}\left(\frac{D}{2}\right)^3$, $\gamma_r = SG\,\gamma_{H_2O}$

Hence, $W = 200\,N = 1.93\left(9.80\times10^3 \frac{N}{m^3}\right)\frac{4\pi}{3}\left(\frac{D}{2}\right)^3$, or $D = 0.272\,m$

Also, $F_B = \rho g \forall = \rho\left(9.81 \frac{m}{s^2}\right)\frac{4\pi}{3}\left(\frac{0.272}{2}m\right)^3 = 0.103\,\rho\ N$, where $\rho \sim \frac{kg}{m^3}$

Thus, Eq. (1) becomes

$0.103\,\rho + C_D \frac{1}{2}\rho U^2 \frac{\pi}{4}(0.272)^2 = 200$, or $0.103\rho + 0.0291\rho C_D U^2 = 200$, where $U \sim \frac{m}{s}$ (2)

a) For falling through air, $\rho = 1.23 \frac{kg}{m^3}$ and Eq. (2) becomes

$0.0358\, C_D U^2 = 199.9$ or $C_D U^2 = 5580$ (3)

Also, $Re = \frac{UD}{\nu} = \frac{(0.272m)\,U}{1.46\times10^{-5}\frac{m^2}{s}} = 1.86\times10^4\,U$, where $U \sim \frac{m}{s}$ (4)

and from Fig. 9.25, 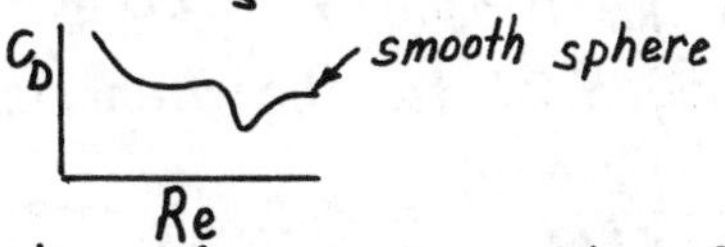

Note: Is the rock a smooth sphere? (5)

Trial and error solution: Assume C_D; obtain U from Eq. (3), Re from Eq. (4); check C_D from Eq. (5), the graph.

Assume $C_D = 0.5 \rightarrow U = 106 \frac{m}{s} \rightarrow Re = 1.97\times10^6 \rightarrow C_D = 0.16 \neq 0.5$

Assume $C_D = 0.16 \rightarrow U = 187 \frac{m}{s} \rightarrow Re = 3.47\times10^6 \rightarrow C_D = 0.18 \neq 0.16$

Assume $C_D = 0.18 \rightarrow U = 176 \frac{m}{s} \rightarrow Re = 3.27\times10^6 \rightarrow C_D = 0.18$ (checks)

Thus, $\underline{\underline{U \approx 176 \frac{m}{s}}}$ in air.

(b) For falling through water, $\rho = 999 \frac{kg}{m^3}$ and Eq. (2) becomes

$C_D U^2 = 3.34$ (6)

Also, $Re = \frac{UD}{\nu} = \frac{(0.272m)\,U}{1.12\times10^{-6}\frac{m^2}{s}} = 2.43\times10^5\,U$ (7)

Trial and error solution with Eqs. (6), (7), and (5).

Assume $C_D = 0.2 \rightarrow U = 4.09 \frac{m}{s} \rightarrow Re = 9.94\times10^5 \rightarrow C_D = 0.10 \neq 0.2$

Assume $C_D = 0.12 \rightarrow U = 5.28 \frac{m}{s} \rightarrow Re = 1.28\times10^6 \rightarrow C_D = 0.12$ (checks)

Thus, $U \approx \underline{\underline{5.28 \frac{m}{s}}}$ in water.

9.14R

9.14R (Drag—composite body) A shortwave radio antenna is constructed from circular tubing, as is illustrated in Fig. P9.14R. Estimate the wind force on the antenna in a 100 km/hr wind.

(ANS: 180 N)

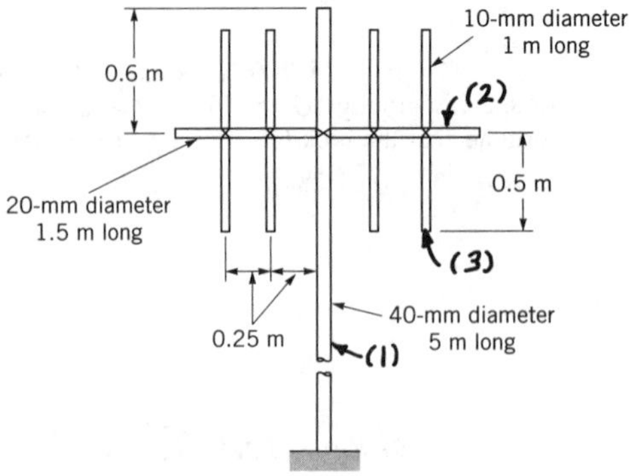

■ FIGURE P9.14R

The antenna is a composite body consisting of one main pole, one horizontal bar, and four vertical rods. Thus,

$$\mathcal{D} = \mathcal{D}_1 + \mathcal{D}_2 + 4\mathcal{D}_3$$

$$= \frac{1}{2}\rho U^2\left[C_{D_1}A_1 + C_{D_2}A_2 + C_{D_3}A_3\right]$$

where $U = 100\,\frac{km}{hr}\left(\frac{10^3 m}{km}\right)\left(\frac{1\,h}{3600\,s}\right) = 27.8\,\frac{m}{s}$

Obtain C_{D_i} from Fig. 9.21 for the given $Re_i = \frac{UD_i}{\nu}$.

Thus, $Re_1 = \frac{(27.8\frac{m}{s})(0.04m)}{1.46\times10^{-5}\frac{m^2}{s}} = 7.62\times10^4 \longrightarrow C_{D_1} = 1.4$

$Re_2 = \frac{(27.8\frac{m}{s})(0.02m)}{1.46\times10^{-5}\frac{m^2}{s}} = 3.81\times10^4 \longrightarrow C_{D_2} = 1.4$

and

$Re_3 = \frac{(27.8\frac{m}{s})(0.01m)}{1.46\times10^{-5}\frac{m^2}{s}} = 1.90\times10^4 \longrightarrow C_{D_3} = 1.4 = C_{D_2} = C_{D_1}$

so that

$$\mathcal{D} = \frac{1}{2}(1.23\,\frac{kg}{m^3})(27.8\,\frac{m}{s})^2(1.4)\left[(5m)(0.04m) + (1.5m)(0.02m) + 4(1m)(0.01m)\right]$$

or

$\mathcal{D} = \underline{\underline{180\text{ N}}}$

9.15R

9.15R (Lift) Show that for level flight the drag on a given airplane is independent of altitude if the lift and drag coefficients remain constant. Note that with C_L constant the airplane must fly faster at a higher altitude.

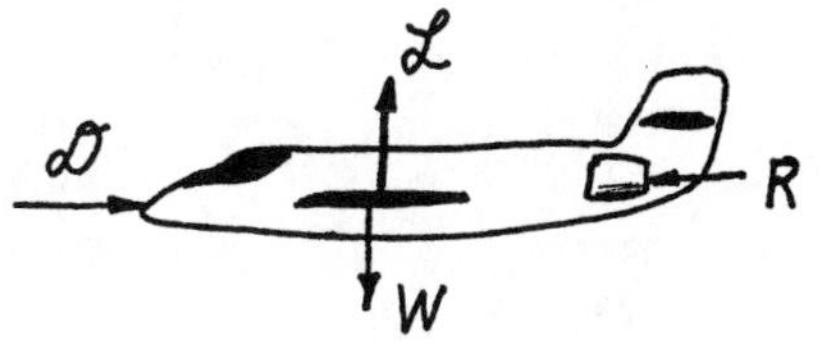

For level flight $\mathscr{L} = W$, where W = airplane weight = constant

and $\mathscr{L} = C_L \frac{1}{2}\rho U^2 A$ (1)

Also, $\mathscr{D} = C_D \frac{1}{2}\rho U^2 A$ (2)

Thus, if C_L, A, and $\mathscr{L}$ are constant (independent of altitude), it follows from Eq. (1) that ρU^2 is also constant. Hence, for constant C_D it follows from Eq. (2) that $\mathscr{D}$ is constant (independent of altitude). (Since ρ decreases with increasing altitude the idea that ρU^2 is constant implies that U increases with altitude.)

9.16R

9.16R (Lift) The wing area of a small airplane weighing 6.22 kN is 10.2 m^2. **(a)** If the cruising speed of the plane is 210 km/hr, determine the lift coefficient of the wing. **(b)** If the engine delivers 150 kW at this speed, and if 60% of this power represents propeller loss and body resistance, what is the drag coefficient of the wing.

(ANS: 0.292; 0.0483)

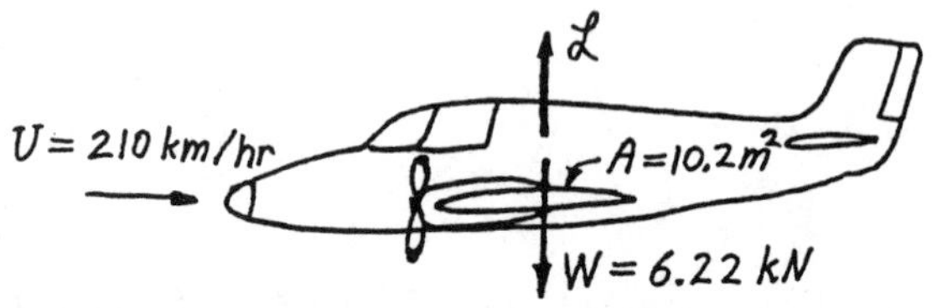

For equilibrium, $\mathcal{L} = W = 6.22\text{ kN}$, where $\mathcal{L} = C_L \frac{1}{2}\rho U^2 A$.

Thus, with $U = (210\,\frac{km}{hr})(\frac{1\,hr}{3600\,s})(\frac{1000\,m}{1\,km}) = 58.3\,\frac{m}{s}$

$$C_L = \frac{\mathcal{L}}{\frac{1}{2}\rho U^2 A} = \frac{6.22\times10^3\,N}{\frac{1}{2}(1.23\,\frac{kg}{m^3})(58.3\,\frac{m}{s})^2(10.2\,m^2)} = \underline{\underline{0.292}}$$

Also,

$$\mathcal{P}_{total} = \mathcal{P}_{body} + \mathcal{P}_{\substack{loss\\prop}} + \mathcal{P}_{wing}, \quad \text{where } \mathcal{P}_{body} + \mathcal{P}_{\substack{loss\\prop}} = 0.6\,\mathcal{P}_{total} = 0.6(150\,kW) = 90\,kW$$

Thus,

$$\mathcal{P}_{wing} = 150\,kW - 90\,kW = 60\,kW$$

where

$$\mathcal{P}_{wing} = \mathcal{D}U = C_D \frac{1}{2}\rho U^2 A U$$

or

$$C_D = \frac{2\,\mathcal{P}_{wing}}{\rho U^3 A} = \frac{2\,(60\times10^3\,\frac{N\cdot m}{s})}{(1.23\,\frac{kg}{m^3})(58.3\,\frac{m}{s})^3(10.2\,m^2)} = \underline{\underline{0.0483}}$$

10
Open-Channel Flow

10.1 General Characteristics of Open-Channel Flow

- Open channel flows have a free surface and are driven by gravity.
- For uniform open channel flows the flow depth is constant; for nonuniform (varied) flows the depth varies with distance along the channel.
- The character of the open channel flow depends on the value of the Froude number, Fr.

10.2 Surface Waves

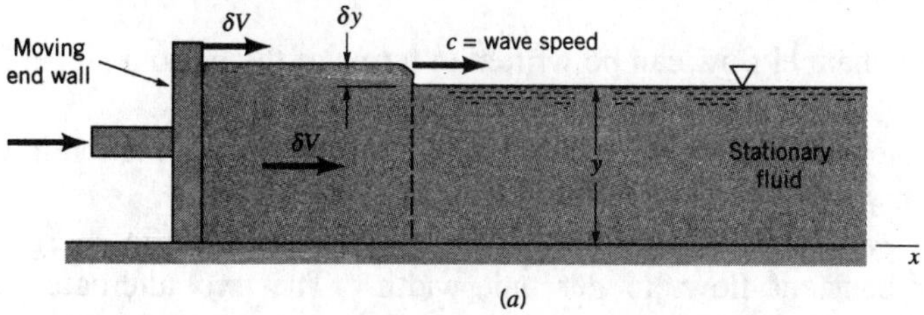

■ **FIGURE 10.2** ***(a)* Production of a single elementary wave in a channel as seen by a stationary observer.**

$$c = \sqrt{gy} \qquad \textbf{(10.3)}$$

- As shown by Fig. 10.2*a*, a small amplitude surface wave travels with a wave speed c relative to the water. The wave speed is given by Eq. 10.3.
- The *Froude number*, $\mathrm{Fr} = V/(gy)^{1/2}$, is the ratio of the flow velocity to the surface wave speed.
- Flows are termed *subcritical*, *critical*, or *supercritical* depending on whether the Froude number is less than, equal to, or greater than one.
- *Caution:* Equation 10.3 is valid only for small amplitude surface waves having a wave length much greater than the liquid depth.

EXAMPLE: Small amplitude surface waves are observed to travel across a shallow pond with a speed of 3 m/s. Determine the depth of the pond.
SOLUTION:

From Eq. 10.3, $c = \sqrt{gy}$ or $y = c^2/g$

Thus, $y = (3\,m/s)^2/(9.81\,m/s^2) = \underline{\underline{0.917\,m}}$

10.3 Energy Considerations

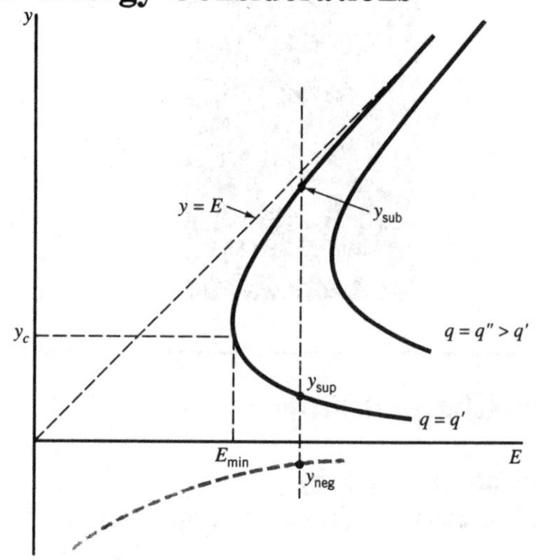

■ **FIGURE 10.7 Specific energy diagram.**

$$y_1 + \frac{V_1^2}{2g} + S_0\ell = y_2 + \frac{V_2^2}{2g} + h_L \quad \textbf{(10.6)}$$

$$E = y + \frac{q^2}{2gy^2} \quad \textbf{(10.10)}$$

- The energy equation for open-channel flow can be written in terms of the bottom slope, S_0, as shown in Eq. 10.6.
- The *specific energy*, E, involves the elevation and velocity heads and can be written in terms of the flowrate per unit width, q, as shown in Eq. 10.10.
- As shown in Fig. 10.7, the *specific energy diagram* is a graph of the specific energy as a function of flow depth for constant flowrate per unit width. The two alternate depths correspond to subcritical and supercritical flow conditions.
- *Caution:* Flows with different flowrates per unit width correspond to different curves on the specific energy diagram.

EXAMPLE: The flowrate in a 20-ft-wide open-channel is 160 ft^3/s. Plot the specific energy diagram for this channel and determine the flow depths possible if the specific energy is 3 ft.

SOLUTION:

From Eq. 10.10, the specific energy diagram is given by $E = y + q^2/(2gy^2)$, where

$q = Q/b = 160\ ft^3/s/(20 ft) = 8\ ft^2/s$

Thus, $E = y + (8\ ft^2/s)^2/[2(32.2\ ft/s)y^2]$, or

$E = y + 0.994/y^2$, where E and y are in ft.

This curve is plotted in the figure. As shown, for E = 3 ft the two positive solutions for y are $\underline{\underline{y = 2.88\ ft}}$ and $\underline{\underline{y = 0.651\ ft}}$

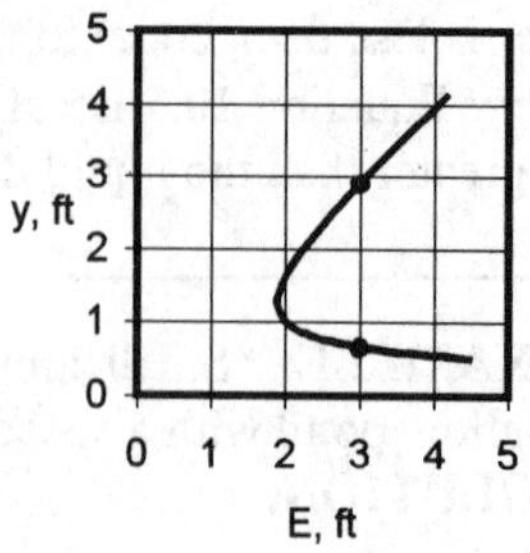

10.4 Uniform Depth Channel Flow

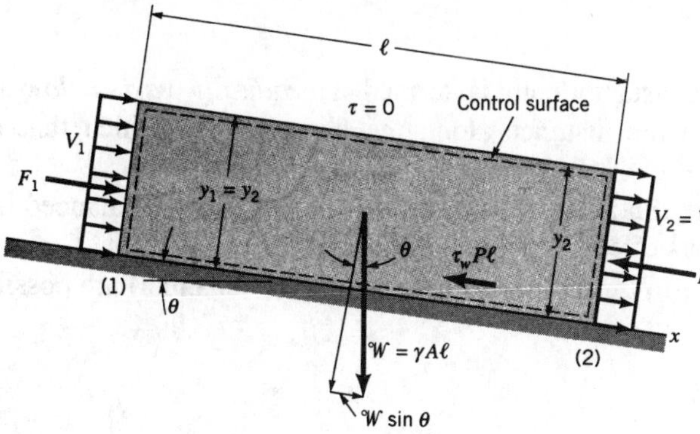

■ **FIGURE 10.10**
Control volume for uniform flow in an open channel.

$$Q = \frac{\kappa}{n} A R_h^{2/3} S_0^{1/2} \qquad \textbf{(10.20)}$$

- Uniform open channel flows occur in channels of constant cross-sectional size and shape having constant flow depth and bottom slope along the channel.
- As shown in Fig. 10.10, uniform open channel flows are a result of the balance between gravity and frictional forces. The component of the water weight down the hill is balanced by the shear stress force up the hill.
- The uniform flow flowrate, Q, can be determined by the *Manning equation*, Eq. 10.20, which involves the channel cross-sectional area, A, the hydraulic radius, R_h, the slope of the channel, S_o, and the Manning coefficient, n.
- The Manning coefficient is a function of the channel surface material; the rougher the surface, the larger the Manning coefficient value.
- The hydraulic radius is equal to the cross-sectional area divided by the wetted perimeter, $R_h = A/P$.
- *Caution:* The parameter κ in the Manning equation equals 1 if SI units are used; it equals 1.49 if BG units are used.
- *Caution:* The wetted perimeter used to determine the hydraulic radius does *not* include the portion of the interface between the free surface of the channel and the air above it.

EXAMPLE: Water is to flow with a velocity of 3 m/s and a depth of 2 m in a 10-m-wide channel. The channel surface is unfinished concrete with a Manning coefficient value of 0.014. Determine the change in elevation needed along a 1 km stretch of this channel to produce this flow.

SOLUTION:

From Eq. 10.20, $Q = AV = \frac{\kappa}{n} A R_h^{2/3} S_o^{1/2}$ so that

$V = \frac{\kappa}{n} R_h^{2/3} S_o^{1/2}$, where $\kappa = 1$ and $n = 0.014$. Also, $R_h = \frac{A}{P}$ or

$R_h = 2\,m\,(10\,m)/(2\,m + 10\,m + 2\,m) = 1.43\,m$

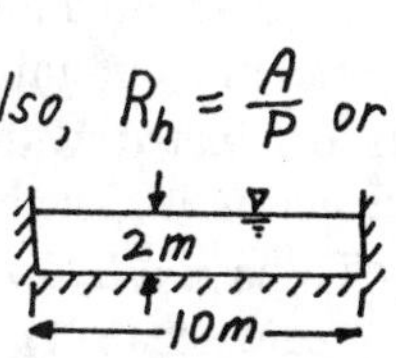

Thus, with $V = 3\,m/s$

$3 = \frac{1}{0.014}(1.43)^{2/3} S_o^{1/2}$, or $S_o = 0.00110$

Hence, since $S_o = \Delta z/\ell$, $\Delta z = S_o \ell = 0.00110\,(1000\,m) = \underline{\underline{1.10\,m}}$

10.5 Gradually Varied Flow

- An open channel flow with non-constant depth is termed a *gradually varied flow* if the change in depth is much less than the distance along the channel over which that depth change occurs.
- If the loss of potential energy of the fluid as it flows downhill is *not* exactly balanced by the dissipation of energy through viscous effects, the flow depth will not be constant.
- A variety of surface shapes (depth as a function of distance along the channel) are possible in gradually varied flows.

10.6 Rapidly Varied Flow

- Open channel flows for which the flow depth changes over relatively short distances (comparable to the water depth) are termed *rapidly varied flows*.

The Hydraulic Jump

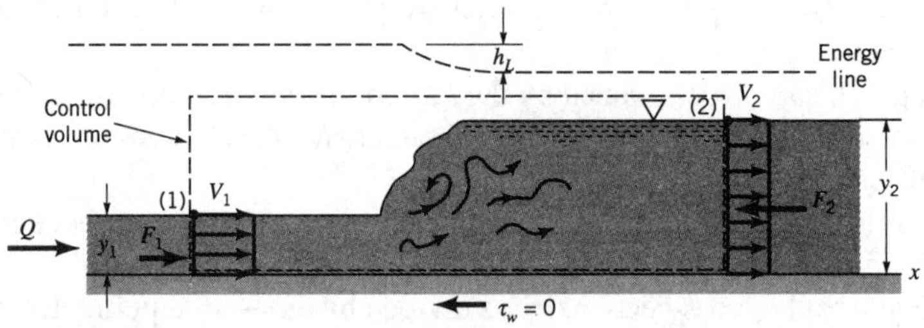

■ FIGURE 10.19 **Hydraulic jump geometry.**

$$\frac{y_2}{y_1} = \frac{1}{2}\left(-1 + \sqrt{1 + 8\text{Fr}_1^2}\right) \qquad \textbf{(10.24)}$$

$$\frac{h_L}{y_1} = 1 - \frac{y_2}{y_1} + \frac{\text{Fr}_1^2}{2}\left[1 - \left(\frac{y_1}{y_2}\right)^2\right] \qquad \textbf{(10.25)}$$

- As shown in Fig. 10.19, a hydraulic jump is a sudden change in flow depth that occurs without any change in the channel configuration. Across a hydraulic jump the water depth increases.
- The depth ratio across a hydraulic jump, given by Eq. 10.24, is obtained by application of the continuity and momentum equations to a control volume that surrounds the jump.
- Energy is dissipated within a hydraulic jump because of the considerable mixing and turbulence within it. The head loss across a hydraulic jump is given by Eq. 10.25.
- *Caution:* To have a hydraulic jump, the flow must be supercritical (i.e., the Froude number must be greater than 1). However, the flow may be supercritical without having the occurrence of a hydraulic jump.

EXAMPLE: A hydraulic jump occurs in a channel where the upstream water depth is 0.5 ft and the upstream velocity is 12 ft/s. **(a)** Determine the water depth downstream of the jump. **(b)** Determine the head loss for this flow.

SOLUTION:

(a) With $y_1 = 0.5$ ft and $V_1 = 12$ ft/s the Froude number is

$$Fr_1 = V_1/\sqrt{gy_1} = 12\ \text{ft/s}/\left[32.2\ \text{ft/s}^2 (0.5\ \text{ft})\right]^{1/2} = 2.99 > 1$$

Thus, if there is a hydraulic jump, then from Eq. 10.24

$$\frac{y_2}{y_1} = \frac{1}{2}\left(-1 + \sqrt{1 + 8Fr_1^2}\right) = \frac{1}{2}\left(-1 + \sqrt{1 + 8(2.99)^2}\right) = 3.76$$

Hence, $y_2 = 3.76\,y_1 = 3.76(0.5\ \text{ft}) = \underline{\underline{1.88\ \text{ft}}}$

(b) From Eq. 10.25, $h_L = y_1\left\{1 - \frac{y_2}{y_1} + \frac{Fr_1^2}{2}\left[1 - \left(\frac{y_1}{y_2}\right)^2\right]\right\}$ so that with $y_2/y_1 = 3.76$, then

$$h_L = 0.5\ \text{ft}\left\{1 - 3.76 + \frac{2.99^2}{2}\left[1 - \left(\frac{1}{3.76}\right)^2\right]\right\} = \underline{\underline{0.697\ \text{ft}}}$$

Sharp-Crested Weirs

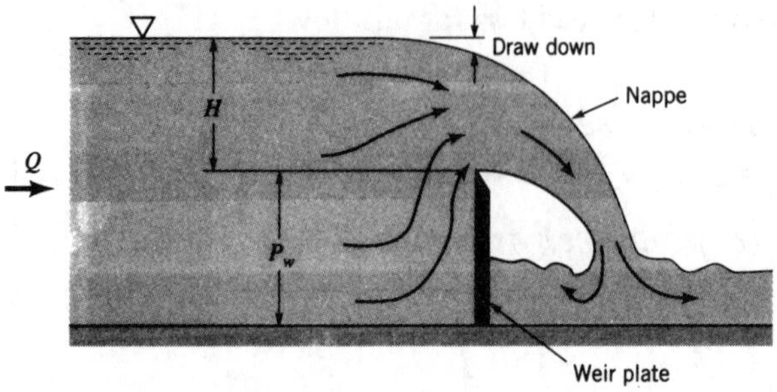

■ **FIGURE 10.22**
Sharp-crested weir geometry.

$$Q = C_{wr} \tfrac{2}{3} \sqrt{2g}\, b\, H^{3/2} \qquad \textbf{(10.30)}$$

$$Q = C_{wt} \frac{8}{15} \tan\left(\frac{\theta}{2}\right) \sqrt{2g}\, H^{5/2} \qquad \textbf{(10.32)}$$

- The flowrate in an open channel can be determined by use of a *sharp-crested weir* as shown in Fig. 10.22.
- The flowrate, Q, over a sharp-crested weir depends on the weir head, H, and the shape of the weir. Eqs. 10.30 and 10.32 give the flowrate for rectangular and triangular sharp-crested weirs, respectively.
- The weir coefficient, C_w, is used to account for the non-ideal conditions that exist in the actual flow over a weir.

EXAMPLE: Water is to flow over a sharp-crested rectangular weir at a rate of 0.5 m³/s when the weir head is 0.2 m. If the weir coefficient is 0.70, determine the width of the weir.

SOLUTION:

From Eq. 10.30, $Q_{wr} = C_{wr} \frac{2}{3} \sqrt{2g}\, b H^{3/2}$, where $Q_{wr} = 0.5\ m^3/s$, $H = 0.2\,m$, and $C_{wr} = 0.70$. Thus,

$$0.5\ m^3/s = 0.70\left(\tfrac{2}{3}\right)\sqrt{2(9.81\ m/s^2)}\,(b\,m)(0.2\,m)^{3/2}$$

or $b = \underline{\underline{2.70\,m}}$

Broad-Crested Weirs

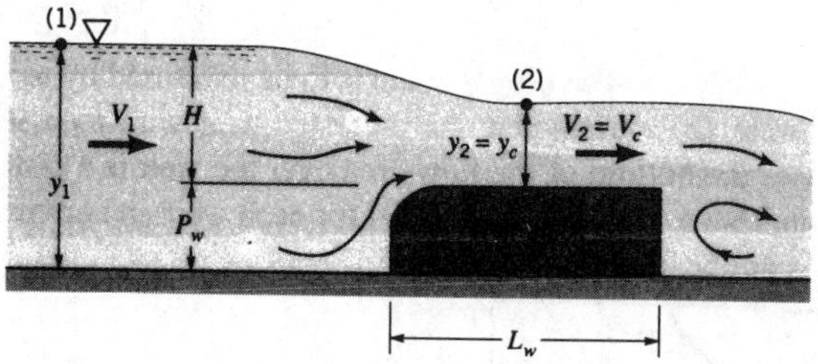

■ FIGURE 10.27
Broad-crested weir geometry.

$$Q = C_{wb}\, b \sqrt{g} \left(\frac{2}{3}\right)^{3/2} H^{3/2} \qquad \textbf{(10.33)}$$

- The flowrate in an open channel can be determined by use of a *broad-crested weir* as shown in Fig. 10.27. The operation of a broad-crested weir is based on the fact that nearly uniform critical flow is achieved above the weir block.
- The flowrate, Q, for a broad-crested weir is a function of the weir head, H, as shown by Eq.10.33.
- The weir coefficient, C_w, is used to account for the non-ideal conditions that exist in the actual flow over a weir.
- *Caution:* To ensure proper operation, broad-crested weirs are generally restricted to the range $0.08 < H/L_w < 0.50$, where L_w is the length of the weir block (see Fig. 10.27).

EXAMPLE: Water flows over a broad-crested weir with a flowrate of 3 ft^3/s when the weir head is 0.5 ft. Determine the flowrate expected when the head is increased to 0.8 ft. Assume that the weir coefficient remains the same for each case.

SOLUTION:

From Eq. 10.33, $Q = C_{wb}\, b\sqrt{g}\left(\frac{2}{3}\right)^{3/2} H^{3/2}$, or with $Q = 3\ ft^3/s$, $H = 0.5\ ft$,

$3\ ft^3/s = C_{wb}\, b \sqrt{32.2\ ft/s^2}\left(\frac{2}{3}\right)^{3/2}(0.5\ ft)^{3/2}$

Thus, $C_{wb}\, b = 2.75\ ft$ Hence, with $H = 0.8\ ft$,

$Q = 2.75\ ft \sqrt{32.2\ ft/s^2}\left(\frac{2}{3}\right)^{3/2}(0.8\ ft)^{3/2} = \underline{6.08\ ft^3/s}$

Underflow Gates

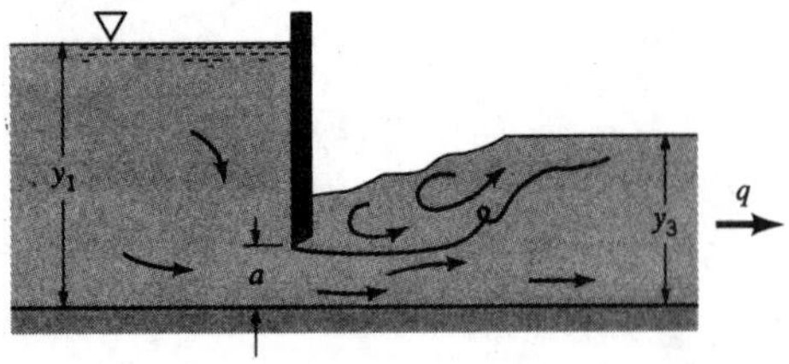

$$q = C_d a \sqrt{2gy_1} \qquad (10.35)$$

■ **FIGURE 10.30 Drowned outflow from a sluice gate.**

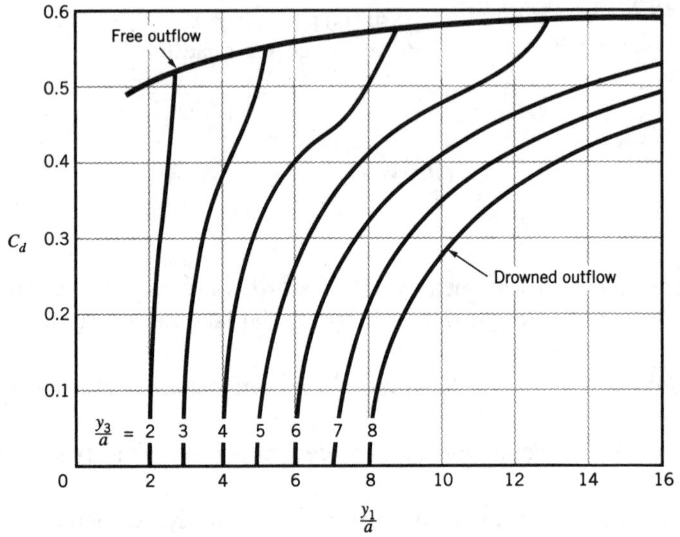

■ **FIGURE 10.29 Typical discharge coefficients for underflow gates (Ref. 3).**

- The flowrate in an open channel can be determined by use of an *underflow gate* as shown in Fig. 10.30.
- As indicated by Eq. 10.35, the flowrate per unit width, q, for an underflow gate depends on the distance, a, between the channel bottom and the bottom of the gate.
- Depending on the conditions downstream of the gate, the stream of water issuing from under the gate may be overlaid by a mass of turbulent water (*drowned outflow*). In other situations the water from under the gate may be a free stream (*free outflow*).
- The discharge coefficient, C_d, for typical underflow gates is as shown in Fig. 10.29. Its value depends on whether the flow involves a free outflow or a drowned outflow.

EXAMPLE: The water depth upstream of the sluice gate shown in Fig. 10.30 is $y_1 = 4$ ft. When $a = 0.5$ ft the water depth downstream of the sluice gate is $y_3 = 2$ ft. Determine the flowrate per unit width.

SOLUTION:

From Eq. 10.35, $q = C_d a\sqrt{2gy_1}$, where $a = 0.5$ ft and $y_1 = 4$ ft.

With $y_3 = 2$ ft it follows that $y_3 > a$ so that the flow is drowned outflow.

Thus, $y_3/a = 2\text{ ft}/0.5\text{ ft} = 4.0$ and $y_1/a = 4\text{ ft}/0.5\text{ ft} = 8.0$

From Fig. 10.29 with $y_3/a = 4$ and $y_1/a = 8$ it follows that $C_d = 0.51$.

Hence, $q = 0.51\,(0.5\text{ ft})\sqrt{2(32.2\text{ ft/s}^2)(4\text{ ft})} = 4.09\text{ ft}^2/\text{s} = \underline{\underline{4.09(\text{ft}^3/\text{s})/\text{ft}}}$

Solutions

to Review Problems

Chapter 10

10.1R

10.1R (Surface waves) If the water depth in a pond is 15 ft, determine the speed of small amplitude, long wavelength ($\lambda \gg y$) waves on the surface.

(ANS: 22.0 ft/s)

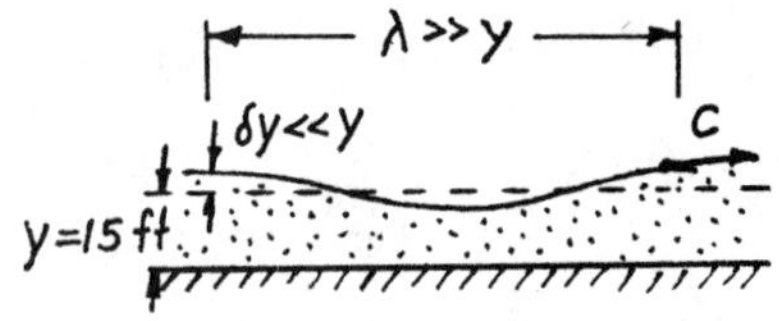

For $\lambda \gg y$, $\tanh \frac{2\pi y}{\lambda} \approx \frac{2\pi y}{\lambda}$ so that (see Eq 10.4)

$$c = \left[\frac{g\lambda}{2\pi} \tanh\left(\frac{2\pi y}{\lambda}\right)\right]^{\frac{1}{2}} \approx \left[\frac{g\lambda}{2\pi}\left(\frac{2\pi y}{\lambda}\right)\right]^{\frac{1}{2}} = \sqrt{gy}$$

or

$$c = \left[\left(32.2\,\frac{ft}{s^2}\right)(15\,ft)\right]^{\frac{1}{2}} = \underline{\underline{22.0\,\frac{ft}{s}}}$$

10.2R

10.2R (Surface waves) A small amplitude wave travels across a pond with a speed of 9.6 ft/s. Determine the water depth.

(ANS: $y \geq 2.86$ ft)

For long wavelength waves $c = \sqrt{gy}$ or $y = \frac{c^2}{g} = \frac{(9.6\frac{ft}{s})^2}{32.2\frac{ft}{s^2}} = 2.86\,ft$

In general, from Eq. 10.4:

$$c = \left[\frac{g\lambda}{2\pi} \tanh\left(\frac{2\pi y}{\lambda}\right)\right]^{\frac{1}{2}}, \text{ or } 9.6\,\frac{ft}{s} = \left[\frac{(32.2\frac{ft}{s^2})\lambda}{2\pi} \tanh\left(\frac{2\pi y}{\lambda}\right)\right]^{\frac{1}{2}}$$

Hence: $\tanh\left(\frac{2\pi y}{\lambda}\right) = \frac{18.0}{\lambda}$ Pick various λ and calculate y. (1)

Note: Since $\tanh\left(\frac{2\pi y}{\lambda}\right) \leq 1$, it is not possible to have $\lambda < 18$ ft (for $c = 9.6$ ft/s). Also, as $\lambda \to \infty$, $y \to 2.86$ ft (the long wavelength limit).

λ, ft	y, ft
18	∞
19	5.46
20	4.69
25	3.61
30	3.31
50	3.00
100	2.90
∞	2.86

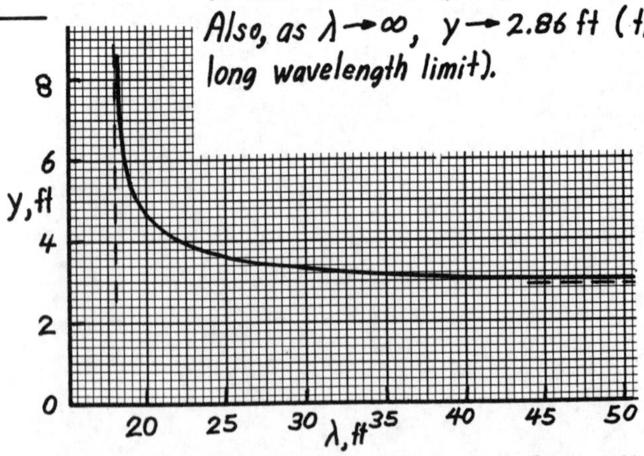

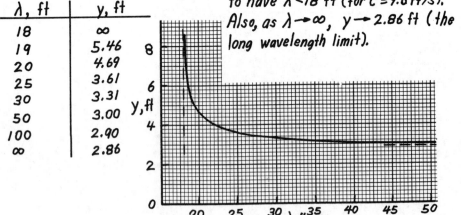

Although we must know the wavelength, λ, to determine the depth, y, we do know that $y \geq \underline{2.86\,ft}$

10.3R

10.3R (Froude number) The average velocity and average depth of a river from its beginning in the mountains to its discharge into the ocean are given in the table below. Plot a graph of the Froude number as a function of distance along the river.

Distance (mi)	Average Velocity (ft/s)	Average Depth (ft)
0	13	1.5
5	10	2.0
10	9	2.3
30	5	3.7
50	4	5.4
80	4	6.0
90	3	6.2

(ANS: Fr = 1.87 at the beginning, 0.212 at the discharge)

$Fr = \frac{V}{c} = \frac{V}{\sqrt{g y}}$, where $g = 32.2 \frac{ft}{s^2}$ and values of V and y are given in the table. The following results are obtained.

Distance, mi	Fr
0	1.87
5	1.25
10	1.05
30	0.458
50	0.303
80	0.288
90	0.212

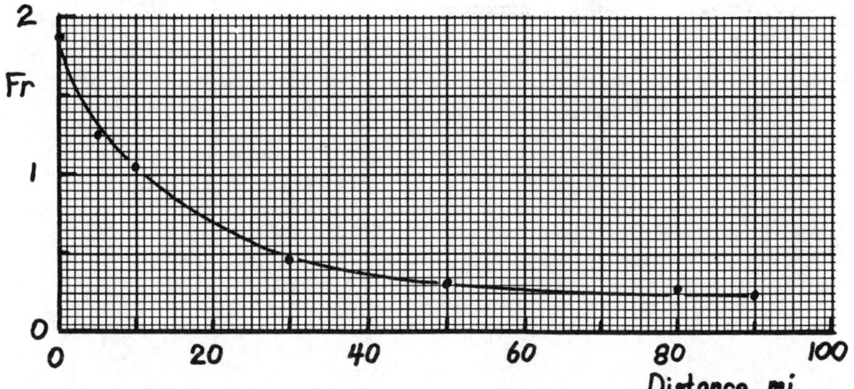

Note that the flow is supercritical for the first 10 miles or so.

10.4R

10.4R (Froude number) Water flows in a rectangular channel at a depth of 4 ft and a flowrate of $Q = 200$ cfs. Determine the minimum channel width if the flow is to be subcritical.

(ANS: 4.41 ft)

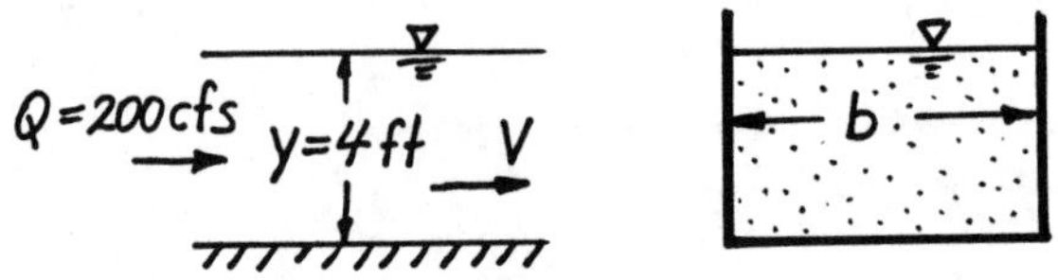

$$V = \frac{Q}{A} = \frac{200\,\frac{ft^3}{s}}{(4ft)\,b} = \frac{50}{b}, \text{ where } b = \text{width} \sim \text{ft and } V \sim \text{ft/s}.$$

Thus, $$Fr = \frac{V}{\sqrt{gy}} = \frac{\left(\frac{50}{b}\right)}{\left[(32.2\frac{ft}{s^2})(4ft)\right]^{1/2}}$$ Note: As b decreases, Fr increases.

Set $Fr = 1$ for minimum width for subcritical flow.

Hence,

$$1 = \frac{50/b}{\left[(32.2)(4)\right]^{1/2}} \quad \text{or} \quad b = \underline{\underline{4.41\text{ ft}}}$$

10.5R (Specific energy) Plot the specific energy diagram for a wide channel carrying $q = 50\ \text{ft}^2/\text{s}$. Determine **(a)** the critical depth, **(b)** the minimum specific energy, **(c)** the alternate depth corresponding to a depth of 2.5 ft, and **(d)** the possible flow velocities if $E = 10$ ft.

(ANS: 4.27 ft; 6.41 ft; 8.12 ft; 5.22 ft/s or 22.3 ft/s)

$$E = y + \frac{q^2}{2g\,y^2}, \quad \text{or} \quad E = y + \frac{(50\,\frac{ft^2}{s})^2}{2(32.2\,\frac{ft}{s^2})y^2}$$

Thus,

$$E = y + \frac{38.8}{y^2}, \quad \text{where } E \sim ft,\ y \sim ft \quad \text{Equation (1) is plotted below.} \qquad (1)$$

Note:

a) $y_c = \left(\frac{q^2}{g}\right)^{\frac{1}{3}} = \left(\frac{(50\,\frac{ft^2}{s})^2}{(32.2\,\frac{ft}{s^2})}\right)^{\frac{1}{3}} = \underline{\underline{4.27\ ft}}$

b) $E_{min} = \frac{3}{2}y_c = \frac{3}{2}(4.27\ ft) = \underline{\underline{6.41\ ft}}$

c) If $y = 2.5$ ft, then $E = 2.5 + \frac{38.8}{(2.5)^2} = 8.71$

Thus, solve for the positive real roots of

$$8.71 = y + \frac{38.8}{y^2}, \quad \text{or} \quad y^3 - 8.71y^2 + 38.8 = 0 \qquad (2)$$

One root is the given $y = 2.5$. Thus, divide $(y - 2.5)$ into Eq. (2) to obtain $y^2 - 6.21y - 15.53 = 0$ which has roots

$$y = \frac{6.21 \pm \sqrt{6.21^2 + 4(15.53)}}{2} = 8.12\ ft, \text{ or } y = -1.91\ ft$$

Thus, the alternate depth corresponding to $y = 2.5$ ft is $y = \underline{\underline{8.12\ ft}}$

d) For $E = 10$ ft Eq. (1) is $10 = y + \frac{38.8}{y^2}$, or $y^3 - 10y^2 + 38.8 = 0$ (3)

which has roots: $y = 9.57$, 2.24, and -1.81. Thus, $y = 9.57$ ft or $y = 2.24$ ft

The corresponding velocities are $V = \frac{q}{y} = \frac{50\,\frac{ft^2}{s}}{9.57\,ft} = \underline{\underline{5.22\,\frac{ft}{s}}}$

$$V = \frac{50\,\frac{ft^2}{s}}{2.24\,ft} = \underline{\underline{22.3\,\frac{ft}{s}}}$$

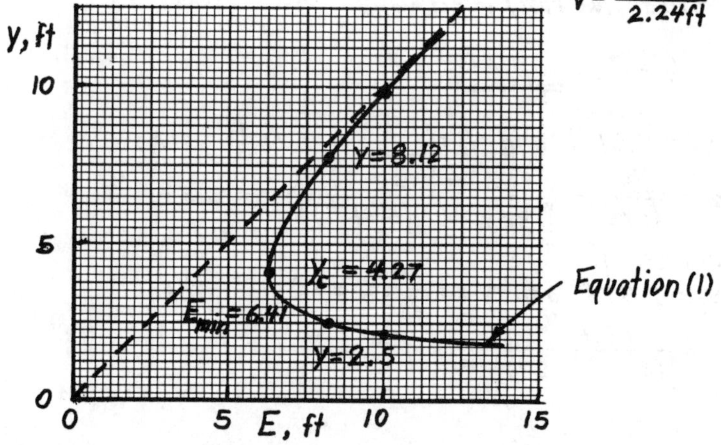

The specific energy diagram is as shown above.

10.6R

10.6R (Specific energy) Water flows at a rate of 1000 ft^3/s in a horizontal rectangular channel 30 ft wide with a 2-ft depth. Determine the depth if the channel contracts to a width of 25 ft. Explain.
(ANS: 2.57 ft)

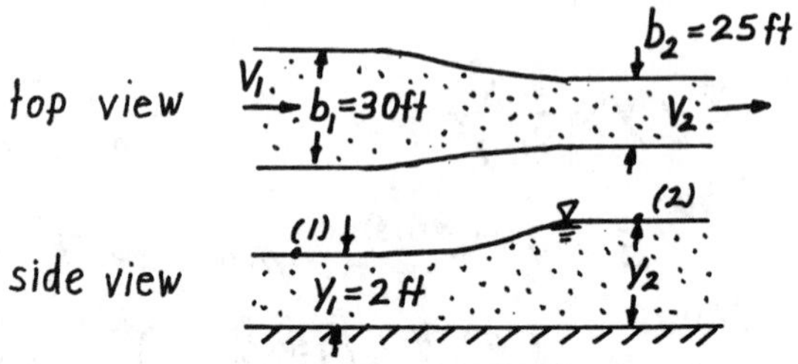

$$\frac{p_1}{\gamma} + \frac{V_1^2}{2g} + z_1 = \frac{p_2}{\gamma} + \frac{V_2^2}{2g} + z_2, \text{ where } z_1 = y_1 = 2\text{ ft},\ z_2 = y_2,\ p_1 = p_2 = 0,$$

$$V_1 = \frac{Q}{A_1} = \frac{1000\frac{ft^3}{s}}{(2ft)(30ft)} = 16.7\frac{ft}{s}, \text{ and } V_2 = \frac{Q}{A_2} = \frac{1000\frac{ft^3}{s}}{(25ft)y_2} = \frac{40}{y_2}\frac{ft}{s} \text{ with } y_2 \sim ft$$

Thus,

$$\frac{(16.7\frac{ft}{s})^2}{2(32.2\frac{ft}{s^2})} + 2\text{ ft} = \frac{(\frac{40}{y_2})^2}{2(32.2\frac{ft}{s^2})} + y_2$$

or

$$y_2^3 - 6.33y_2^2 + 24.8 = 0 \text{ which has roots } y_2 = 5.51, 2.57, -1.75 \text{ ft.}$$

$$\text{Note: } Fr_1 = \frac{V_1}{\sqrt{gy_1}} = \frac{16.7\frac{ft}{s}}{[(32.2\frac{ft}{s^2})(2ft)]^{\frac{1}{2}}} = 2.08 > 1$$

If there is no relative minimum area between (1) and (2) where critical flow can occur, it follows that $Fr_2 > 1$ also. Thus, the $y_2 = 5.51$ (subcritical flow root) is not valid.

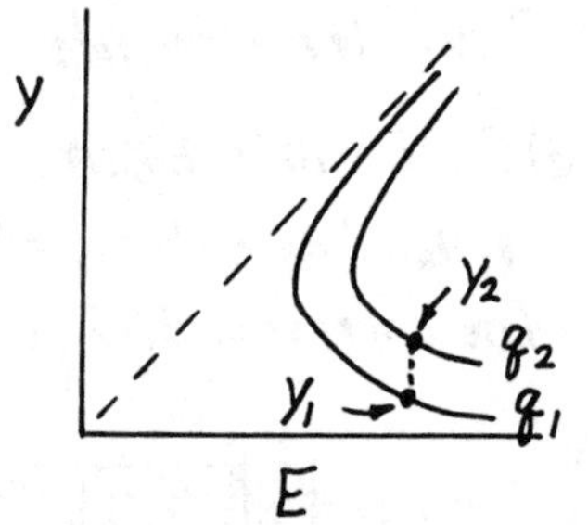

Thus, $y_2 = z_2 = \underline{\underline{2.57 \text{ ft}}}$

Note: If $y_2 = 5.51$ ft and $V_2 = \frac{40}{5.51} = 7.26\frac{ft}{s}$, then

$$Fr_2 = \frac{V_2}{\sqrt{gy_2}} = \frac{7.26 \text{ ft/s}}{[(32.2 \text{ ft/s}^2)(5.51 \text{ ft})]^{1/2}} = 0.545 < 1,$$

whereas if

$y_1 = 2.57$ ft and $V_2 = \frac{40}{2.57} = 15.6\frac{ft}{s}$, then

$$Fr_2 = \frac{15.6}{[(32.2)(2.57)]^{1/2}} = 1.71 > 1$$

10.7R

10.7R (Wall shear stress) Water flows in a 10-ft-wide rectangular channel with a flowrate of 150 cfs and a depth of 3 ft. If the slope is 0.005, determine the Manning coefficient, n, and the average shear stress at the sides and bottom of the channel.

(ANS: 0.0320; 0.585 lb/ft²)

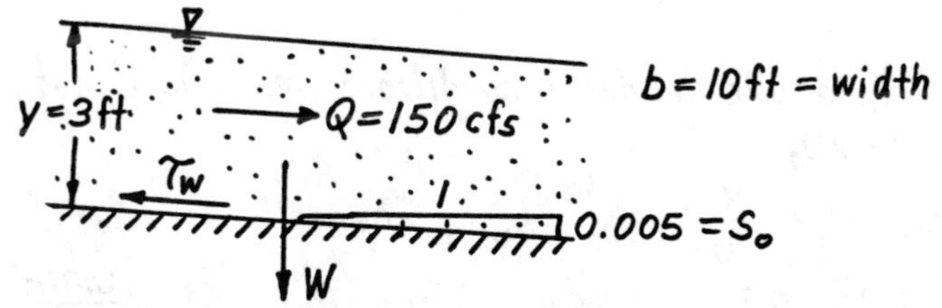

$Q = \frac{\kappa}{n} A R_h^{2/3} S_o^{1/2}$, where $\kappa = 1.49$, $A = by = (10\text{ft})(3\text{ft}) = 30\text{ft}^2$

$P = b + 2y = 10\text{ft} + 2(3\text{ft}) = 16\text{ ft}$, or $R_h = \frac{A}{P} = \frac{30\text{ft}^2}{16\text{ft}} = 1.875\text{ ft}$

Thus,

$$n = \frac{1.49 A R_h^{2/3} S_o^{1/2}}{Q} = \frac{1.49(30)(1.875)^{2/3}(0.005)^{1/2}}{150} = \underline{\underline{0.0320}}$$

Also,

$$\tau_w = \gamma R_h S_o = (62.4\ \tfrac{\text{lb}}{\text{ft}^3})(1.875\text{ft})(0.005) = \underline{\underline{0.585\ \tfrac{\text{lb}}{\text{ft}^2}}}$$

10.8R

10.8R (Manning equation) The triangular flume shown in Fig. P10.8R is built to carry its design flowrate, Q_0, at a depth of 0.90 m as is indicated. If the flume is to be able to carry up to twice its design flowrate, $Q = 2Q_0$, determine the freeboard, ℓ, needed.

(ANS: 0.378 m)

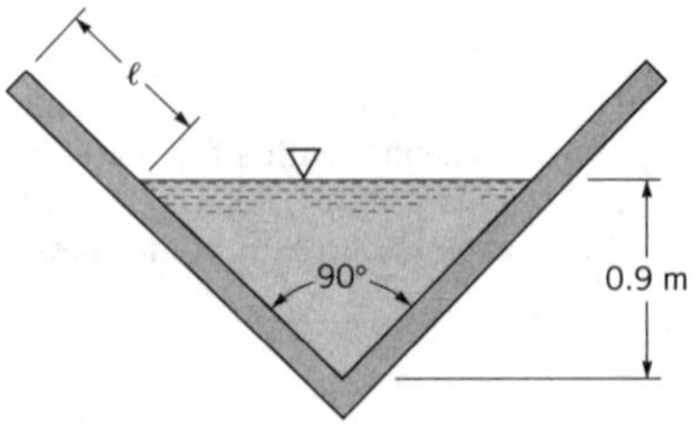

■ FIGURE P10.8R

Let $(\ )_0$ denote the design conditions and $(\ )_2$ denote conditions with $Q = 2Q_0$

Thus, $Q_0 = \frac{K}{n_0} A_0 R_{h_0}^{2/3} S_{00}$, where $K = 1$, $A_0 = (0.9\,m)^2 = 0.81\,m^2$,

$P_0 = 2\sqrt{2}\,(0.9\,m) = 2.55\,m$ Thus, $R_{h_0} = \frac{A_0}{P_0} = \frac{0.81\,m^2}{2.55\,m} = 0.318\,m$

Hence, $Q_0 = \frac{1}{n_0}(0.810)(0.318)^{2/3} S_{00}^{1/2}$

or

$$Q_0 = \frac{0.377\,S_{00}^{1/2}}{n_0} \qquad (1)$$

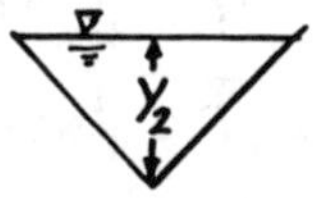

Also, $Q_2 = \frac{K}{n_2} A_2 R_{h_2}^{2/3} S_{02}$, where $A_2 = y_2^2$,

$P_2 = 2\sqrt{2}\,y_2$, or $R_{h_2} = \frac{A_2}{P_2} = \frac{y_2^2}{2\sqrt{2}\,y_2} = 0.354\,y_2$

Hence, with $n_0 = n_2$ and $S_{00} = S_{02}$,

$$Q_2 = \frac{1}{n_0} y_2^2 (0.354\,y_2)^{2/3} S_{00}^{1/2}$$

or

$$Q_2 = \frac{0.500\,S_{00}^{1/2}}{n_0} y_2^{8/3} \qquad (2)$$

From Eqs. (1) and (2) with $Q_2 = 2Q_0$ we obtain

$0.500\,y_2^{8/3} = 2(0.377)$ or $y_2 = 1.167\,m$

However,

$y_2 - 0.9\,m = \ell \sin 45°$ so that

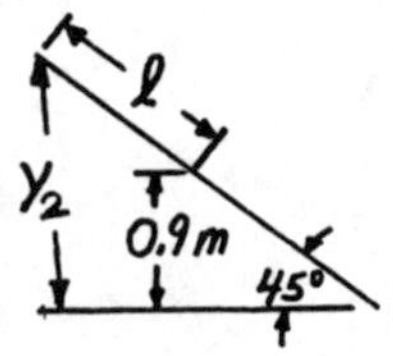

$$\ell = \frac{y_2 - 0.9\,m}{\sin 45°} = \frac{1.167\,m - 0.9\,m}{\sin 45°} = \underline{\underline{0.378\,m}}$$

10.9R

10.9R (Manning equation) Water flows in a rectangular channel of width b at a depth of $b/3$. Determine the diameter of a circular channel (in terms of b) that carries the same flowrate when it is half-full. Both channels have the same Manning coefficient, n, and slope.

(ANS: 0.889 b)

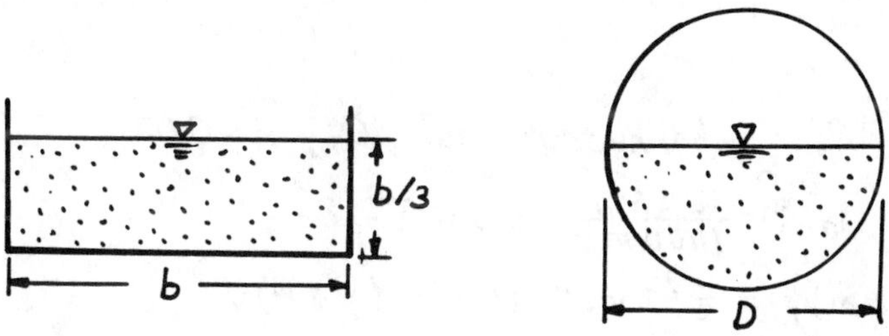

$$Q = \frac{\kappa}{n} A R_h^{2/3} S_o^{1/2} \quad (1)$$

For the rectangle, $A = b\left(\frac{b}{3}\right)$ and $P = b + 2\left(\frac{b}{3}\right) = \frac{5}{3}b$

or $R_h = \frac{A}{P} = \frac{\frac{1}{3}b^2}{\frac{5}{3}b} = \frac{1}{5}b$

For the semi-circle, $A = \frac{\pi}{8}D^2$ and $P = \frac{\pi}{2}D$

or $R_h = \frac{1}{4}D$

Thus, from Eq.(1) with equal Q, κ, n, and S_o for either case:

$$\left(\frac{b^2}{3}\right)\left(\frac{1}{5}b\right)^{2/3} = \left(\frac{\pi}{8}D^2\right)\left(\frac{1}{4}D\right)^{2/3}$$

or $\underline{\underline{D = 0.889\,b}}$

10.10R

10.10R (Manning equation) A weedy irrigation canal of trapezoidal cross section is to carry 20 m^3/s when built on a slope of 0.56 m/km. If the sides are at a 45° angle and the bottom is 8 m wide, determine the width of the waterline at the free surface.

(ANS: 12.0 m)

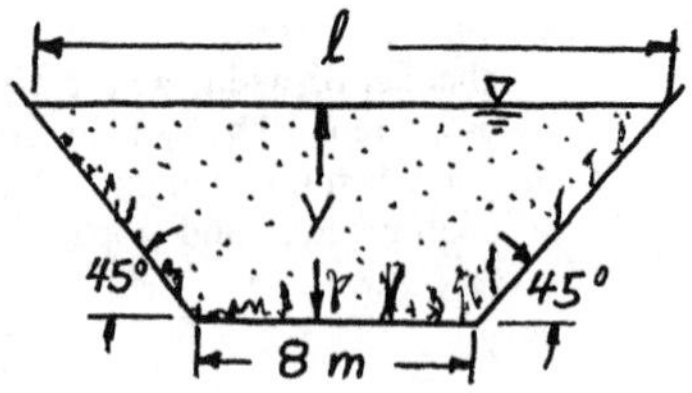

$$Q = \frac{\kappa}{n} A R_h^{2/3} S_o^{1/2}, \text{ where from Table 10.1, } n = 0.03 \quad (1)$$

Also, $\kappa = 1$ and $S_o = \frac{0.56\,m}{1000\,m} = 5.6\times10^{-4}$

$A = \frac{1}{2}(\ell + 8m)y = \frac{1}{2}(2y+8+8)y = (y+8)y$

and

$P = 8m + 2\frac{y}{\sin 45°} = 8 + 2.83y$

so that

$R_h = \frac{A}{P} = \frac{(y+8)y}{(8+2.83y)}$

Thus, Eq. (1) becomes

$$20 = \frac{1}{0.03}(y+8)y\left[\frac{(y+8)y}{(8+2.83y)}\right]^{2/3}(5.6\times10^{-4})^{1/2}$$

or

$$25.4 = \frac{(y^2+8y)^{5/3}}{(8+2.83y)^{2/3}}$$

or

$$y^2 + 8y - 6.96(8+2.83y)^{0.4} = 0 \equiv F(y) \quad (2)$$

Solve for $F(y) = 0$

From the graph, $F = 0$ when

$y \approx 1.98\,m$

Thus, $\ell = 2y + 8\,m$

or

$\ell = 2(1.98m) + 8m = \underline{\underline{12.0m}}$

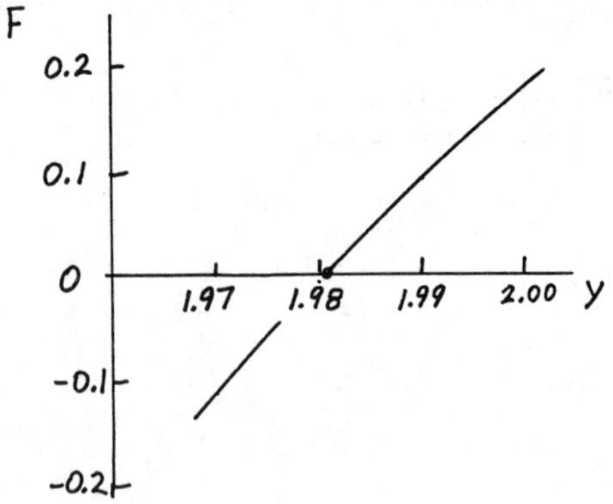

10.11R

10.11R (Manning equation) Determine the maximum flowrate possible for the creek shown in Fig. P10.11R if it is not to overflow onto the floodplain. The creek bed drops an average of 5 ft/half mile of length. Determine the flowrate during a flood if the depth is 8 ft.

(ANS: 182 ft^3/s 1517 ft^3/s)

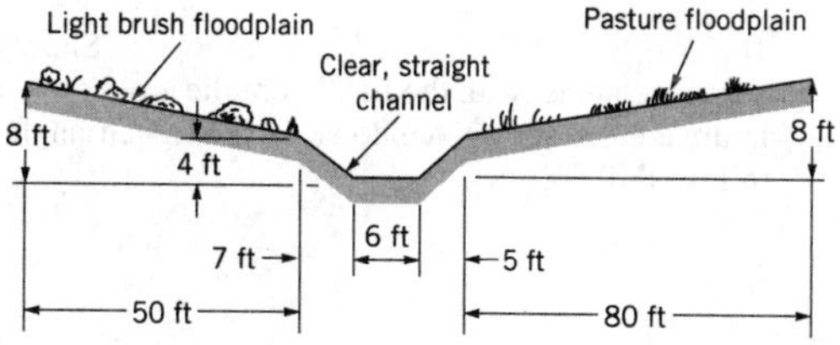

FIGURE P10.11R

(a) Maximum Q without overflowing:

$$Q = \frac{\kappa}{n} A R_h^{2/3} S_o^{1/2}, \text{ where } \kappa = 1.49,\ S_o = \frac{5\text{ ft}}{\left(\frac{5280}{2}\text{ft}\right)} = 0.00189, \qquad (1)$$

and from Table 10.1 $n = 0.030$

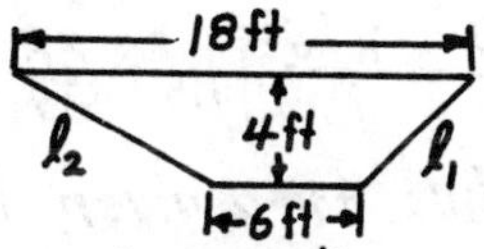

Also, $A = \frac{1}{2}(6\text{ft} + 18\text{ft})(4\text{ft}) = 48\text{ft}^2$

and $P = 6.40\text{ ft} + 6\text{ ft} + 8.06\text{ ft} = 20.5\text{ ft}$

so that $R_h = \frac{A}{P} = \frac{48\text{ ft}^2}{20.5\text{ ft}} = 2.34\text{ ft}$

$\ell_1 = (5^2 + 4^2)^{1/2} = 6.40\text{ft}$

$\ell_2 = (7^2 + 4^2)^{1/2} = 8.06\text{ ft}$

Thus, from Eq. (1)

$$Q = \frac{1.49}{0.030}(48)(2.34)^{2/3}(0.00189)^{1/2} = \underline{\underline{182\ \frac{ft^3}{s}}}$$

(b) At a flood depth of 8 ft:

$Q = Q_1 + Q_2 + Q_3$, where

$Q_i = \frac{\kappa}{n_i} A_i R_{h_i}^{2/3} S_o^{1/2}$ with

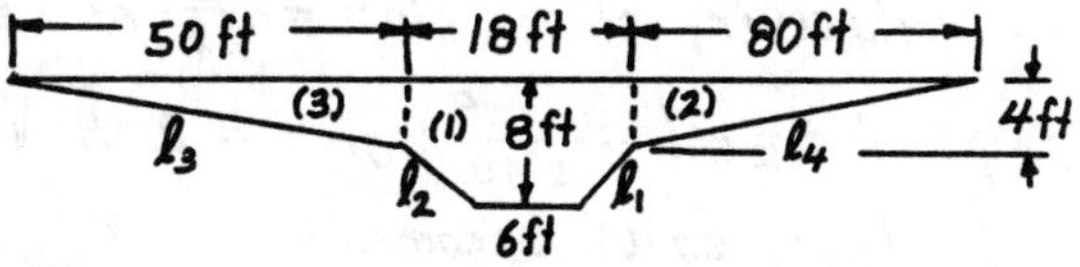

$\kappa = 1.49$, $S_o = 0.00189$, and from Table 10.1: $n_1 = 0.030$, $n_2 = 0.035$, $n_3 = 0.050$

From part (a), $\ell_1 = 6.40\text{ft}$ and $\ell_2 = 8.06\text{ft}$

while $\ell_3 = (50^2 + 4^2)^{1/2} = 50.2\text{ ft}$ and $\ell_4 = (80^2 + 4^2)^{1/2} = 80.1\text{ ft}$

Thus, $A_1 = (4\text{ft})(18\text{ft}) + \frac{1}{2}(6\text{ft} + 18\text{ft})(4\text{ft}) = 120\text{ft}^2$,

$A_2 = \frac{1}{2}(4\text{ft})(80\text{ft}) = 160\text{ft}^2$, and $A_3 = \frac{1}{2}(4\text{ft})(50\text{ft}) = 100\text{ft}^2$

and $P_1 = (6.40 + 6 + 8.06)\text{ft} = 20.5\text{ ft}$, $P_2 = 80.1\text{ ft}$, and $P_3 = 50.2\text{ ft}$

Hence, since $R_{h_i} = \frac{A_i}{P_i}$ we obtain $A_i R_{h_i}^{2/3} = A_i^{5/3}/P_i^{2/3}$, or

$$Q = 1.49(0.00189)^{1/2}\left[\frac{(120)^{5/3}}{(20.5)^{2/3}(0.03)} + \frac{(160)^{5/3}}{(80.1)^{2/3}(0.035)} + \frac{(100)^{5/3}}{(50.2)^{2/3}(0.05)}\right]$$

$$= (842 + 470 + 205)\ \frac{ft^3}{s}$$

or

$$\underline{\underline{Q = 1517\ \frac{ft^3}{s}}}$$

10.12 R

10.12R (Best hydraulic cross section) Show that the triangular channel with the best hydraulic cross section (i.e., minimum area, A, for a given flowrate) is a right triangle as is shown in Fig. E10.8*b*.

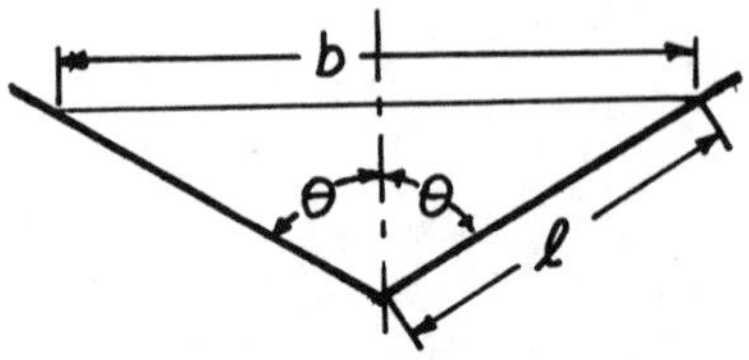

$Q = \frac{K}{n} A R_h^{2/3} S_o^{1/2}$, or with K, n, S_o constant

$$\frac{dQ}{d\theta} = \frac{K S_o^{1/2}}{n}\left[R_h^{2/3}\frac{dA}{d\theta} + A\left(\frac{2}{3}\right)R_h^{-\frac{1}{3}}\frac{dR_h}{d\theta}\right] \qquad (1)$$

Thus, for a given flowrate $\frac{dQ}{d\theta} = 0$ and for the minimum area $\frac{dA}{d\theta} = 0$

Eq. (1) gives

$$\frac{dR_h}{d\theta} = 0$$

Also, $R_h = \frac{A}{P}$, where $P = 2\ell$ and $A = \frac{1}{2}b(\ell\cos\theta)$

or

$$R_h = \frac{\frac{1}{2}b(\ell\cos\theta)}{2\ell} = \frac{1}{4}b\cos\theta \qquad (2)$$

However, since $\sin\theta = \frac{b}{2\ell}$ or $\ell = \frac{b}{2\sin\theta}$ it follows that

$$A = \frac{1}{2}b\left(\frac{b\cos\theta}{2\sin\theta}\right) \text{ or } b = 2\sqrt{A}\sqrt{\tan\theta}$$

Thus, Eq. (2) becomes:

$$R_h = \frac{\cos\theta}{4}\left(2\sqrt{A}\sqrt{\tan\theta}\right) = \frac{1}{2}\sqrt{A}\left(\sin\theta\cos\theta\right)^{\frac{1}{2}}$$

so that

$$\frac{dR_h}{d\theta} = \frac{1}{2}(\sin\theta\cos\theta)^{\frac{1}{2}}\frac{1}{2}A^{-\frac{1}{2}}\frac{dA}{d\theta} + \frac{1}{2}\sqrt{A}\left(\frac{1}{2}\right)(\sin\theta\cos\theta)^{-\frac{1}{2}}(\cos^2\theta - \sin^2\theta)$$

Thus, with $\frac{dA}{d\theta} = 0$ (i.e. minimum area), $\frac{dR_h}{d\theta} = 0$ when $\cos^2\theta - \sin^2\theta = 0$

or

$\underline{\underline{\theta = 45^\circ}}$ (i.e. the best hydraulic cross-section occurs with a right triangle)

10.13R

10.13R (Hydraulic jump) At the bottom of a water ride in an amusement park, the water in the rectangular channel has a depth of 1.2 ft and a velocity of 15.6 ft/s. Determine the height of the "standing wave" (a hydraulic jump) that the boat passes through for its final "splash."

(ANS: 2.50 ft)

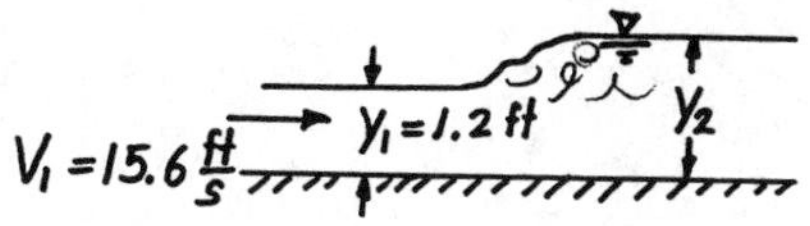

The depth ratio for a hydraulic jump is

$$\frac{y_2}{y_1} = \frac{1}{2}\left[-1 + \sqrt{1 + 8Fr_1^2}\right], \text{ where } Fr_1 = \frac{V_1}{(g\,y_1)^{1/2}} = \frac{15.6\,\frac{ft}{s}}{\left[(32.2\frac{ft}{s^2})(1.2\,ft)\right]^{1/2}} = 2.51$$

Hence,

$$\frac{y_2}{y_1} = \frac{1}{2}\left[-1 + \sqrt{1 + 8(2.51)^2}\right] = 3.08 \text{ so that } y_2 = (3.08)(1.2\,ft) = 3.70\,ft$$

Thus, $y_2 - y_1 = (3.70 - 1.20)\,ft = \underline{\underline{2.50\,ft}}$

10.14R

10.14R (Hydraulic jump) Water flows in a rectangular channel with velocity $V = 6$ m/s. A gate at the end of the channel is suddenly closed so that a wave (a moving hydraulic jump) travels upstream with velocity $V_w = 2$ m/s as is indicated in Fig. P10.14R. Determine the depths ahead of and behind the wave. Note that this is an unsteady problem for a stationary observer. However, for an observer moving to the left with velocity V_w, the flow appears as a steady hydraulic jump.

(ANS: 0.652 m; 2.61 m)

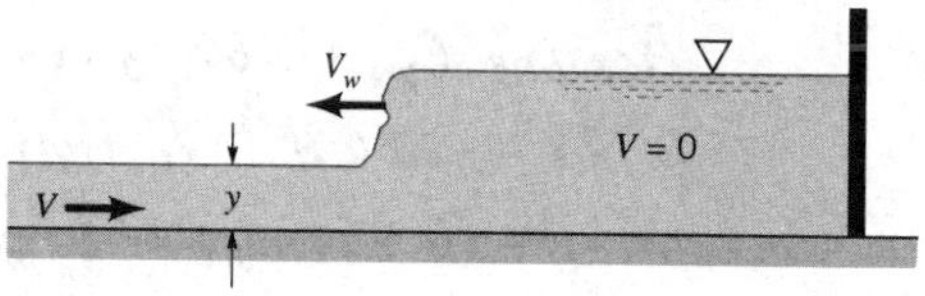

FIGURE P10.14R

For an observer moving to the left with speed $V_w = 2\frac{m}{s}$ the flow appears as shown below.

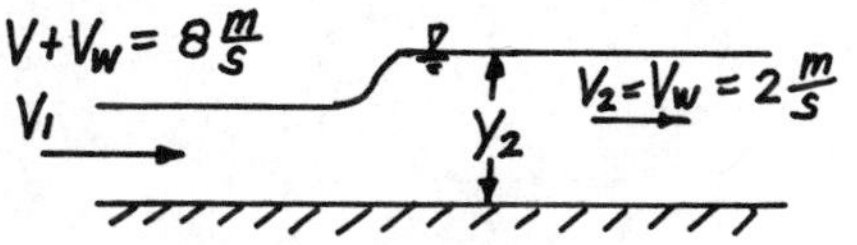

Thus, treat as a jump with

$V_1 = 8\frac{m}{s}$, $V_2 = 2\frac{m}{s}$

Since $A_1V_1 = A_2V_2$ or $\frac{y_2}{y_1} = \frac{V_1}{V_2} = \frac{8\frac{m}{s}}{2\frac{m}{s}} = 4$ it follows that

$$\frac{y_2}{y_1} = \frac{1}{2}\left[-1 + \sqrt{1 + 8Fr_1^2}\right] = 4 \quad \text{Hence, } Fr_1 = 3.16$$

However, $Fr_1 = \frac{V_1}{(g\,y_1)^{1/2}}$ so that

$$y_1 = \frac{V_1^2}{g\,Fr_1^2} = \frac{(8\frac{m}{s})^2}{(9.81\frac{m}{s^2})(3.16)^2} = \underline{\underline{0.652\,m}}$$

and

$y_2 = 4y_1 = 4(0.652\,m) = \underline{\underline{2.61\,m}}$

10.15R

10.15R (Sharp-crested weir) Determine the head, H, required to allow a flowrate of 600 m^3/hr over a sharp-crested triangular weir with $\theta = 60°$.

(ANS: 0.536 m)

$Q = C_{wt} \frac{8}{15} \tan\frac{\theta}{2} \sqrt{2g}\, H^{5/2}$, where $Q = 600 \frac{m^3}{hr}\left(\frac{1hr}{3600s}\right) = 0.167 \frac{m^3}{s}$ and $\theta = 60°$

Thus, $0.167 \frac{m^3}{s} = C_{wt} \frac{8}{15}(\tan 30°)\left[2(9.81\frac{m}{s^2})\right]^{1/2} H^{5/2}$

or

$0.122 = C_{wt} H^{5/2}$ (1)

The value of C_{wt} is obtained from Fig. 10.25 as a function of H.

Trial and error solution of Eqs. (1), (2):

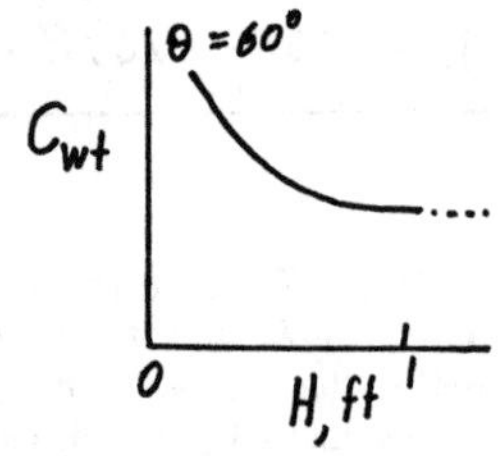

Fig. 10.25 (2)

Assume $C_{wt} = 0.60$, or

$0.122 = 0.60\, H^{5/2}$ so that $H = 0.529\, m$

or $H = (0.529m)(3.28 \frac{ft}{m}) = 1.73\, ft$

From Fig. 10.25 this gives $C_{wt} \approx 0.58$ (extrapolate for $H > 1 ft$)

Assume $C_{wt} = 0.58$, or

$0.122 = 0.58\, H^{5/2}$ so that $H = (0.536m)(3.28 \frac{ft}{m}) = 1.76\, ft$

which checks with $C_{wt} = 0.58$, the assumed value.

Thus,

$H = \underline{\underline{0.536\, m}}$

10.16R (Broad-crested weir) The top of a broad-crested weir block is at an elevation of 724.5 ft, which is 4 ft above the channel bottom. If the weir is 20-ft wide and the flowrate is 400 cfs, determine the elevation of the reservoir upstream of the weir.
(ANS: 730.86 ft)

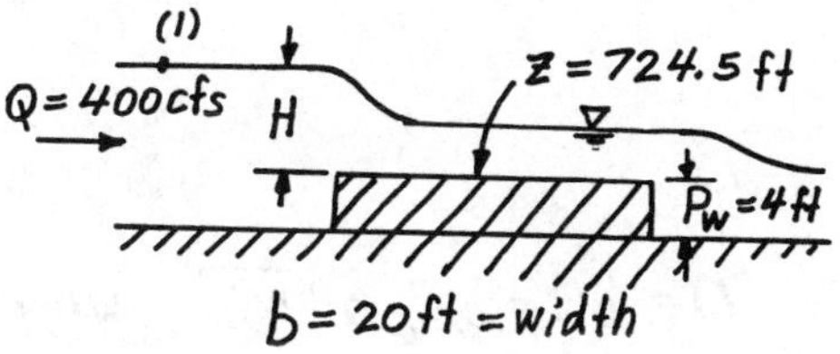

$$Q = \frac{0.65}{(1+\frac{H}{P_w})^{1/2}}\, b\sqrt{g}\left(\frac{2}{3}\right)^{3/2} H^{3/2}$$, or with $P_w = 4\,ft$, $b = 20\,ft$, and $Q = 400\,cfs$,

$$400\,\frac{ft^3}{s} = \frac{0.65}{(1+\frac{H}{4})^{1/2}}\,(20\,ft)\sqrt{32.2\,\frac{ft}{s^2}}\left(\frac{2}{3}\right)^{3/2} H^{3/2}$$, where $H \sim ft$

or

$$9.96\,(1+\frac{H}{4})^{1/2} = H^{3/2}$$

Thus,

$$H^3 - 24.8\,H - 99.2 = 0$$

The roots of this cubic equation are $H = 6.36$ and two complex roots.
Thus, $H = 6.36\,ft$ so that

$$z_1 = 724.5\,ft + 6.36\,ft = \underline{\underline{730.86\,ft}}$$

10.17R

10.17R (Underflow gate) Water flows under a sluice gate in a 60-ft-wide finished concrete channel as is shown in Fig. P10.17R. Determine the flowrate. If the slope of the channel is 2.5 ft/100 ft, will the water depth increase or decrease downstream of the gate? Assume $C_c = y_2/a = 0.65$. Explain.

(ANS: 1670 ft^3/s; decrease)

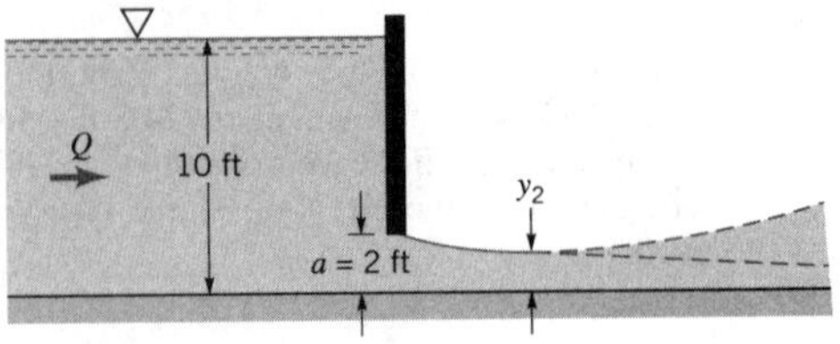

■ FIGURE P10.17R

$Q = bq = b C_d a \sqrt{2 g y_1}$, where $b = 60$ ft, $a = 2$ ft, and from Fig. 10.29

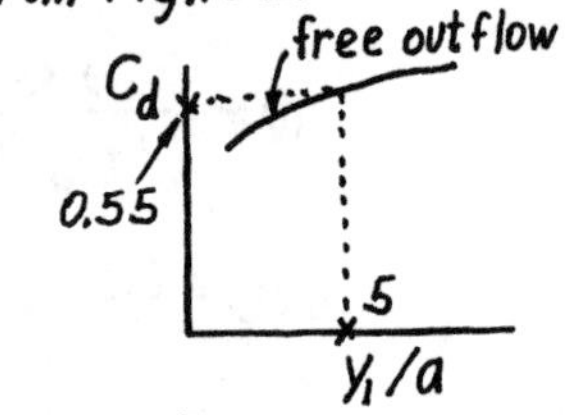

since $\frac{y_1}{a} = \frac{10\,ft}{2\,ft} = 5$, it follows that $C_d = 0.55$

Hence,

$$Q = (60\,ft)(0.55)(2\,ft)\left[2\left(32.2\,\frac{ft}{s^2}\right)(10\,ft)\right]^{1/2} = \underline{\underline{1670\,\frac{ft^3}{s}}}$$

Determine the slope needed to maintain uniform flow downstream of the gate:

$$Q = \frac{\kappa}{n} A R_h^{2/3} S_o^{1/2}, \text{ where } \kappa = 1.49 \text{ and from Table 10.1 } n = 0.012 \quad (1)$$

Also, $y_2 = C_c a = 0.65\,(2\,ft) = 1.3\,ft$

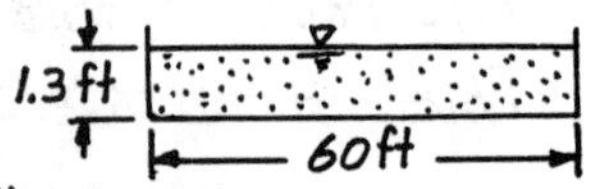

so that

$A = (1.3\,ft)(60\,ft) = 78\,ft^2$, $P = (60 + 2(1.3))\,ft = 62.6\,ft$

and

$R_h = \frac{A}{P} = \frac{78\,ft^2}{62.6\,ft} = 1.245\,ft$

Thus, from Eq. (1):

$$1670 = \frac{1.49}{0.012}(78)(1.245)^{2/3} S_o^{1/2}, \text{ or } S_o = 0.0222$$

Hence, the required slope for uniform flow is $S_o = 0.0222$, but the actual slope is $S_o = \frac{2.5\,ft}{100\,ft} = 0.0250$, more than required. The fluid will speed up and the <u>depth decrease</u>.

11 Compressible Flow

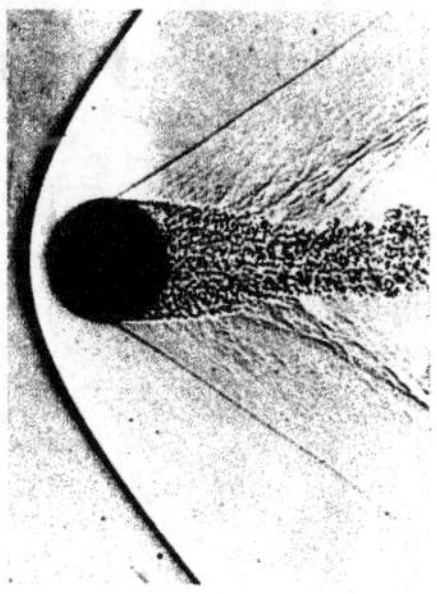

11.1 Ideal Gas Relationships

$$p = \rho RT \qquad \textbf{(11.1)}$$

$$\check{u}_2 - \check{u}_1 = c_v(T_2 - T_1) \qquad \textbf{(11.5)} \qquad \check{h}_2 - \check{h}_1 = c_p(T_2 - T_1) \qquad \textbf{(11.9)}$$

$$c_p = \frac{Rk}{k-1} \qquad \textbf{(11.14)} \qquad c_v = \frac{R}{k-1} \qquad \textbf{(11.15)}$$

- The equation of state for an ideal gas, Eq. 11.1, expresses the relationship among pressure, temperature, and density.
- Changes in internal energy, $\check{u}$, and enthalpy, $\check{h} = \check{u} + p/\rho$, for an ideal gas are directly related to temperature changes as shown by Eqs. 11.5 and 11.9, respectively.
- The specific heats at constant volume, c_v, and constant pressure, c_p, are related to the gas constant, R, and the specific heat ratio, $k = c_p/c_v$, as shown in Eqs. 11.14 and 11.15.
- *Caution:* The pressure and temperature used in Eq. 11.1 must be the absolute pressure and absolute temperature.
- *Caution:* Eqns. 11.5 and 11.9 (as well as all problems and examples in this book) are valid only for constant specific heats.

EXAMPLE: A tank contains helium at a temperature of 80 °F and a pressure of 60 psi above the standard 14.7 psia pressure in the room containing the tank. **(a)** Determine the density of the helium. **(b)** Determine the specific heat at constant pressure.

SOLUTION:

(a) From Eq. 11.1, $\rho = p/RT$,

where $p = (60 + 14.7)\text{psia} = 74.7\ \text{lb/in.}^2(\text{abs})(144\ \text{in.}^2/\text{ft}^2)$
$= 1.08 \times 10^4\ \text{lb/ft}^2\ (\text{abs})$

Also, $T = (80 + 460)°R = 540°R$ Hence,

$\rho = 1.08\times10^4\ \text{lb/ft}^2 / [1.242\times10^4\ \text{ft}\cdot\text{lb/slug}\cdot°R\,(540°R)] = \underline{\underline{0.00161\ \text{slugs/ft}^3}}$

(b) From Eq. 11.14, $c_p = kR/(k-1) = 1.66(1.242\times10^4\ \text{ft}\cdot\text{lb/slug}\cdot°R)/(1.66-1)$
$= \underline{\underline{3.12\times10^4\ \text{ft}\cdot\text{lb/slug}\cdot°R}}$

11.2 Mach Number and Speed of Sound

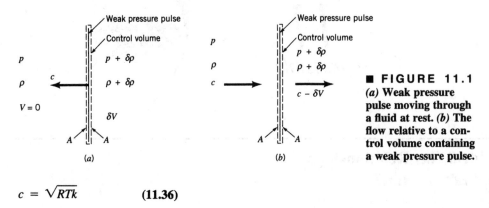

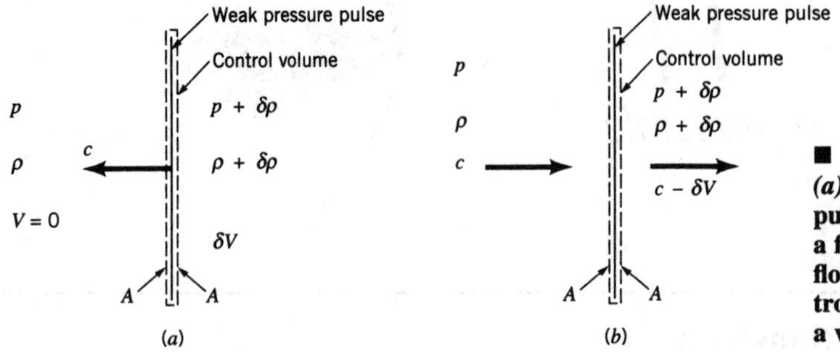

■ FIGURE 11.1 **(a) Weak pressure pulse moving through a fluid at rest. (b) The flow relative to a control volume containing a weak pressure pulse.**

$$c = \sqrt{RTk} \qquad \textbf{(11.36)}$$

- What we perceive as sound consists of weak pressure pulses that move through the fluid at the speed of sound, c, as shown in Fig. 11.1.
- The Mach number, $\text{Ma} = V/c$, is the ratio of the flow velocity to the speed of sound.
- As shown by Eq. 11.36, the speed of sound for an ideal gas is a function of the temperature only, independent of pressure and density.
- *Caution:* The temperature used in Eq. 11.36 must be the absolute temperature.

EXAMPLE: Air at -20 °C flows past an airplane with a speed of 600 m/s. Determine the Mach number of this flow.
SOLUTION:

$Ma = V/c$, where $c = \sqrt{kRT}$

Thus, with $k = 1.4$, $R = 286.9\ \text{J/kg}\cdot\text{K}$, and $T = (273-20)\text{K} = 253\text{K}$,

$c = \sqrt{1.4\,(286.9\ \text{J/kg}\cdot\text{K})(253\text{K})} = 319\,(\text{N}\cdot\text{m/kg})^{1/2} = 319\ \text{m/s}$

Recall $\text{N}\cdot\text{m/kg} = (\text{kg}\cdot\text{m/s}^2)\text{m/kg} = \text{m}^2/\text{s}^2$

Hence, $Ma = V/c = 600\text{m/s} / 319\ \text{m/s} = \underline{\underline{1.88}}$

11.3 Categories of Compressible Flow

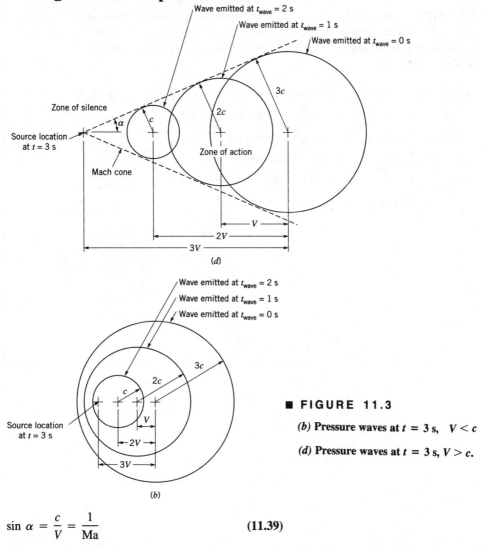

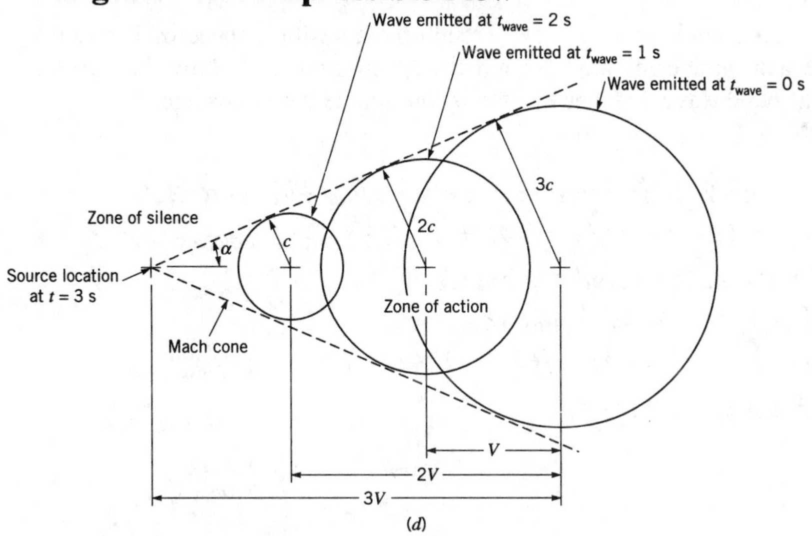

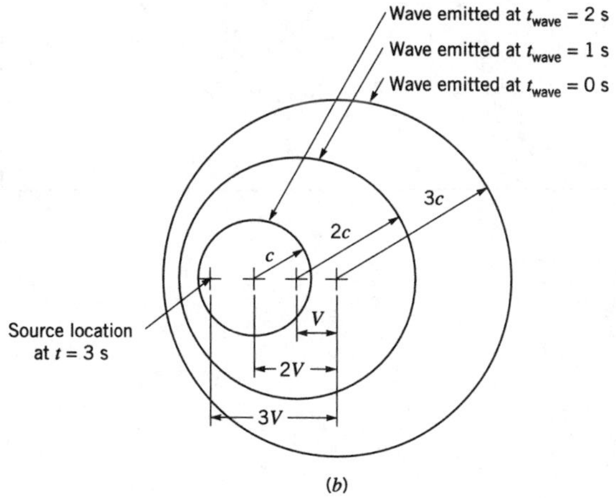

■ **FIGURE 11.3**

(b) **Pressure waves at *t* = 3 s, $V < c$**

(d) **Pressure waves at *t* = 3 s, $V > c$.**

$$\sin \alpha = \frac{c}{V} = \frac{1}{\text{Ma}} \qquad \textbf{(11.39)}$$

- If a fluid were truly incompressible, the speed of sound in that fluid would be infinite. All fluids are compressible to one extent or another, although compressibility effects are often negligible if the Mach number is less than approximately 0.3 or so.
- A flow with the Mach number less than, equal to, or greater than one is termed a *subsonic*, *sonic*, or *supersonic* flow, respectively.
- As shown in Fig. 11.3*b*, for subsonic flows sound waves (small pressure pulses) can travel outward in all directions from the sound source.
- As indicated in Fig. 11.3*d*, for supersonic flows sound waves are confined to a conical region, denoted the Mach cone, whose half angle is given by Eq. 11.39.

EXAMPLE: A sound source travels through the air with a speed of $V = 2000$ ft/s and a Mach number of 2.0. Sketch the Mach cone and show the current location of a sound wave that was emitted by the source 2 seconds ago.
SOLUTION:

With $Ma = V/c = 2$ it follows that $\sin\alpha = 1/Ma = \frac{1}{2}$ or $\alpha = 30°$ and $c = V/2 = (2000\ ft/s)/2 = 1000\ ft/s$. Thus, in 2 seconds the sound wave has traveled a distance $d = 1000\ ft/s\ (2\ s) = 2000\ ft$ and the sound source has traveled $\ell = 2000\ ft/s\ (2\ s) = 4000\ ft$. This is shown in the figure.

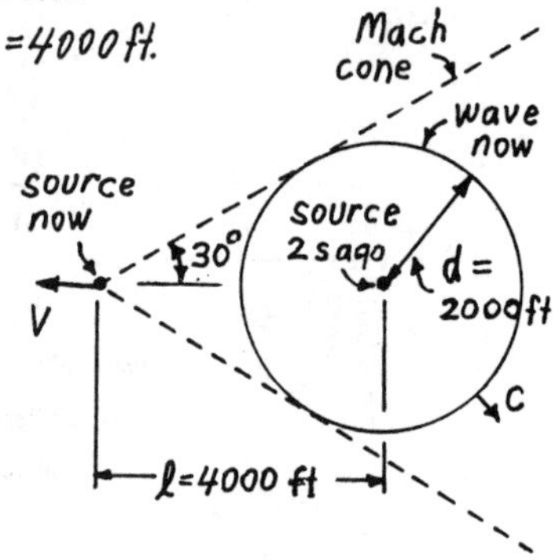

11.4 Isentropic Flow of an Ideal Gas

Effect of Variations in Flow Cross-Sectional Area

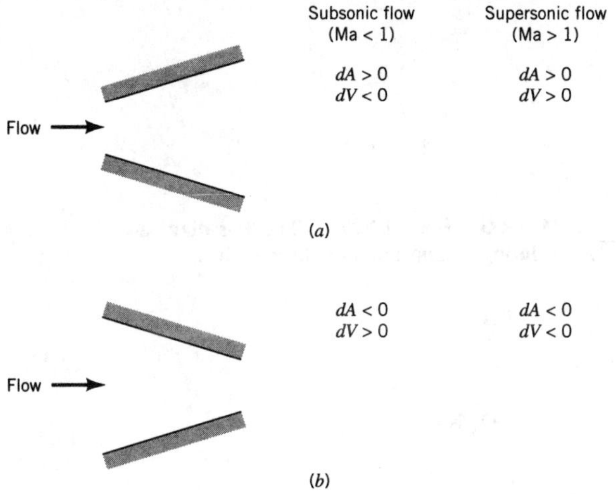

■ **FIGURE 11.5**
***(a)* A diverging duct. *(b)* A converging duct.**

$$\frac{dA}{dV} = -\frac{A}{V}(1 - \mathrm{Ma}^2) \qquad \textbf{(11.50)}$$

- An *isentropic flow* is a flow for which the entropy remains constant. An adiabatic, frictionless flow is an isentropic flow.
- The relationship between an area change and the corresponding velocity change for isentropic flow in a variable duct is given by Eq. 11.50 and shown graphically in Fig. 11.5.
- For steady isentropic flow of an ideal gas, the sonic condition (Ma =1) can be attained at the throat (minimum area) of a converging-diverging duct. To accelerate an isentropic subsonic flow in a duct to supersonic conditions, a converging-diverging section is required.
- *Caution:* Some flow characteristics for subsonic flows are exactly opposite of those for supersonic flows. For example, a *converging* duct accelerates a subsonic flow and decelerates a supersonic flow. A *diverging* duct decelerates a subsonic flow and accelerates a supersonic flow.

EXAMPLE: Air flows isentropically through the variable area duct shown in the figure. **(a)** Is it possible for the flow at section (2) to be supersonic if the flow at section (1) is subsonic? Explain **(b)** Is it possible to have $V_2 > V_1$? Explain.

SOLUTION:

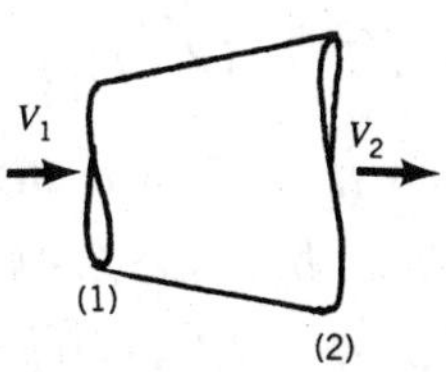

(a) For an isentropic flow to go from subsonic to supersonic conditions, a converging-diverging section is required. Thus, with the diverging section shown, if $Ma_1 < 1$, then $Ma_2 < 1$ also. Not possible.

(b) Since a diverging section accelerates a supersonic flow, it is possible to have $V_2 > V_1$, provided $Ma_1 > 1$.

Pressure, Temperature, and Density as Functions of Mach Number

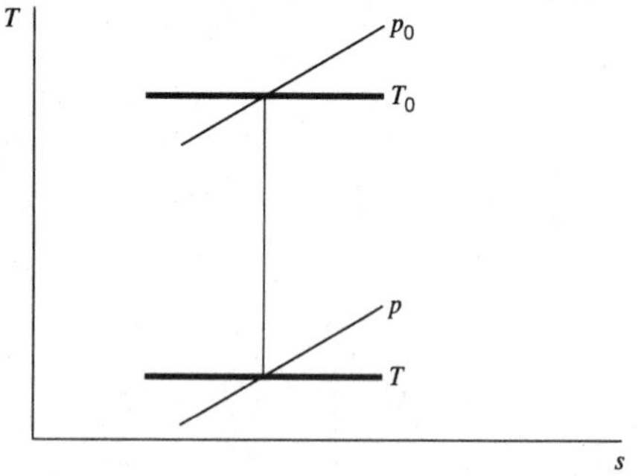

■ **FIGURE 11.7** The *T–s* diagram relating stagnation and static states.

$$\frac{T}{T_0} = \frac{1}{1 + [(k - 1)/2]\mathrm{Ma}^2} \qquad \textbf{(11.56)}$$

$$\frac{p}{p_0} = \left\{\frac{1}{1 + [(k - 1)/2]\mathrm{Ma}^2}\right\}^{k/(k-1)} \qquad \textbf{(11.59)}$$

$$\frac{\rho}{\rho_0} = \left\{\frac{1}{1 + [(k - 1)/2]\mathrm{Ma}^2}\right\}^{1/(k-1)} \qquad \textbf{(11.60)}$$

- The *stagnation state* is the state the fluid would have if the velocity were zero with the entropy equal to that of the flowing fluid.
- For steady, isentropic flow of an ideal gas with constant specific heats, the temperature, pressure, and density (relative to the stagnation state) are related to the Mach number of the flow as given by Eqs. 11.56, 11.59, and 11.60.
- Isentropic flow from the stagnation state where Ma = 0 [denoted $( )_0$] to another state with the fluid flowing (Ma > 0) is represented as the vertical line on the temperature-entropy diagram shown in Fig. 11.7. Such flows involve decreases in temperature, pressure, and density.
- *Caution:* The temperatures and pressures involved in Eqs. 11.56 and 11.59 are absolute temperatures and absolute pressures.

EXAMPLE: Air flows isentropically from its stagnation state in a tank where the temperature is 70 °F and the pressure is 200 psia. Determine the temperature and pressure at a location in the flow where the Mach number is Ma = 2.0.

SOLUTION:

The stagnation conditions within the tank are $p_0 = 200$ psia and $T_0 = 70°F = (70+460)°R = 530°R$. Thus, from Eq. 11.56 with Ma = 2.0,

$T/T_0 = 1/\{1+[(k-1)/2]\mathrm{Ma}^2\} = 1/\{1+[(1.4-1)/2](2.0)^2\} = 0.556$

so that $T = 0.556\,T_0 = 0.556(530°R) = 295°R = (295-460)°F = \underline{\underline{-165°F}}$

Also, from Eq. 11.59

$p/p_0 = 1/\{1+[(k-1)/2]\mathrm{Ma}^2\}^{k/(k-1)} = 1/\{1+[(1.4-1)/2](2.0)^2\}^{1.4/(1.4-1)} = 0.128$

so that $p = 0.128\,p_0 = 0.128(200\text{ psia}) = \underline{\underline{25.6\text{ psia}}}$

Converging-Diverging Duct Flow

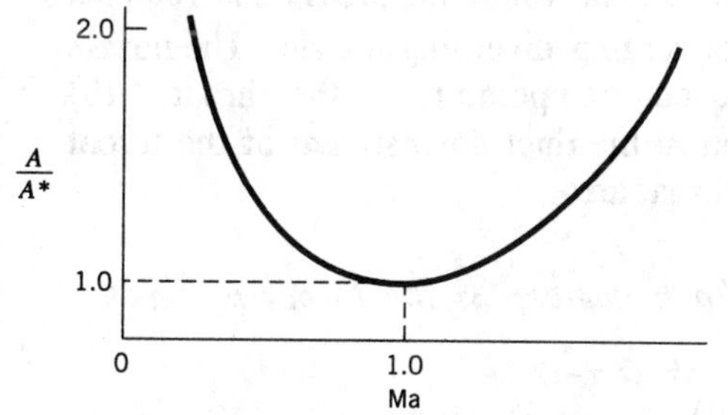

■ FIGURE 11.10 **The variation of area ratio with Mach number for isentropic flow of an ideal gas (k = 1.4, linear coordinate scales).**

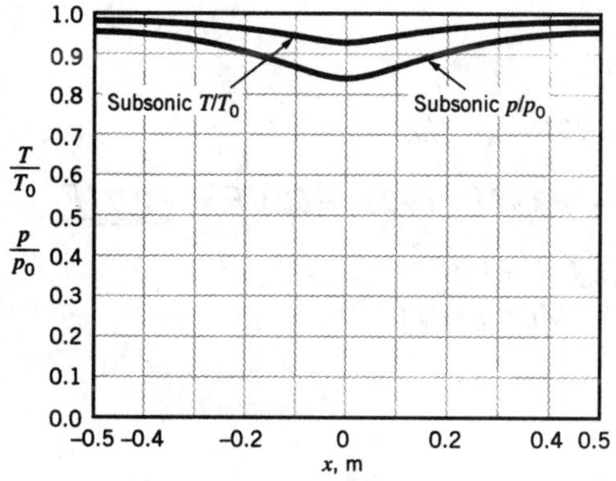

■ FIGURE E11.10 (c)

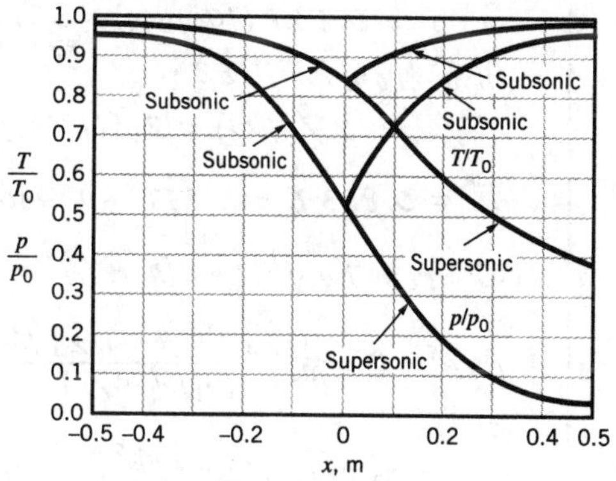

■ FIGURE E11.8 (c)

$$\frac{p^*}{p_0} = \left(\frac{2}{k+1}\right)^{k/(k-1)} \quad \textbf{(11.61)}$$

$$\frac{T^*}{T_0} = \frac{2}{k+1} \quad \textbf{(11.63)}$$

$$\frac{A}{A^*} = \frac{1}{\text{Ma}}\left\{\frac{1 + [(k-1)/2]\text{Ma}^2}{1 + [(k-1)/2]}\right\}^{(k+1)/[2(k-1)]} \quad \textbf{(11.71)}$$

- When Ma = 1 at the throat of a converging-diverging duct, the flow is termed *choked flow*. A decrease in the pressure downstream of the throat cannot increase the flowrate.
- The state associated with Ma = 1 and the same entropy level as the flowing fluid is termed the *critical state*. The ratio of the duct area to the critical area, A/A^*, (where A^* is the area where Ma = 1) is a function of the Mach number as given by Eq. 11.71 and shown in Fig. 11.10.
- The critical pressure ratio, p^*/p_0 , and critical temperature ratio, T^*/T_0 , are given by Eqs. 11.61 and 11.63, respectively.
- Depending on the imposed pressure at the end of a converging-diverging duct connected to a tank, the flow may either: (1) remain subsonic throughout the duct (Fig. E11.10*c*), (2) accelerate from subsonic conditions upstream of the throat, to sonic at the throat, and back to subsonic downstream of the throat (Fig. E11.8*c*), or (3) accelerate to sonic conditions at the throat and continue to accelerate as supersonic flow downstream of the throat (Fig. E11.8*c*).

EXAMPLE: Air flows isentropically from a tank, where the pressure is 100 psia and the temperature is 60 °F, through a converging-diverging nozzle. The nozzle is choked. **(a)** Determine the pressure and temperature at the throat. **(b)** Determine the Mach number at a location in the duct downstream of the throat where the duct area is 2 times that of the throat area.

SOLUTION:

(a) Since the nozzle is choked, the Mach number at the throat is $Ma = 1$.

From Eq. 11.61

$$p^*/p_0 = (2/(k+1))^{k/(k-1)} = (2/(1.4+1))^{1.4/(1.4-1)} = 0.528$$

or $p^* = 0.528\,(100\text{psia}) = \underline{\underline{52.8\text{ psia}}}$

Also, from Eq. 11.63

$T^*/T_0 = 2/(k+1) = 2/(1.4+1) = 0.833$ so that

$$T^* = 0.833\,T_0 = 0.833\,(60+460)^\circ R = 433^\circ R = (433-460)^\circ F = \underline{\underline{-27^\circ F}}$$

(b) From Eq. 11.71, with $A/A^* = 2$ and $k = 1.4$,

$$A/A^* = 2 = \frac{1}{Ma}\left\{\frac{1+[(1.4-1)/2]\,Ma^2}{1+[(1.4-1)/2]}\right\}^{(1.4+1)/[2(1.4-1)]}$$

or

$$2 = \frac{1}{Ma}\left\{\frac{1+0.2\,Ma^2}{1.2}\right\}^3$$ which has solutions $Ma = \underline{\underline{0.305}}$ and $Ma = \underline{\underline{2.20}}$

11.5 Nonisentropic Flow of an Ideal Gas

Adiabatic Constant-Area Duct Flow with Friction (Fanno Flow)

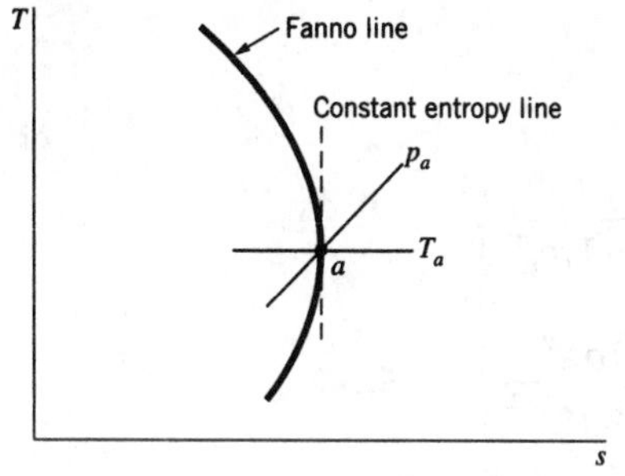

■ **FIGURE 11.16 The *T–s* diagram for Fanno flow.**

$$\frac{1}{k}\frac{(1-\text{Ma}^2)}{\text{Ma}^2} + \frac{k+1}{2k}\ln\left\{\frac{[(k+1)/2]\text{Ma}^2}{1+[(k-1)/2]\text{Ma}^2}\right\} = \frac{f(\ell^* - \ell)}{D} \qquad \textbf{(11.98)}$$

$$\frac{T}{T^*} = \frac{(k+1)/2}{1+[(k-1)/2]\text{Ma}^2} \qquad \textbf{(11.101)}$$

$$\frac{V}{V^*} = \left\{\frac{[(k+1)/2]\text{Ma}^2}{1+[(k-1)/2]\text{Ma}^2}\right\}^{1/2} \qquad \textbf{(11.103)}$$

$$\frac{p}{p^*} = \frac{1}{\text{Ma}}\left\{\frac{(k+1)/2}{1+[(k-1)/2]\text{Ma}^2}\right\}^{1/2} \qquad \textbf{(11.107)}$$

- Flow of an ideal gas in a constant-area duct without heat transfer (adiabatic), but with friction, is termed *Fanno flow*.
- As shown in Fig. 11.16, the Fanno line is the locus of states on the temperature-entropy diagram for a flow having a given, constant stagnation temperature, T_0 , density-velocity product, ρV, and inlet temperature, pressure, and entropy.
- For Fanno flow to be consistent with the second law of thermodynamics, entropy must increase, and flow can proceed along the Fanno line only toward the critical state where Ma = 1, denoted $(\ )_a$ in Fig. 11.16.
- For subsonic (supersonic) Fanno flow the Mach number increases (decreases) in the flow direction. That is, friction accelerates (decelerates) a subsonic (supersonic) flow
- The length of duct, $\ell^* - \ell$, needed to reach the critical (choked) Ma = 1 condition from a given state with Mach number Ma can be obtained from Eq. 11.98.
- For a given Fanno flow (constant specific heat ratio, duct diameter, and friction factor) the temperature, velocity, and pressure relative to values at the critical state (Ma = 1) are given as a function of Ma by Eqs. 11.101, 11.103, and 11.107.
- *Caution:* For given conditions, if a duct is long enough the flow will be choked with Ma = 1 at the exit. For such conditions, lowering the exit pressure cannot increase the flowrate.

EXAMPLE: At one section in a 0.1-m-diameter duct the Mach number of the flowing air is 0.3. The friction factor is 0.02. Determine the length of duct necessary to accelerate the flow to the critical state of Ma = 1.
SOLUTION:

From Eq. 11.98 with $Ma = 0.3$ and $k = 1.4$,

$$\frac{f(\ell^* - \ell)}{D} = \frac{1}{k}\frac{(1 - Ma^2)}{Ma^2} + \frac{k+1}{2k}\ln\left\{\frac{[(k+1)/2]Ma^2}{1 + [(k-1)/2]Ma^2}\right\}$$

$$= \frac{1}{1.4}\frac{(1 - 0.3^2)}{0.3^2} + \frac{(1.4+1)}{2(1.4)}\ln\left\{\frac{[(1.4+1)/2]0.3^2}{1 + [(1.4-1)/2]0.3^2}\right\} = 5.30$$

Thus, $\ell^* - \ell = 5.30D/f = 5.30(0.1\text{m})/0.02 = \underline{\underline{26.5\text{m}}}$

Frictionless Constant-Area Duct Flow with Heat Transfer (Rayleigh Flow)

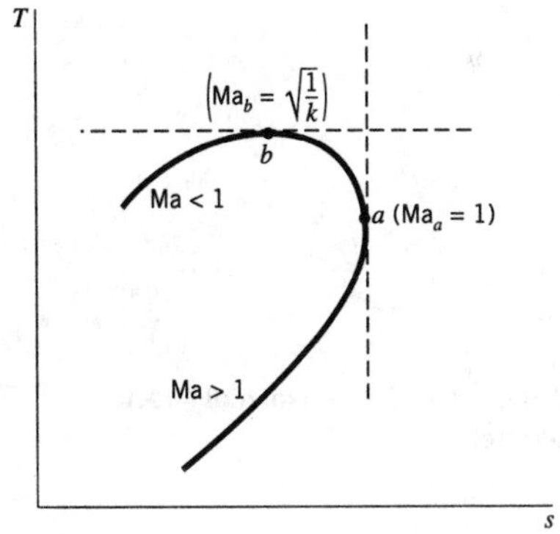

■ FIGURE 11.22 **Rayleigh line.**

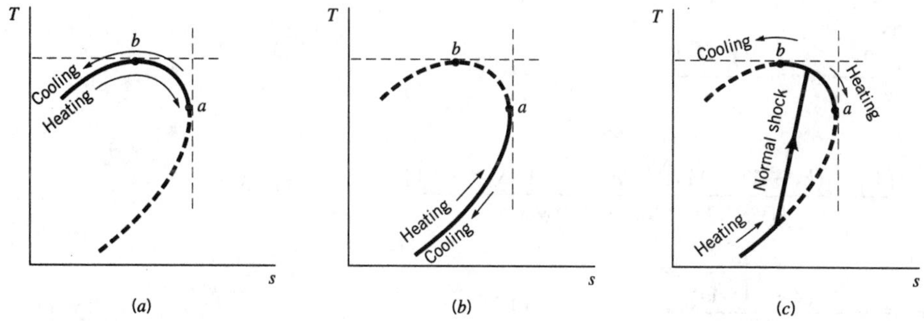

■ FIGURE 11.23 **(a) Subsonic Rayleigh flow. (b) Supersonic Rayleigh flow. (c) Normal shock in a Rayleigh flow.**

- Frictionless flow of an ideal gas through a constant-area duct with heat transfer is termed *Rayleigh flow*. The Rayleigh line is the locus of states on the temperature-entropy diagram for such flows (see Fig. 11.22).
- As shown in Fig. 11.23, for subsonic (supersonic) Rayleigh flow heating the fluid accelerates (decelerates) the fluid. Conversely, for subsonic (supersonic) Rayleigh flow cooling decelerates (accelerates) the fluid.
- *Caution:* Depending on the Ma number involved, adding heat to the flow can result in a temperature decrease and removing heat from the fluid can result in a temperature increase.

Normal Shock Waves

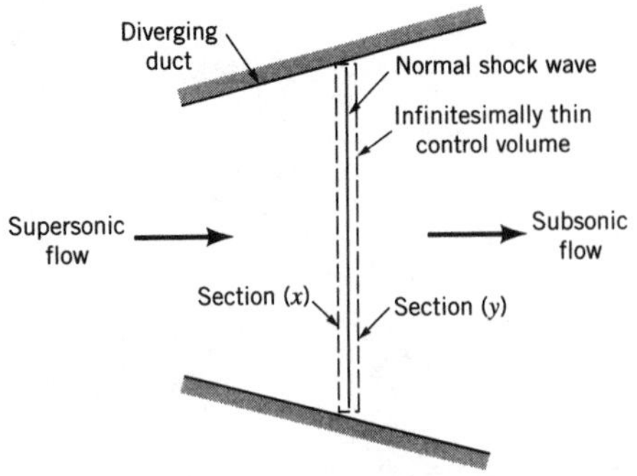

■ FIGURE 11.24 **Normal shock control volume.**

$$\mathrm{Ma}_y^2 = \frac{\mathrm{Ma}_x^2 + [2/(k-1)]}{[2k/(k-1)]\mathrm{Ma}_x^2 - 1} \qquad \textbf{(11.149)}$$

$$\frac{p_y}{p_x} = \frac{2k}{k+1}\mathrm{Ma}_x^2 - \frac{k-1}{k+1} \qquad \textbf{(11.150)}$$

$$\frac{T_y}{T_x} = \frac{\{1 + [(k-1)/2]\mathrm{Ma}_x^2\}\{[2k/(k-1)]\mathrm{Ma}_x^2 - 1\}}{\{(k+1)^2/[2(k-1)]\}\mathrm{Ma}_x^2} \qquad \textbf{(11.151)}$$

$$\frac{\rho_y}{\rho_x} = \frac{V_x}{V_y} = \frac{(k+1)\mathrm{Ma}_x^2}{(k-1)\mathrm{Ma}_x^2 + 2} \qquad \textbf{(11.154)}$$

- As shown in Fig. 11.24, a *normal shock wave* can be treated as a discontinuity in the flow across which the flow decelerates from supersonic to subsonic conditions and the pressure, temperature, density, and entropy increase.
- The relationship between the supersonic Mach number upstream of the normal shock, Ma_x, and the subsonic Mach number downstream of the shock, Ma_y, is as given in Eq. 11.149.
- The pressure, temperature, density, and velocity ratios across a normal shock are given in terms of the upstream Mach number by Eqs. 11.150, 11.151, and 11.154.
- Although the stagnation temperature remains constant across a normal shock wave, the stagnation pressure decreases.
- *Caution:* The pressures and temperatures used in Eqs. 11.150 and 11.151 are absolute values.

EXAMPLE: Upstream of a normal shock wave the air velocity, pressure, and temperature are 1000 mph (1467 ft/s), 14.7 psia, and 80 °F. Determine the velocity, pressure, and temperature downstream of the normal shock wave.

SOLUTION:

With $c_x = \sqrt{kRT_x} = \left[1.4(1716\ \text{ft}\cdot\text{lb/slug}\cdot{}^\circ\text{R})(80+460)^\circ\text{R}\right]^{1/2} = 1139\ \text{ft/s}$

it follows that $Ma_x = V_x/c_x = (1467\ \text{ft/s})/(1139\ \text{ft/s}) = 1.29$

Thus, from Eq. 11.154, $V_x/V_y = (k+1)Ma_x^2/[(k-1)Ma_x^2+2]$

$$= (1.4+1)(1.29)^2/[(1.4-1)(1.29)^2+2] = 1.50$$

so that $V_y = V_x/1.50 = (1467\ \text{ft/s})/1.50 = \underline{\underline{978\ \text{ft/s}}}$

From Eq. 11.150, $p_y/p_x = [2k/(k+1)]Ma_x^2 - (k-1)/(k+1)$

$$= [2(1.4)/(1.4+1)](1.29)^2 - (1.4-1)/(1.4+1) = 1.78$$

so that $p_y = 1.78\,p_x = 1.78\,(14.7\text{psia}) = \underline{\underline{26.2\ \text{psia}}}$

Finally, from Eq. 11.151,

$$\frac{T_y}{T_x} = \frac{\{1+[(k-1)/2]Ma_x^2\}\{[2k/(k-1)]Ma_x^2-1\}}{\{(k+1)^2/[2(k-1)]\}Ma_x^2}$$

$$= \frac{\{1+[(1.4-1)/2](1.29)^2\}\{[2(1.4)/(1.4-1)](1.29)^2-1\}}{\{(1.4+1)^2/[2(1.4-1)]\}(1.29)^2} = 1.19$$

so that $T_y = 1.19\,T_x = 1.19\,(80+460)^\circ\text{R} = 643^\circ\text{R} = (640-460)^\circ\text{F} = \underline{\underline{180^\circ\text{F}}}$

Solutions

to Review Problems

Chapter 11

11.1R

11.1R (Speed of sound) Determine the speed of sound in air for a hot summer day when the temperature is 100 °F; for a cold winter day when the temperature is −20 °F.

(ANS: 1160 ft/s; 1028 ft/s)

For a perfect gas $c = \sqrt{kRT}$, where for air $k = 1.4$ and $R = 1716 \frac{ft \cdot lb}{slug \cdot {}^\circ R}$ (see Table 1.7).

Hence,

$$c_{100^\circ F} = \sqrt{1.4\left(1716 \frac{ft\cdot lb}{slug\cdot{}^\circ R}\right)(460+100)^\circ R} = 1160\sqrt{\frac{ft\cdot lb}{slug}} = 1160\sqrt{\frac{ft(slug\cdot ft/s^2)}{slug}}$$

or

$$c_{100^\circ F} = \underline{\underline{1160 \frac{ft}{s}}} = 791 \text{ mph}$$

Similarly,

$$c_{-20^\circ F} = \sqrt{1.4\left(1716 \frac{ft\cdot lb}{slug\cdot{}^\circ R}\right)(460-20)^\circ R} = \underline{\underline{1028 \frac{ft}{s}}} = 701 \text{ mph}$$

11.2R

11.2R (Speed of sound) Compare values of the speed of sound in ft/s in the following liquids at 68 °F: **(a)** ethyl alcohol, **(b)** glycerin, **(c)** mercury.

(ANS: 3810 ft/s; 6220 ft/s; 4760 ft/s)

For a liquid $c = \sqrt{\frac{E_v}{\rho}}$ where the values of E_v (bulk modulus) and ρ (density) are given in Table 1.5.

Thus, for ethyl alcohol

$$c_{al} = \left[\frac{\left(1.54\times10^5 \frac{lb}{in.^2}\right)\left(144\frac{in.^2}{ft^2}\right)}{1.53 \frac{slugs}{ft^3}}\right]^{\frac{1}{2}} = 3810\left[\frac{ft\cdot lb}{slug}\right]^{\frac{1}{2}} = 3810\left[\frac{ft\cdot(slug\cdot ft/s^2)}{slug}\right]^{\frac{1}{2}}$$

or

$$c_{al} = \underline{\underline{3810 \frac{ft}{s}}} = 2600 \text{ mph}$$

For glycerin

$$c_{gl} = \left[\frac{\left(6.56\times10^5 \frac{lb}{in.^2}\right)\left(144\frac{in.^2}{ft^2}\right)}{2.44 \frac{slugs}{ft^3}}\right]^{\frac{1}{2}} = \underline{\underline{6\,220 \frac{ft}{s}}} = 4240 \text{ mph}$$

and for mercury

$$c_m = \left[\frac{\left(4.14\times10^6 \frac{lb}{in.^2}\right)\left(144\frac{in.^2}{ft^2}\right)}{26.3 \frac{slugs}{ft^3}}\right]^{\frac{1}{2}} = \underline{\underline{4760 \frac{ft}{s}}} = 3250 \text{ mph}$$

11.3R

11.3R (Sound waves) A stationary point source emits weak pressure pulses in a flow that moves uniformly from left to right with a Mach number of 0.5. Sketch the instantaneous outline at time = 10s of pressure waves emitted earlier at time = 5s and time = 8s. Assume the speed of sound is 1000 ft/s.

Since $Ma = \frac{V}{c}$ it follows that the fluid speed is

$V = c\,Ma = 1000\,\frac{ft}{s}(0.5) = 500\,\frac{ft}{s}$.

Assume that the sound source is located at $x = y = 0$.

At time $t = 10s$, the pulse that was emitted at $t = 8s$ has traveled a distance $R_8 = c\,\Delta t = 1000\,\frac{ft}{s}(10s - 8s) = 2000\,ft$ relative to the moving fluid that has moved a distance $x_8 = V\Delta t = 500\,\frac{ft}{s}(10s - 8s) = 1000\,ft$. Similarly, for the pulse that was emitted at $t = 5s$, $R_5 = 1000\,\frac{ft}{s}(10s - 5s) = 5000\,ft$ and $x_5 = 500\,\frac{ft}{s}(10s - 5s) = 2500\,ft$.

These pulses at $t = 10s$ are shown below. They are circles.

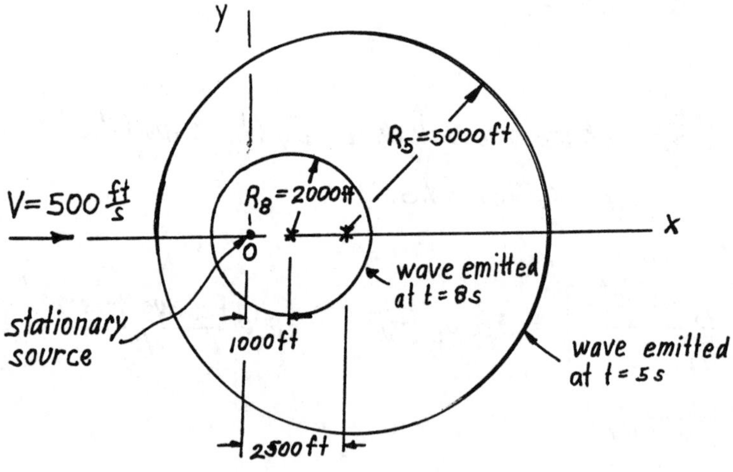

11.4R

11.4R (Mach number) An airplane moves forward in air with a speed of 500 mph at an altitude of 40,000 ft. Determine the Mach number involved if the air is considered as U.S. standard atmosphere (see Table C.1).

(ANS: 0.757)

$Ma = \frac{V}{c}$, where $c = \sqrt{kRT}$

From Table C.1, the temperature at 40,000 ft is −69.7°F.

Thus,

$$c = \left[1.4\left(1716 \frac{ft \cdot lb}{slug \cdot °R}\right)(460 - 69.7)°R\right]^{\frac{1}{2}} = 968 \frac{ft}{s}$$

so that

$$Ma = \frac{\left(500 \frac{mi}{hr}\right)\left(\frac{1\ hr}{3600\ s}\right)\left(\frac{5280\ ft}{1\ mi}\right)}{968 \frac{ft}{s}} = \underline{\underline{0.757}}$$

11.5R

11.5R (Isentropic flow) At section (1) in the isentropic flow of carbon dioxide, $p_1 = 40$ kPa(abs), $T_1 = 60$ °C, and $V_1 = 350$ m/s. Determine the flow velocity, V_2, in m/s, at another section, section (2), where the Mach number is 2.0. Also calculate the section area ratio, A_2/A_1.

(ANS: 500 m/s; 1.71)

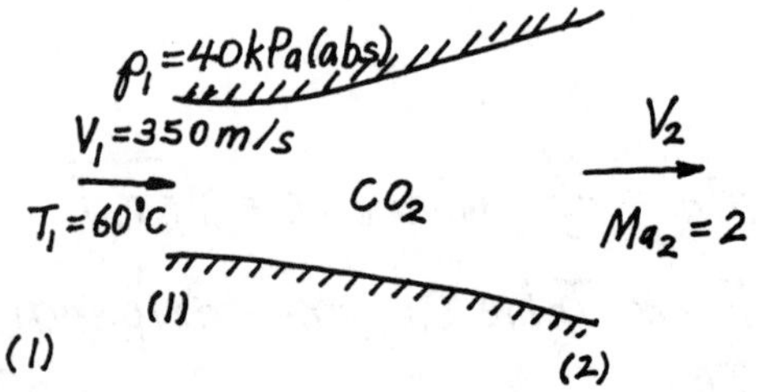

$$V_2 = Ma_2\, c_2 = 2\, c_2 = 2\sqrt{kRT_2} \qquad (1)$$

where for CO_2, $k = 1.30$ and $R = 188.9 \frac{N\cdot m}{kg\cdot K}$ (see Table 1.8)

Also, from Eq. 11.56

$$\frac{T_2}{T_{0,2}} = \frac{1}{1+\left(\frac{k-1}{2}\right)Ma_2^2} \text{ and } \frac{T_1}{T_{0,1}} = \frac{1}{1+\left(\frac{k-1}{2}\right)Ma_1^2}$$, where for isentropic flow, $T_{0,1} = T_{0,2}$. Thus, by dividing these equations we obtain

$$\frac{T_2}{T_1} = \frac{1+\left(\frac{k-1}{2}\right)Ma_1^2}{1+\left(\frac{k-1}{2}\right)Ma_2^2} \qquad (2)$$

However,

$$Ma_1 = \frac{V_1}{\sqrt{kRT_1}} = \frac{350\,\frac{m}{s}}{\sqrt{188.9\frac{N\cdot m}{kg\cdot K}(1.3)(273+60)K}} = 1.22$$

Thus, from Eq. (2),

$$T_2 = (273+60)K\,\frac{\left[1+\left(\frac{1.3-1}{2}\right)(1.22)^2\right]}{\left[1+\left(\frac{1.3-1}{2}\right)(2)^2\right]} = 255\,K, \text{ so from Eq. (1)}$$

$$V_2 = 2\left[1.3\left(188.9\frac{N\cdot m}{kg\cdot K}\right)(255)\right]^{1/2} = \underline{\underline{500\,\frac{m}{s}}}$$

Also, (see Eq. 11.71)

$$\frac{A_2}{A_1} = \frac{\left(\frac{A_2}{A^*}\right)}{\left(\frac{A_1}{A^*}\right)} = \frac{\frac{1}{Ma_2}\left[\frac{1+\left(\frac{k-1}{2}\right)Ma_2^2}{1+\left(\frac{k-1}{2}\right)}\right]^{\frac{k+1}{2(k-1)}}}{\frac{1}{Ma_1}\left[\frac{1+\left(\frac{k-1}{2}\right)Ma_1^2}{1+\left(\frac{k-1}{2}\right)}\right]^{\frac{k+1}{2(k-1)}}} = \frac{Ma_1}{Ma_2}\left[\frac{1+\left(\frac{k-1}{2}\right)Ma_2^2}{1+\left(\frac{k-1}{2}\right)Ma_1^2}\right]^{\frac{k+1}{2(k-1)}}$$

so that with

$$\frac{k-1}{2} = \frac{1.3-1}{2} = 0.15,\quad \frac{k+1}{2(k-1)} = \frac{1.3+1}{2(1.3-1)} = 3.83,\quad Ma_1 = 1.22, \text{ and } Ma_2 = 2$$

we obtain

$$\frac{A_2}{A_1} = \frac{1.22}{2}\left[\frac{1+0.15(2)^2}{1+0.15(1.22)^2}\right]^{3.83} = \underline{\underline{1.71}}$$

11.6R

11.6R (Isentropic flow) An ideal gas in a large storage tank at 100 psia and 60 °F flows isentropically through a converging duct to the atmosphere. The throat area of the duct is 0.1 ft^2. Determine the pressure, temperature, velocity, and mass flowrate of the gas at the duct throat if the gas is **(a)** air, **(b)** carbon dioxide, **(c)** helium.

(ANS: **(a)** 52.8 psia; 433 °R; 1020 ft/s; 1.04 slugs/s; **(b)** 54.6 psia; 452 °R; 815 ft/s; 1.25 slugs/s; **(c)** 48.8 psia; 391 °R; 2840 ft/s; 0.411 slugs/s)

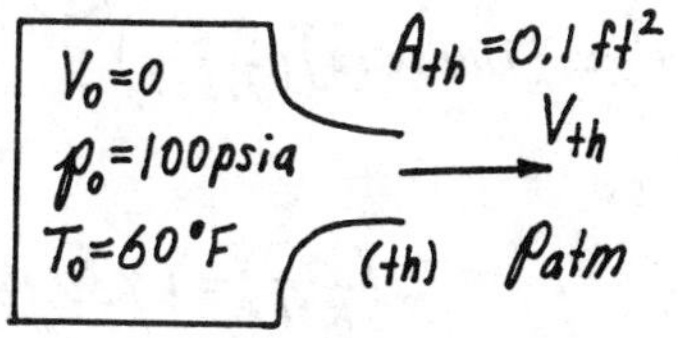

If the nozzle is choked, then the fluid speed at the exit is equal to the speed of sound there and the pressure is the critical pressure, p^*, where from Eq. 11.61

$$\frac{p^*}{p_0} = \left(\frac{2}{k+1}\right)^{k/(k-1)}$$

If $p^* < p_{atm}$, then the nozzle is not choked and $p_{th} = p_{atm}$. If $p^* > p_{atm}$, then the nozzle is choked and $p_{th} = p^*$. As shown below, the nozzle is choked in each case considered.

a) For air $k=1.4$ so that

$$p^* = (100\,psia)\left(\frac{2}{1.4+1}\right)^{1.4/(1.4-1)} = 52.8\,psia > p_{atm} = 14.7\,psia$$

Thus, $p_{th} = p^* = \underline{\underline{52.8\,psia}}$

From Eq. 11.63, since $Ma_{th}=1$ we have $T_{th} = T^* = T_0\left(\frac{2}{k+1}\right)$

or

$$T_{th} = (460+60)\left(\frac{2}{1.4+1}\right) = \underline{\underline{433\,°R}} \text{ so that}$$

$$V_{th} = Ma_{th}\,c_{th} = \sqrt{kRT_{th}} = \left[1.4\left(1716\,\frac{ft\cdot lb}{slug\cdot °R}\right)(433\,°R)\right]^{1/2} = \underline{\underline{1020\,\frac{ft}{s}}}$$

Finally,

$$\dot{m} = \rho AV = \left(\frac{p_{th}}{RT_{th}}\right)A_{th}V_{th} = \frac{\left(52.8\,\frac{lb}{in.^2}\right)\left(144\,\frac{in.^2}{ft^2}\right)(0.1\,ft^2)\left(1020\,\frac{ft}{s}\right)}{\left(1716\,\frac{ft\cdot lb}{slug\cdot °R}\right)(433\,°R)}$$

or

$$\dot{m} = \underline{\underline{1.04\,\frac{slugs}{s}}}$$

(continued)

b) Similarly for carbon dioxide $k = 1.3$ so that

$$p^* = (100\,\text{psia})\left(\frac{2}{1.3+1}\right)^{1.3/(1.3-1)} = \underline{\underline{54.6\,\text{psia}}} > p_{atm} \text{ so } p_{th} = p^*$$

Also

$$T_{th} = T^* = T_0\left(\frac{2}{k+1}\right) = (460+60)\left(\frac{2}{1.3+1}\right) = \underline{\underline{452\,°R}} \text{ so that}$$

$$V_{th} = \sqrt{kRT_{th}} = \left[1.3\left(1130\,\frac{\text{ft}\cdot\text{lb}}{\text{slug}\cdot°\text{R}}\right)(452\,°\text{R})\right]^{1/2} = \underline{\underline{815\,\frac{\text{ft}}{\text{s}}}}$$

and

$$\dot{m} = \left(\frac{p_{th}}{R\,T_{th}}\right)A_{th}V_{th} = \frac{\left(54.6\,\frac{\text{lb}}{\text{in.}^2}\right)\left(144\,\frac{\text{in.}^2}{\text{ft}^2}\right)(0.1\,\text{ft}^2)\left(815\,\frac{\text{ft}}{\text{s}}\right)}{\left(1130\,\frac{\text{ft}\cdot\text{lb}}{\text{slug}\cdot°\text{R}}\right)(452\,°\text{R})} = \underline{\underline{1.25\,\frac{\text{slugs}}{\text{s}}}}$$

c) Finally, for helium $k = 1.66$ so that

$$p^* = (100\,\text{psia})\left(\frac{2}{1.66+1}\right)^{1.66/(1.66-1)} = \underline{\underline{48.8\,\text{psia}}} > p_{atm} \text{ so that } p_{th} = p^*$$

Also,

$$T_{th} = T^* = T_0\left(\frac{2}{k+1}\right) = (460+60)\left(\frac{2}{1.66+1}\right) = \underline{\underline{391\,°R}} \text{ so that}$$

$$V_{th} = \sqrt{kRT_{th}} = \left[1.66\left(12{,}420\,\frac{\text{ft}\cdot\text{lb}}{\text{slug}\cdot°\text{R}}\right)(391\,°\text{R})\right]^{1/2} = \underline{\underline{2840\,\frac{\text{ft}}{\text{s}}}}$$

and

$$\dot{m} = \left(\frac{p_{th}}{R\,T_{th}}\right)A_{th}V_{th} = \frac{\left(48.8\,\frac{\text{lb}}{\text{in.}^2}\right)\left(144\,\frac{\text{in.}^2}{\text{ft}^2}\right)(0.1\,\text{ft}^2)\left(2840\,\frac{\text{ft}}{\text{s}}\right)}{\left(12{,}420\,\frac{\text{ft}\cdot\text{lb}}{\text{slug}\cdot°\text{R}}\right)(391\,°\text{R})} = \underline{\underline{0.411\,\frac{\text{slugs}}{\text{s}}}}$$

11.7R

11.7R (Fanno flow) A long, smooth wall pipe ($f = 0.01$) is to deliver 8000 ft^3/min of air at 60 °F and 14.7 psia. The inside diameter of the pipe is 0.5 ft and the length of the pipe is 100 ft. Determine the static temperature and pressure required at the pipe entrance if the flow through the pipe is adiabatic.
(ANS: 539 °R; 23.4 psia)

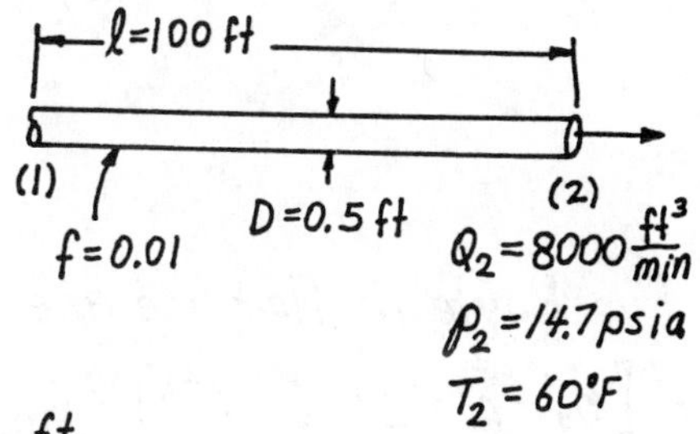

At the exit

$$V_2 = \frac{Q_2}{A_2} = \frac{4Q_2}{\pi D^2} = \frac{4\left(8000\,\frac{ft^3}{min}\right)\left(\frac{1\,min}{60\,s}\right)}{\pi\,(0.5\,ft)^2} = 679\,\frac{ft}{s}$$

so that

$$Ma_2 = \frac{V_2}{c_2} = \frac{V_2}{\sqrt{kRT_2}} = \frac{679\,\frac{ft}{s}}{\left[1.4\left(1716\,\frac{ft\cdot lb}{slug\cdot{}^\circ R}\right)(460+60)^\circ R\right]^{1/2}} = 0.608$$

Thus, from Fig. D.2 (Fanno flow) with $Ma_2 = 0.608$ we obtain

$$\frac{f(\ell^* - \ell_2)}{D} = 0.45 \text{ so that from}$$

$$\frac{f(\ell_2 - \ell_1)}{D} = \frac{(0.01)(100\,ft)}{0.5\,ft} = 2.00 = \frac{f(\ell^* - \ell_1)}{D} - \frac{f(\ell^* - \ell_2)}{D}$$

we obtain

$$\frac{f(\ell^* - \ell_1)}{D} = 2.00 + 0.45 = 2.45$$

Hence, from Fig. D.2 we find $Ma_1 = 0.39$, $\frac{T_1}{T^*} = 1.16$, and $\frac{p_1}{p^*} = 2.75$

Thus, with

$T_1 = T_2\left(\frac{T^*}{T_2}\right)\left(\frac{T_1}{T^*}\right)$ and $p_1 = p_2\left(\frac{p^*}{p_2}\right)\left(\frac{p_1}{p^*}\right)$ we can obtain T_1 and p_1.

From Fig. D.2 with $Ma_2 = 0.608$ we have

$$\frac{T_2}{T^*} = 1.12 \text{ and } \frac{p_2}{p^*} = 1.73$$

Thus,

$$T_1 = (460+60)^\circ R\,\frac{(1.16)}{(1.12)} = \underline{\underline{539^\circ R}}$$

and

$$p_1 = 14.7\,psia\,\frac{(2.75)}{(1.73)} = \underline{\underline{23.4\,psia}}$$

11.8R

11.8R (Rayleigh flow) Air enters a constant-area duct that may be considered frictionless with $T_1 = 300$ K and $V_1 = 300$ m/s. Determine the amount of heat transfer in kJ/kg required to choke the Rayleigh flow involved.

(ANS: 5020 J/kg)

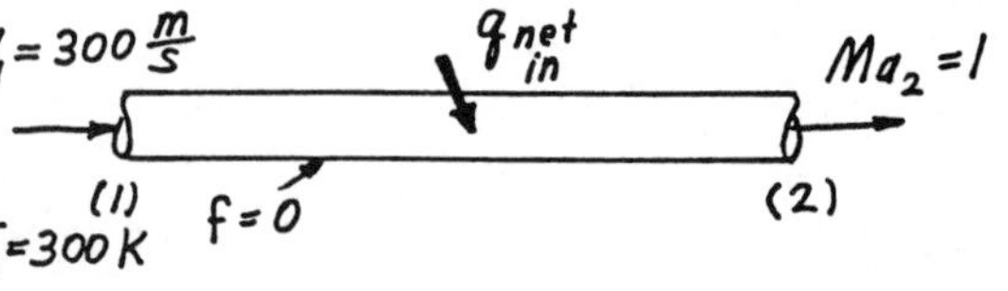

For Rayleigh flow the energy equation (Eq. 5.69) is

$$\dot{m}\left[\check{h}_2 - \check{h}_1 + \tfrac{1}{2}(V_2^2 - V_1^2)\right] = \dot{Q}_{net\ in} ,$$

or

$$q_{net\ in} = \frac{\dot{Q}_{net\ in}}{\dot{m}} = h_{o_2} - h_{o_1} \text{ where } h_o = \check{h} + \tfrac{1}{2}V^2 = c_p T_o \text{ is}$$

the stagnation enthalpy. Thus,

$$q_{net\ in} = c_p\,(T_{o_2} - T_{o_1}) \qquad (1)$$

At section (1), $Ma_1 = \frac{V_1}{c_1} = \frac{V_1}{\sqrt{kRT_1}} = \frac{300\ \frac{m}{s}}{\left[1.4\left(286.9\ \frac{N\cdot m}{kg\cdot K}\right)(300K)\right]^{1/2}} = 0.864$

Thus, from Eq. (11.56),

$$\frac{T_1}{T_{o_1}} = \frac{1}{1 + \frac{(k-1)}{2}Ma_1^2} , \text{ or } T_{o_1} = (300K)\left[1 + \frac{(1.4-1)}{2}(0.864)^2\right] = 345\ K$$

Since the flow is choked (i.e. $Ma_2 = 1$), $T_{o_2} = T_{oa}$ where from Eq. (11.131),

$$\frac{T_{o1}}{T_{oa}} = \frac{2(k+1)\left(1 + \frac{k-1}{2}Ma_1^2\right)Ma_1^2}{(1 + k\,Ma_1^2)^2} , \text{ or}$$

$$T_{oa} = (345\ K)\,\frac{\left(1 + 1.4(0.864)^2\right)^2}{2(1.4+1)\left[1 + \frac{(1.4-1)}{2}(0.864)^2\right](0.864)^2} = 350K$$

Thus,

$T_{o2} = T_{oa} = 350\ K$ so that with $c_p = \frac{Rk}{k-1} = \frac{1.4\left(286.9\ \frac{N\cdot m}{kg\cdot K}\right)}{(1.4-1)} = 1004\ \frac{J}{kg\cdot K}$

Eq. (1) gives

$$q_{net\ in} = \left(1004\ \frac{J}{kg\cdot K}\right)(350K - 345K) = 5020\ \frac{J}{kg}$$

11.9R

11.9R (Normal shock waves) Standard atmospheric air enters subsonically and accelerates isentropically to supersonic flow in a duct. If the ratio of duct exit to throat cross-section areas is 3.0, determine the ratio of back pressure to inlet stagnation pressure that will result in a standing normal shock at the duct exit. Determine also the stagnation pressure loss across the normal shock in kPa.

(ANS: 0.375: 56.1 kPa)

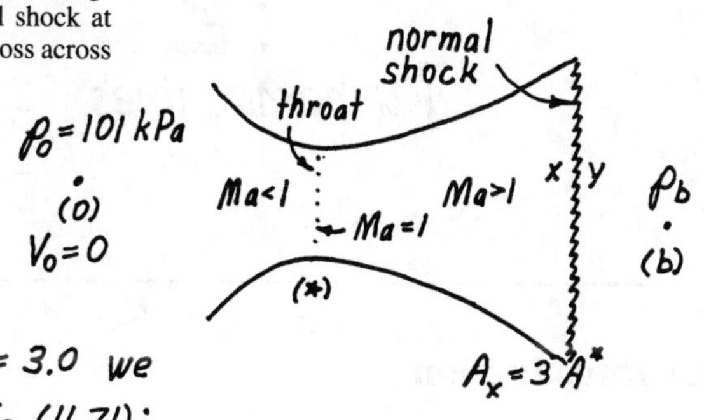

For isentropic flow with $\frac{A_x}{A^*} = 3.0$ we obtain Ma_x from Fig. D.1 or Eq. (11.71):

$Ma_x = 2.64$

Then from Eq. (11.59) we obtain

$$\frac{p_x}{p_{0x}} = \left[\frac{1}{1+\left(\frac{k-1}{2}\right)Ma_x^2}\right]^{k/(k-1)} = \left[\frac{1}{1+\left(\frac{1.4-1}{2}\right)(2.64)^2}\right]^{1.4/(1.4-1)} = 0.0471$$

while

$$\frac{p_b}{p_{0_1}} = \frac{p_y}{p_{0x}} = \frac{p_y}{p_x}\frac{p_x}{p_{0x}},$$ where from Eq. (11.150) (or Fig. D.4)

$$\frac{p_y}{p_x} = \frac{2k}{(k+1)}Ma_x^2 - \frac{k-1}{k+1} = \frac{2(1.4)}{(1.4+1)}(2.64)^2 - \frac{(1.4-1)}{(1.4+1)} = 7.96$$

Thus, since $p_0 = p_{0x} = p_{01}$

$$\frac{p_b}{p_0} = (7.96)(0.0471) = \underline{\underline{0.375}},$$ or $p_b = 101\ kP(0.375) = 37.9\ kPa$

Finally, from Eq. (11.156) (or Fig. D.4)

$$\frac{p_{0y}}{p_{0x}} = \frac{\left(\frac{k+1}{2}Ma_x^2\right)^{k/(k-1)}\left(1+\frac{k-1}{2}Ma_x^2\right)^{k/(1-k)}}{\left(\frac{2k}{k+1}Ma_x^2 - \frac{k-1}{k+1}\right)^{1/(k-1)}}$$

$$= \frac{\left[\left(\frac{1.4+2}{2}\right)(2.64)^2\right]^{[1.4/0.4]}\left[1+\frac{0.4}{2}(2.64)^2\right]^{-\frac{1.4}{0.4}}}{\left[\frac{2.8}{2.4}(2.64)^2 - \frac{0.4}{2.4}\right]^{1/0.4}} = 0.445$$

Thus,

$$p_{0x} - p_{0y} = p_{0x}\left[1 - \frac{p_{0y}}{p_{0x}}\right] = 101\ kPa\ (abs)\left[1 - 0.445\right] = \underline{\underline{56.1\ kPa}}$$

12
Turbomachines

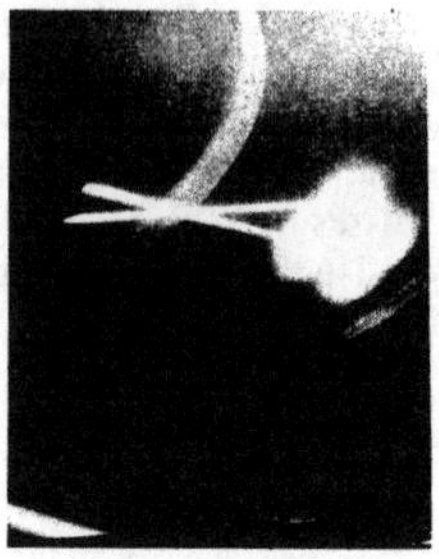

12.1 Introduction

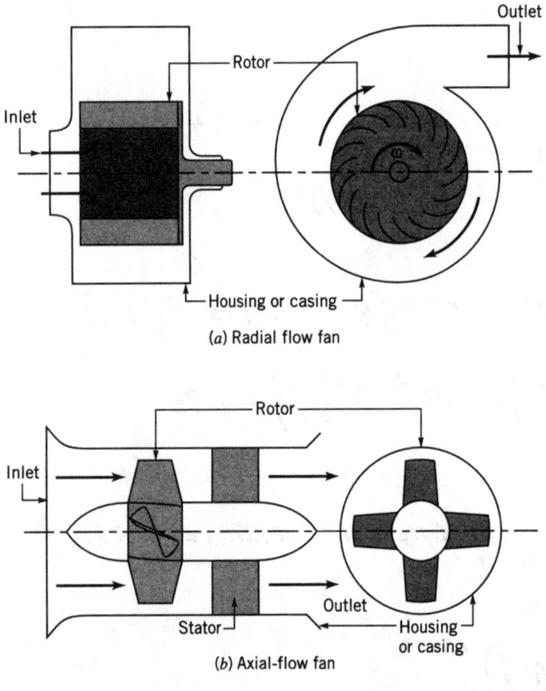

■ FIGURE 12.2 ***(a)* A radial-flow turbomachine, *(b)* An axial-flow turbomachine.**

- Turbomachines are mechanical devices that either extract energy from a fluid (a turbine) or add energy to a fluid (a pump) as a result of dynamic interactions between the device and the fluid.
- Turbomachines are classified as *axial-flow*, *mixed-flow*, or *radial-flow* machines depending on the predominant direction of the fluid motion relative to the rotor's axis as the fluid passes the blades (see Fig. 12.2).

12.2 Basic Energy Considerations

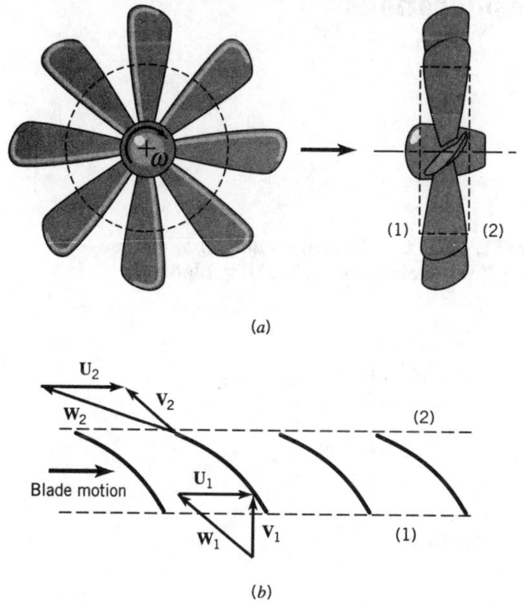

■ FIGURE 12.4 **Idealized flow through a windmill: *(a)* Windmill blade geometry; *(b)* Absolute velocity, V; relative velocity, W; and blade velocity, U; at the inlet and exit of the windmill blade section.**

$$\mathbf{V} = \mathbf{W} + \mathbf{U} \tag{12.1}$$

- Flow through a turbomachine is often described in terms of the *absolute* fluid velocity, **V**, the *relative* velocity, **W**, and the *blade* velocity, **U**, which are related as shown in Eq.12.1.
- The *velocity triangles* at the inlet and outlet of a turbomachine, as shown in Fig. 12.4, are graphical representations of the relationship among the absolute, relative, and blade velocities.
- As a fluid flows through a *turbine*, the blades push on the fluid in a direction opposite to the direction of the blade, causing the fluid to change direction. For the case shown in Fig. 12.4, the tangential component of absolute velocity of the fluid entering the turbine is zero, but the tangential component of the fluid leaving the turbine is nonzero and in the direction opposite to the blade motion. The fluid does work on the blade.
- For a *pump* or fan the blades push on the fluid in the same direction as the motion of the blades and the blade does work on the fluid.

EXAMPLE: The velocity triangles at the inlet and outlet of a turbomachine are as shown in Fig. 12.4*b*. Is the device a pump or a turbine? Explain.

SOLUTION:

The fluid enters the turbomachine in the axial direction with zero tangential component of absolute velocity. To produce the flow at the exit, the blade had to push against the fluid in the direction opposite to the blade motion (i.e. the fluid is "turned back" in the negative tangential direction). Conversely, the fluid pushes on the blade in the direction of the blade motion. The fluid does work on the blade. The device is a turbine.

12.3 Basic Angular Momentum Considerations

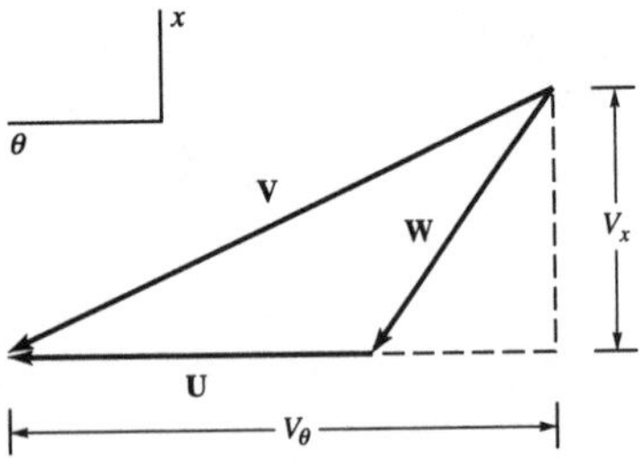

■ FIGURE 12.5 **Velocity triangle: V = absolute velocity, W = relative velocity, U = blade velocity.**

$$T_{\text{shaft}} = -\dot{m}_1(r_1V_{\theta 1}) + \dot{m}_2(r_2V_{\theta 2}) \quad \textbf{(12.2)}$$

$$\dot{W}_{\text{shaft}} = -\dot{m}_1(U_1V_{\theta 1}) + \dot{m}_2(U_2V_{\theta 2}) \quad \textbf{(12.4)}$$

$$w_{\text{shaft}} = -U_1V_{\theta 1} + U_2V_{\theta 2} \quad \textbf{(12.5)}$$

$$w_{\text{shaft}} = \frac{V_2^2 - V_1^2 + U_2^2 - U_1^2 - (W_2^2 - W_1^2)}{2} \quad \textbf{(12.8)}$$

- All turbomachines involve the rotation of an impeller or rotor about a central axis. Their performance can be discussed in terms of torque, angular momentum, and the axial component of the moment-of-momentum equation.
- As shown by Eq. 12.2, the *shaft torque*, T_{shaft}, is a function of the mass flowrate, $\dot{m}$, the radii at the rotor inlet and outlet, r_1, and r_2, and the tangential components of the absolute velocity at the inlet and outlet, $V_{\theta 1}$ and $V_{\theta 2}$.
- A typical velocity triangle illustrating the tangential component of the absolute velocity is shown in Fig. 12.5.
- The *shaft power*, $\dot{W}_{\text{shaft}} = T_{\text{shaft}}\ \omega$, can be written in terms of the mass flowrate, the blade speed, and the tangential component of absolute velocity as shown by Eq. 12.4.
- The *work per unit mass*, $w_{\text{shaft}} = \dot{W}_{\text{shaft}} / \dot{m}$, can be written in terms of the blade speed and the tangential component of absolute velocity as shown by Eq. 12.5. Alternatively, it can be written in terms of the blade speed, the absolute fluid speed, and the relative fluid speed as shown by Eq. 12.8.
- *Caution:* The tangential component of the absolute velocity, V_θ, is positive if V_θ and U are in the same direction. The shaft torque, T_{shaft}, is positive if it is in the same direction as the direction of rotation.

EXAMPLE: Water flows through a turbomachine as shown in the Fig. 12.4*b* with a mass flowrate of 2 slugs/s. The radii at the inlet and outlet are both 1.5 ft and the blade speed is 10 ft/s at both locations. The magnitudes of the tangential components of the absolute velocity at the inlet and outlet are 0 and 20 ft/s, respectively. **(a)** Determine the shaft power for this turbomachine. **(b)** Is the device a pump or a turbine? Explain.

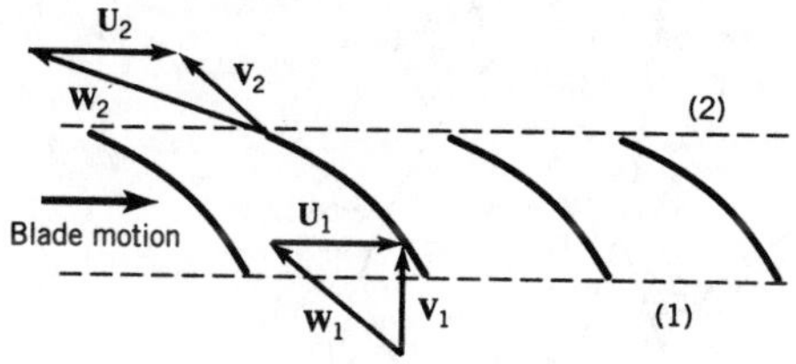

■ **FIGURE 12.4** (*b*)

SOLUTION:

(a) From Eq. 12.4, $\dot{W}_{shaft} = -\dot{m}_1 (U_1 V_{\theta 1}) + \dot{m}_2 (U_2 V_{\theta 2})$

where $\dot{m}_1 = \dot{m}_2 = 2\ slugs/s$, $U_1 = U_2 = 10\ ft/s$, $V_{\theta 1} = 0$, and $V_{\theta 2} = -20\ ft/s$

Thus, $\dot{W}_{shaft} = -2\ slugs/s\,(10\ ft/s)(0) + 2\ slugs/s\,(10\ ft/s)(-20\ ft/s)$

$= -400 (slug \cdot ft/s^2) \cdot ft/s = \underline{-400\ lb \cdot ft/s}$

(b) Since $\dot{W}_{shaft} < 0$ the device is a <u>turbine.</u>

12.4 The Centrifugal Pump

Theoretical Considerations

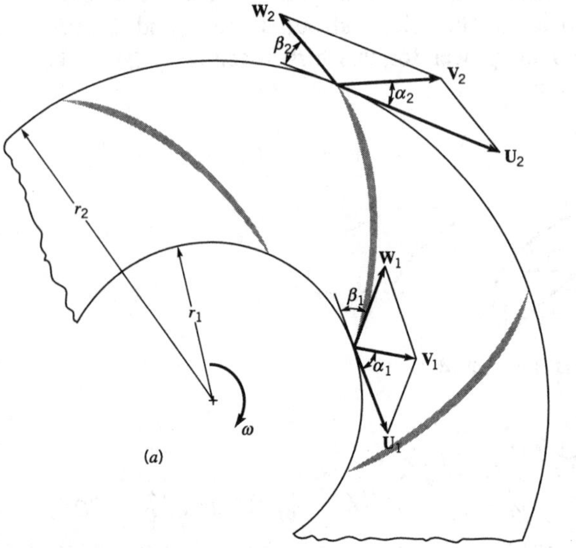

■ FIGURE 12.8 Velocity diagrams at the inlet and exit of a centrifugal pump impeller.

$$T_{\text{shaft}} = \dot{m}(r_2 V_{\theta 2} - r_1 V_{\theta 1}) \qquad \textbf{(12.9)}$$

$$\dot{W}_{\text{shaft}} = \rho Q(U_2 V_{\theta 2} - U_1 V_{\theta 1}) \qquad \textbf{(12.11)}$$

$$w_{\text{shaft}} = \frac{\dot{W}_{\text{shaft}}}{\rho Q} = U_2 V_{\theta 2} - U_1 V_{\theta 1} \qquad \textbf{(12.12)}$$

$$h_i = \frac{1}{g}(U_2 V_{\theta 2} - U_1 V_{\theta 1}) \qquad \textbf{(12.13)}$$

- Inlet and exit velocity diagrams for the impeller of a centrifugal pump are shown in Fig. 12.8*a*.
- The shaft torque, T_{shaft}, shaft power, $\dot{W}_{\text{shaft}}$, shaft power per mass flowrate, w_{shaft}, and ideal head rise are given by Eqs. 12.9, 12.11, 12.12, and 12.13.
- *Caution:* The actual head rise realized by fluid passing through a centrifugal pump is less than the ideal head rise by an amount equal to the head loss suffered.

EXAMPLE: The velocity triangles for water flow through a centrifugal pump are as shown in the figure. The tangential component of the absolute velocity at the inlet to the pump is -10 m/s. **(a)** Determine the shaft power per mass flowrate. **(b)** Determine the ideal head rise.

SOLUTION:

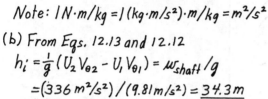

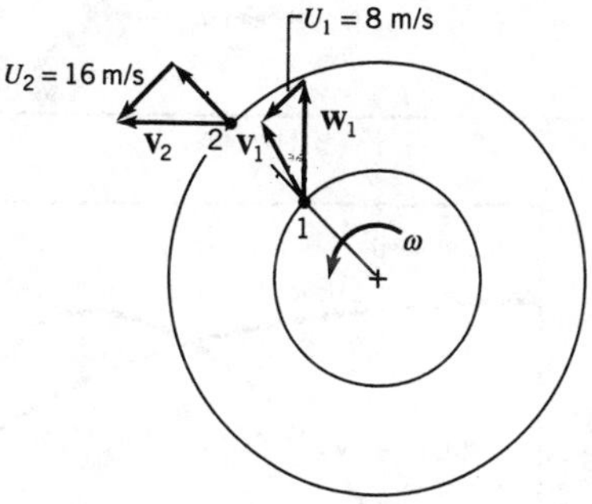

(a) From Eq. 12.12, $w_{shaft} = U_2 V_{\theta 2} - U_1 V_{\theta 1}$

where $U_1 = 8\,m/s$, $U_2 = 16\,m/s$, $V_{\theta 1} = -10\,m/s$, and from the exit velocity triangle $V_{\theta 2} = U_2 = 16\,m/s$. Thus,

$$w_{shaft} = 16\,m/s\,(16\,m/s) - 8\,m/s\,(-10\,m/s)$$
$$= 336\,m^2/s^2 = \underline{\underline{336\,N\cdot m/kg}}$$

Note: $1\,N\cdot m/kg = 1\,(kg\cdot m/s^2)\cdot m/kg = m^2/s^2$

(b) From Eqs. 12.13 and 12.12

$$h_i = \frac{1}{g}(U_2 V_{\theta 2} - U_1 V_{\theta 1}) = w_{shaft}/g$$
$$= (336\,m^2/s^2)/(9.81\,m/s^2) = \underline{\underline{34.3\,m}}$$

Pump Performance Characteristics

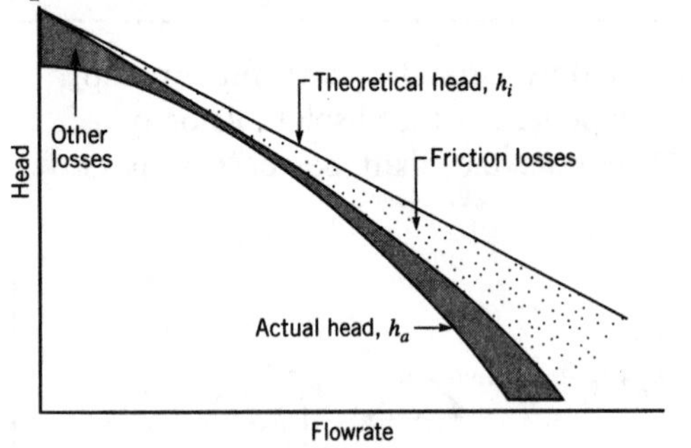

■ FIGURE 12.9 **Effect of losses on the pump head-flowrate curve.**

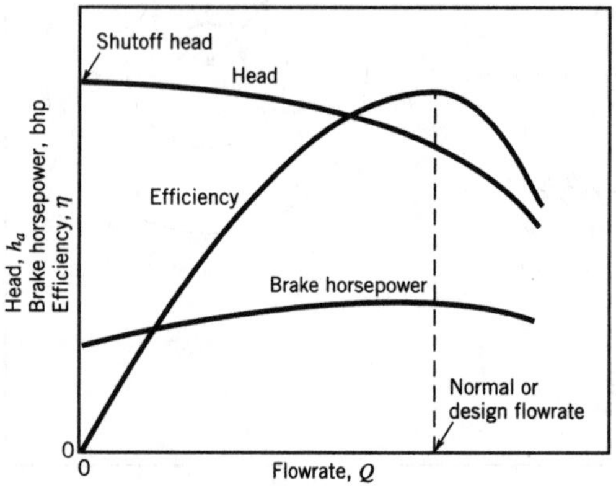

■ FIGURE 12.11 **Typical performance characteristics for a centrifugal pump of a given size operating at a constant impeller speed.**

$$h_a = \frac{p_2 - p_1}{\gamma} + z_2 - z_1 + \frac{V_2^2 - V_1^2}{2g} \qquad \textbf{(12.19)}$$

$$\eta = \frac{\gamma Q h_a / 550}{\text{bhp}} \qquad \textbf{(12.23)}$$

- The *actual head rise*, h_a, gained by a fluid passing through a pump is less than the theoretical (ideal) head rise, h_i, (see Fig. 12.9) and can be determined experimentally by use of the energy equation, Eq. 12.19.
- The *overall efficiency*, η, of a pump is the ratio of the power gained by the fluid, $\gamma Q h_a$, to the shaft power driving the pump (referred to as brake horsepower, bhp), as given by Eq. 12.23.
- Typical performance characteristics for a centrifugal pump of a given size operating at a constant impeller speed are shown in Fig. 12.11. The curve of pump head as a function of flowrate is termed the *pump performance curve*.
- *Caution:* In Eq. 12.23 the units on $\gamma Q h_a$ are ft•lb/s and the units on bhp are horsepower. The conversion factor is 1 hp = 550 ft•lb/s as indicated.

EXAMPLE: If the overall efficiency of a centrifugal pump is 0.60, determine the amount of power added to the fluid when the motor driving the pump supplies 5 hp to the pump.

SOLUTION:

From Eq. 12.23, $\eta = (\gamma Q h_a / 550)/\text{bhp}$ where bhp = 5 hp.

Thus, $\gamma Q h_a / 550 = \eta(\text{bhp}) = 0.6(5 \text{ hp}) = \underline{3 \text{ hp}}$

System Characteristics and Pump Selection

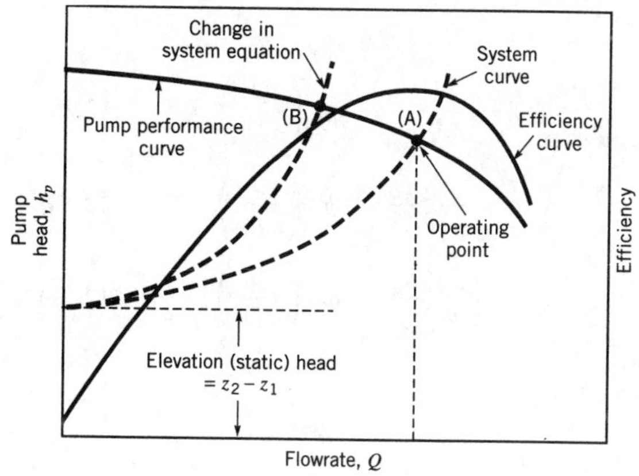

■ **FIGURE 12.15 Utilization of the system curve and the pump performance curve to obtain the operating point for the system.**

$$h_p = z_2 - z_1 + KQ^2 \quad \textbf{(12.27)}$$

- The *system equation*, Eq. 12.27, gives the pump head, h_p , required to produce the flowrate, Q, and elevation change, $z_2 - z_1$, for a given pipe system.
- The pump characteristics can be used to decide if a particular pump is suitable for a given application.
- As shown in Fig. 12.15, the flowrate expected for a given pump and pipe system is represented by the intersection of the pump performance curve and the system curve.
- *Caution:* If the pipe flow is laminar the frictional losses will be proportional to Q rather than Q^2 as indicated in Eq. 12.27 and Fig. 12.15.

EXAMPLE: A pump with a pump performance curve $h_p = 30 - 200\ Q^2$, where h_p is in ft and Q is in ft^3/s, is used to pump a fluid up a 25-ft-tall hill. The system equation is $h_p = 25 + 100\ Q^2$. **(a)** Determine the flowrate expected. **(b)** Is this pump a reasonable choice to use if the fluid is to be pumped up a 35-ft-tall hill so that the system equation is $h_p = 35 + 100\ Q^2$? Explain.

SOLUTION:

(a) The flowrate will be such that the pump provides just the head that the system requires. Thus,

$30 - 200\,Q^2 = 25 + 100\,Q^2$

or $Q = \underline{\underline{0.129\ \text{ft}^3/\text{s}}}$

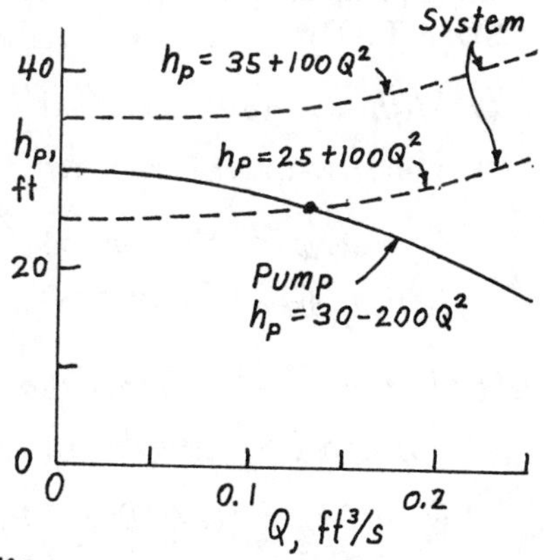

(b) Similarly, for the system with the 35 ft hill,

$30 - 200\,Q^2 = 35 + 100\,Q^2$, or $Q^2 = -0.0167$

Thus, since $Q^2 < 0$ there is no solution!

This pump is <u>not a reasonable choice</u> because it can lift the fluid up only a 30 ft hill (not a 35 ft hill).

12.5 Dimensionless Parameters and Similarity Laws

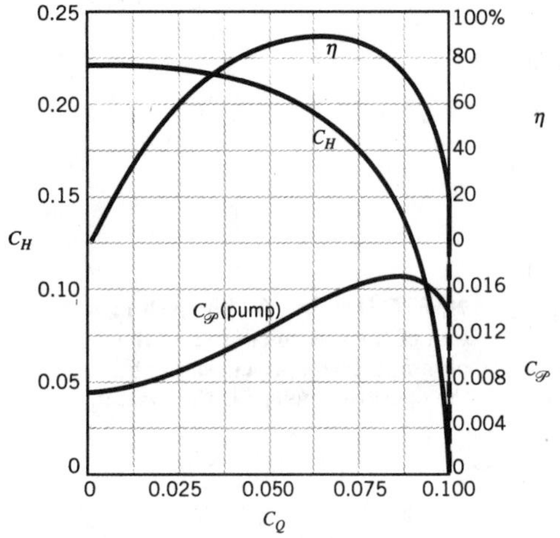

■ FIGURE 12.17 *(b)*

$$C_H = \frac{gh_a}{\omega^2 D^2} = \phi_1\left(\frac{Q}{\omega D^3}\right) \quad \textbf{(12.29)}$$

$$C_{\mathcal{P}} = \frac{\dot{W}_{\text{shaft}}}{\rho\omega^3 D^5} = \phi_2\left(\frac{Q}{\omega D^3}\right) \quad \textbf{(12.30)}$$

$$\eta = \frac{\rho g Q h_a}{\dot{W}_{\text{shaft}}} = \phi_3\left(\frac{Q}{\omega D^3}\right) \quad \textbf{(12.31)}$$

- The main dependent pump variables are the actual head rise, h_a, the shaft power, $\dot{W}_{\text{shaft}}$, and the efficiency, η. These can be written in dimensionless form as the *head rise coefficient*, C_H, the *power coefficient*, $C_{\mathcal{P}}$, and the *efficiency*, η, as shown by the above equations.
- As shown by Eqs.12.29, 12.30, and 12.31, for geometrically similar pumps the head rise coefficient, the power coefficient, and the efficiency are functions of only the *flow coefficient*, $C_Q = Q/\omega D^3$.
- Typical performance data for a centrifugal pump are shown in dimensionless form in Fig. 12.17*b*.

EXAMPLE: Two geometrically similar water pumps with performance characteristics given by Fig. 12.17*b* have diameters of 0.3 m and 0.5 m and operate at 800 rpm and 600 rpm, respectively. **(a)** Determine the efficiency of the smaller pump if its flowrate is 0.1 m/s^3. **(b)** Determine the flowrate of the larger pump operating at the same efficiency under similar conditions.

SOLUTION:

(a) With $\omega = (800\,\text{rev/min})(1\,\text{min}/60\,\text{s})(2\pi\,\text{rad/rev}) = 83.8\,\text{rad/s}$, the flow coefficient for the smaller pump is

$C_Q = Q/\omega D^3 = (0.1\,\text{m}^3/\text{s})/[83.8\,\text{rad/s}\,(0.3\,\text{m})^3] = 0.0442$

Hence, from Fig. 12.17b, $\eta = \underline{81\%}$

(b) Under similar conditions $\eta_\ell = \eta_s$ so that $C_{Q\ell} = C_{Qs}$, where $(\ )_\ell$ and $(\ )_s$ denote the large and small pumps. Thus,

$(Q/\omega D^3)_\ell = (Q/\omega D^3)_s$, or $Q_\ell = (\omega_\ell/\omega_s)(D_\ell/D_s)^3 Q_s$ so that

$Q_\ell = (600\,\text{rpm}/800\,\text{rpm})(0.5\,\text{m}/0.3\,\text{m})^3(0.1\,\text{m}^3/\text{s}) = \underline{0.347\,\text{m}^3/\text{s}}$

Specific Speed

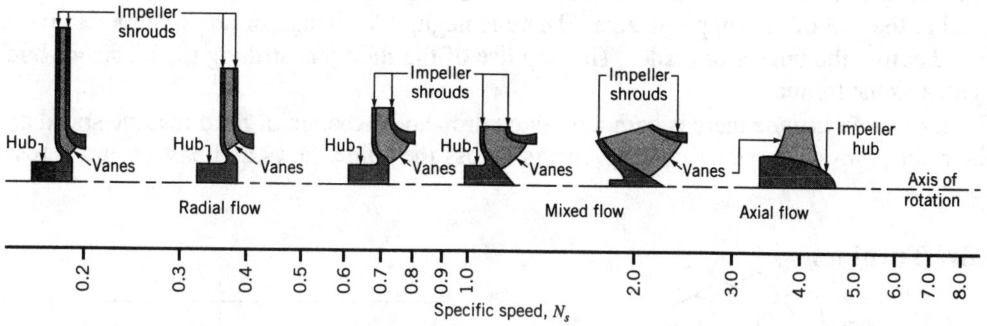

■ **FIGURE 12.18** **Variation in specific speed with type of pump. (Adapted from Ref. 17, used with permission.)**

$$\frac{(Q/\omega D^3)^{1/2}}{(gh_a/\omega^2 D^2)^{3/4}} = \frac{\omega\sqrt{Q}}{(gh_a)^{3/4}} = N_s \qquad \textbf{(12.43)}$$

- The *specific speed, N_s* , which is a function of the pump head, flowrate, and angular velocity as shown by Eq. 12.43, is used to determine which type of pump (radial flow, mixed flow, or axial flow) would be most efficient for a particular application.
- Fig. 12.18 illustrates how the specific speed changes as the configuration of the pump changes from radial-flow pumps to axial-flow pumps.
- Centrifugal pumps typically are low-capacity, high-head pumps with low specific speeds. Axial-flow pumps are essentially high-capacity, low-head pumps with large specific speeds.

EXAMPLE: A pump is to supply a flowrate of 0.05 m³/s with a head of 20 m while operating at 500 rpm. What type of pump should be selected?
SOLUTION:

From Eq. 12.43 with $\omega = (500\,\text{rev/min})(1\,\text{min}/60\,\text{s})(2\pi\,\text{rad/rev}) = 52.4\,\text{rad/s}$ it follows that the specific speed is

$N_s = \omega\sqrt{Q}/(g\,h_a)^{3/4} = (52.4\,\text{rad/s})(0.05\,\text{m}^3/\text{s})^{1/2}/[(9.81\,\text{m/s}^2)\,20\,\text{m}]^{3/4} = 0.224$

From Fig. 12.18 with $N_s = 0.224$ it is seen that a radial flow (centrifugal) pump should be selected.

12.8 Turbines

- For an *impulse turbine* the total head of the incoming fluid is converted into a large velocity head at the exit of the supply nozzle. There is negligible change in pressure or relative fluid speed across the bucket or blade. The impulse of the fluid jets striking the blades or buckets generates the torque.
- For a *reaction turbine* there is both a pressure drop and a change in fluid relative speed across the rotor. Part of the pressure drop occurs across the guide vanes and part occurs across the rotor.

Impulse Turbines

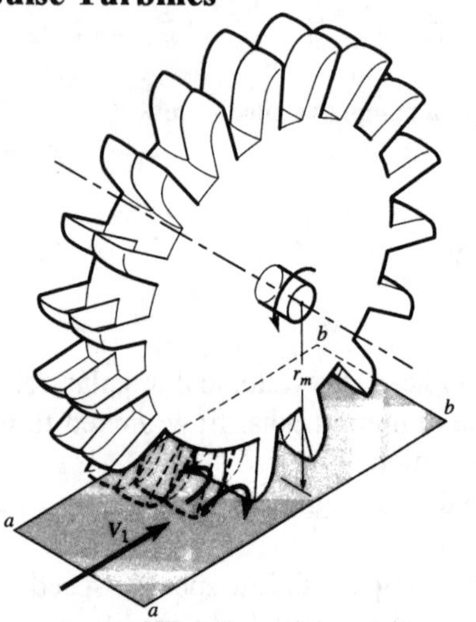

■ FIGURE 12.24 Details of Pelton wheel turbine bucket.

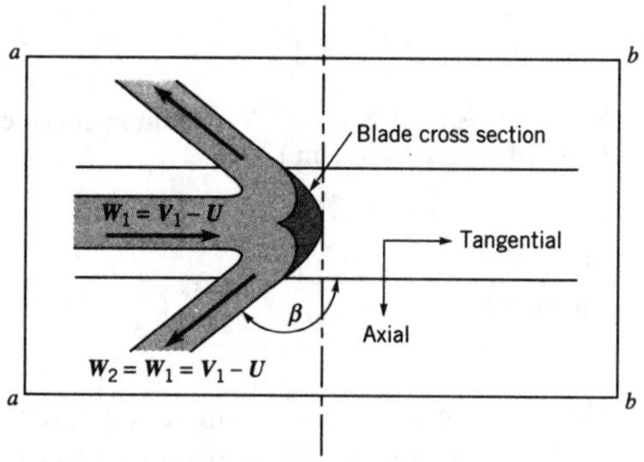

■ FIGURE 12.26 Flow as viewed by an observer riding on the Pelton wheel—relative velocities.

$$T_{\text{shaft}} = \dot{m} r_m (U - V_1)(1 - \cos\beta)$$

$$\dot{W}_{\text{shaft}} = T_{\text{shaft}}\omega = \dot{m}U(U - V_1)(1 - \cos\beta) \qquad \textbf{(12.51)}$$

- As shown in Figs. 12.24 and 12.26, a *Pelton wheel* turbine is an impulse turbine that produces torque and power as a result of the change in direction of the velocity of the fluid as it strikes the turbine wheel buckets.
- The *relative* speed of the fluid does not change as it is deflected by the buckets. The *absolute* speed of the fluid does change. That is, $W_2 = W_1$, but $V_2 \neq V_1$.
- The shaft torque, T_{shaft}, and shaft power, $\dot{W}_{\text{shaft}}$, are given in terms of the pertinent parameters by the above equations.
- The maximum torque occurs when the turbine rotor is stationary ($U = 0$). The minimum torque ($T_{\text{shaft}} = 0$) occurs when the rotor is free-wheeling ($U = V_1$).
- The maximum power occurs when the bucket speed is one half the fluid speed ($U = 0.5\ V_1$).
- *Caution:* For turbines the shaft torque and shaft power are negative, indicating that power is removed from the flowing fluid.

EXAMPLE: A 0.15-ft-diameter water jet with a velocity of 60 ft/s strikes a Pelton wheel at a radial distance of $r_m = 1$ ft and is deflected through an angle of $\beta = 135$ degrees. **(a)** Determine the shaft torque and shaft power if the angular velocity of the turbine rotor is 300 rpm. **(b)** Determine the free-wheeling angular velocity if the shaft torque were zero.

SOLUTION:

(a) From the above equation, $T_{shaft} = \dot{m} r_m (U - V_1)(1 - \cos\beta)$

where $\dot{m} = \rho A_1 V_1 = 1.94 \text{ slugs/ft}^3 (\pi (0.15 \text{ ft})^2/4)(60 \text{ ft/s}) = 2.06 \text{ slugs/s}$

and $U = \omega r_m = (300 \text{ rev/min})(1 \text{ min}/60 \text{ s})(2\pi \text{ rad/rev})(1 \text{ ft}) = 31.4 \text{ ft/s}$

Thus, $T_{shaft} = 2.06 \text{ slugs/s } (1 \text{ ft})(31.4 \text{ ft/s} - 60 \text{ ft/s})(1 - \cos 135°)$

$= \underline{\underline{-101 \text{ ft}\cdot\text{lb}}}$

and

$\dot{W}_{shaft} = T_{shaft}\,\omega = -101 \text{ ft}\cdot\text{lb } (300 \text{ rev/min})(1 \text{ min}/60 \text{ s})(2\pi \text{ rad/rev})$

$= -3170 \text{ ft}\cdot\text{lb/s } (1 \text{ hp}/550 \text{ ft}\cdot\text{lb/s}) = \underline{\underline{-5.76 \text{ hp}}}$

(b) With $T_{shaft} = 0 = \dot{m} r_m (U - V_1)(1 - \cos\beta)$ it follows that $U = V_1$

Thus, $\omega r_m = V_1$ or

$\omega = V_1 / r_m = (60 \text{ ft/s})/1 \text{ ft} = 60 \text{ rad/s } (1 \text{ rev}/2\pi \text{ rad})(60 \text{ s/min})$

$= \underline{\underline{573 \text{ rev/min}}}$

Reaction Turbines

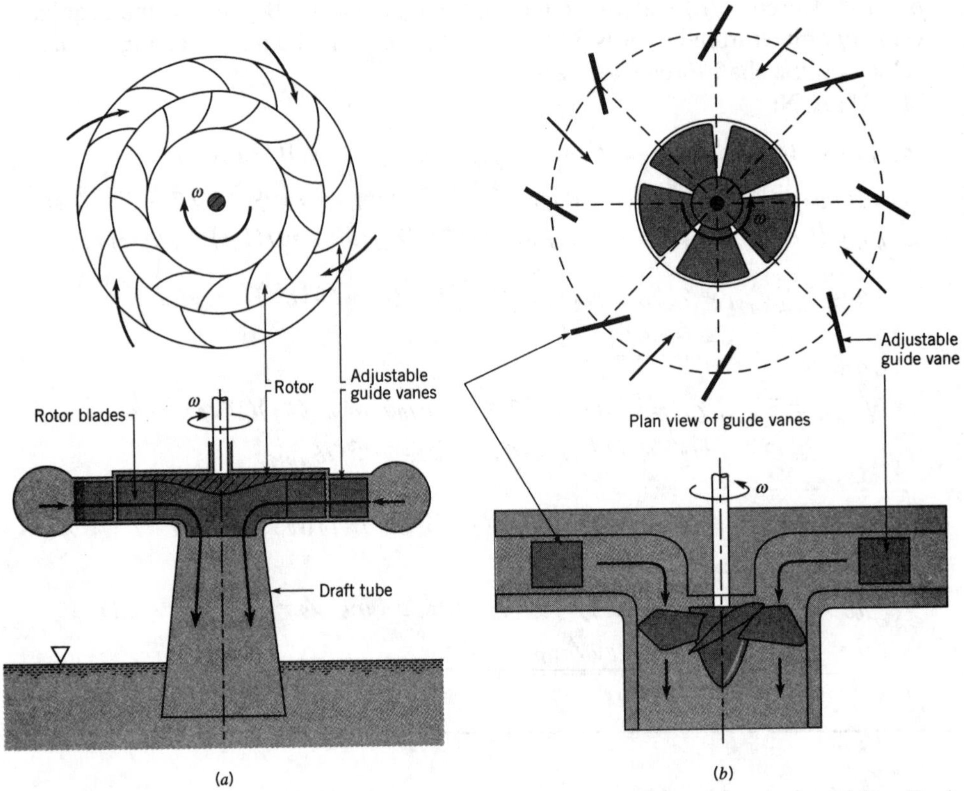

■ **FIGURE 12.31** ***(a)* Typical radial-flow Francis turbine, *(b)* typical axial-flow Kaplan turbine.**

- As shown in Fig. 12.31, in a reaction turbine the fluid completely fills the passageways through which it flows. Angular momentum, pressure, and velocity of the fluid decrease as the fluid flows through the turbine rotor.
- Reaction turbines are best suited for high flowrate, low head situations. Impulse turbines are best suited for low flowrate, high head situations.

Turbine Specific Speed

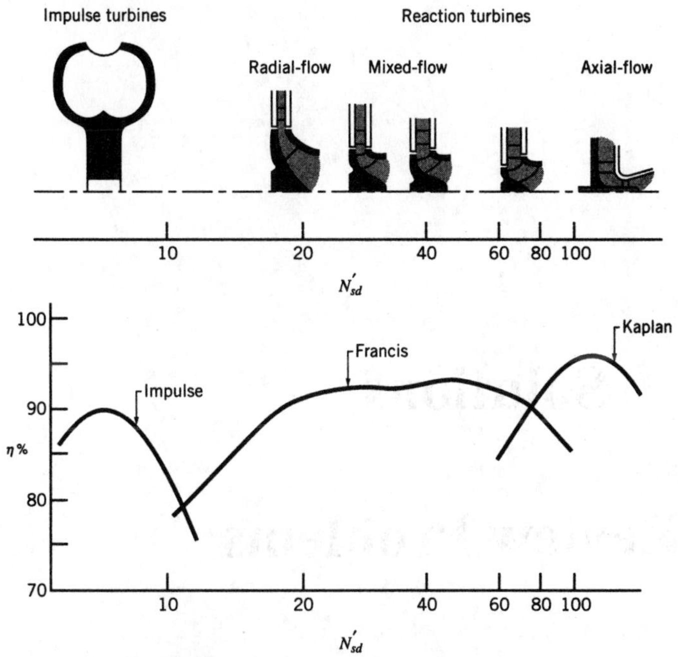

■ **FIGURE 12.32** **Typical turbine cross sections and maximum efficiencies as a function of specific speed.**

$$N'_{sd} = \frac{\omega(\text{rpm})\sqrt{\dot{W}_{\text{shaft}}\,(\text{bhp})}}{[h_T(\text{ft})]^{5/4}} \qquad \textbf{(12.53)}$$

- A *turbine's specific speed*, N'_{sd}, is a function of its angular velocity, shaft power, and turbine head as shown in Eq. 12.53.
- A turbine's efficiency, η, is a function of specific speed. Typical values of the efficiency for various types of turbines are indicated in Fig. 12.32.
- *Caution:* Specific speed, as defined by Eq. 12.53, is not dimensionless. The dimensions for angular velocity, shaft power, and turbine head must be expressed as rpm, bhp, and ft, respectively, as indicated.

EXAMPLE: A small hydraulic turbine is to produce 0.2 bhp when operating at 12 rev/s and a head of 10 ft. What type of turbine should be selected? Explain.
SOLUTION:

From Eq. 12.53, for the conditions given the specific speed is

$N'_{sd} = (12\ \text{rev/s})(60\ \text{s/min})(0.2\ \text{bhp})^{1/2} / (10\ \text{ft})^{5/4} = 18.1$

Thus, from Fig. 12.32, at this specific speed the most efficient type of turbine would be a radial flow Francis turbine.

Solutions

to Review Problems

Chapter 12

12.1R

12.1R (Angular momentum) Water is supplied to a dishwasher through the manifold shown in Fig. P12.1R. Determine the rotational speed of the manifold if bearing friction and air resistance are neglected. The total flowrate of 2.3 gpm is divided evenly among the six outlets, each of which produces a 5/16-in.-diameter stream.

(ANS: 0.378 rev/s)

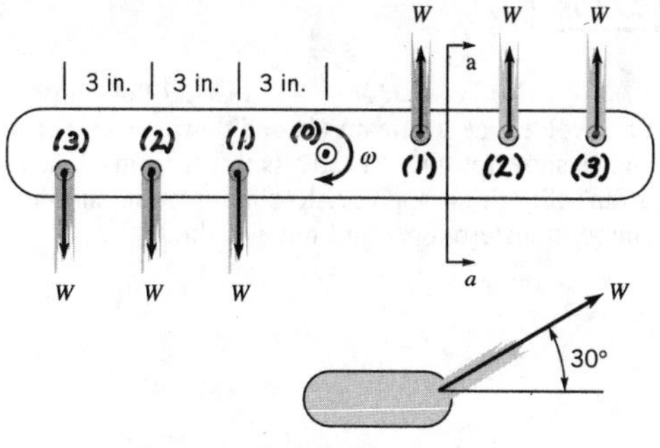

■ FIGURE P12.1R

With points (0), (1), (2), and (3) located as in the diagram above,

$T = $ "$\dot{m}(r_{out}V_{\theta out} - r_{in}V_{\theta in})$" where $V_{\theta in} = 0$. Thus,

$T = 2\dot{m}_1 r_1 V_{\theta 1} + 2\dot{m}_2 r_2 V_{\theta 2} + 2\dot{m}_3 r_3 V_{\theta 3} = 0$ since there is no friction.

But $\dot{m}_1 = \dot{m}_2 = \dot{m}_3$ so that the above becomes

$$r_1 V_{\theta 1} + r_2 V_{\theta 2} + r_3 V_{\theta 3} = 0 \quad (1)$$

But $U_i + V_{\theta i} = W_i \cos 30°\ ,\ i = 1,2,3 \quad (2)$

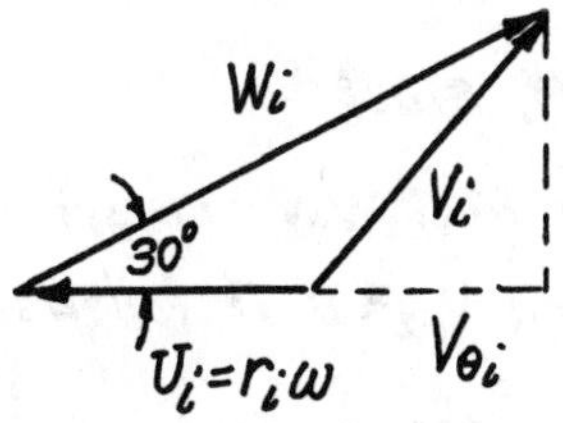

where

$$Q = 6A_iW_i = 6\left[\frac{\pi}{4}\left(\frac{(\frac{5}{16})}{12}\,ft\right)^2\right]W_i = 0.00320\,W_i$$

with

$$Q = \left(2.3\,\frac{gal}{min}\right)\left(\frac{1\,min}{60\,s}\right)\left(231\,\frac{in.^3}{gal}\right)\left(\frac{1\,ft^3}{1728\,in.^3}\right) = 0.00512\,\frac{ft^3}{s}$$

Thus, $W_i = 1.60\ \frac{ft}{s}$ so that from Eq. (2) $V_{\theta i} = W_i \cos 30° - U_i$

or $V_{\theta i} = (1.60\ \frac{ft}{s})\cos 30° - r_i\omega$. With $r_1 = \frac{3}{12}\,ft$, $r_2 = \frac{6}{12}\,ft$, and $r_3 = \frac{9}{12}\,ft$

Eq. (1) becomes

$$\frac{3}{12}\left(1.386 - \frac{3}{12}\omega\right) + \frac{6}{12}\left(1.386 - \frac{6}{12}\omega\right) + \frac{9}{12}\left(1.386 - \frac{9}{12}\omega\right) = 0$$

or $24.9 = 10.5\,\omega$

Thus, $\omega = 2.37\,\frac{rad}{s} \times \frac{1\,rev}{2\pi\,rad} = \underline{\underline{0.378\,\frac{rev}{s}}}$

12.2R

12.2R (Velocity triangles) An axial-flow turbomachine rotor involves the upstream (1) and downstream (2) velocity triangles shown in Fig. P12.2R. Is this turbomachine a turbine or a fan? Sketch an appropriate blade section and determine the energy transferred per unit mass of fluid.

(ANS: turbine; $-36.9\ \text{ft}^2/\text{s}^2$)

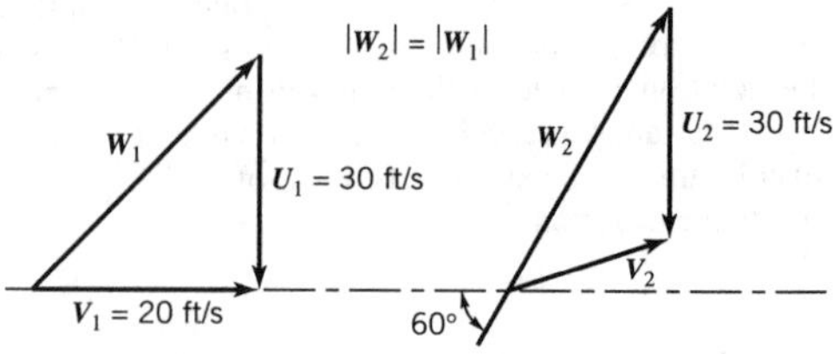

■ FIGURE P12.2R

From the given inlet conditions $W_1 = \sqrt{\left(30\frac{ft}{s}\right)^2 + \left(20\frac{ft}{s}\right)^2} = 36.06\frac{ft}{s}$

Thus, with $W_2 = W_1 = 36.06\frac{ft}{s}$ the outlet velocity triangle is as shown below,

or

$V_{\theta 2} = \left(36.06\frac{ft}{s}\right)\sin 60° - 30\frac{ft}{s} = 1.229\frac{ft}{s}$

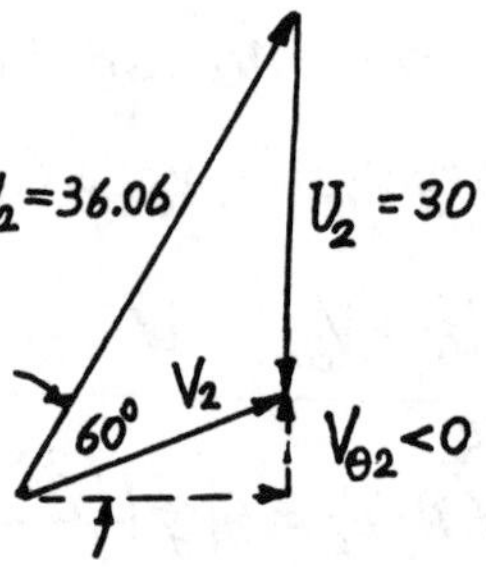

Thus, since

$T = \dot{m}(r_2 V_{\theta 2} - r_1 V_{\theta 1})$ with $V_{\theta 1} = 0$

and $V_{\theta 2} < 0$ it follows that $T < 0$,

i.e., the machine is a turbine.

From the figure it follows that

$\tan\theta = \frac{U_1}{V_1} = \frac{30\frac{ft}{s}}{20\frac{ft}{s}}$ or $\theta = 56.3°$

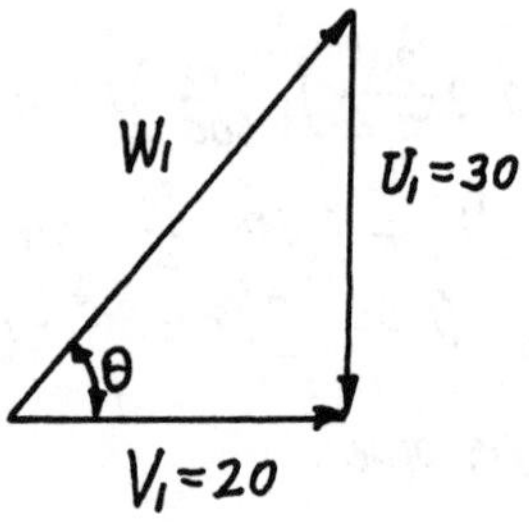

Hence, the blade would be shaped as follows:

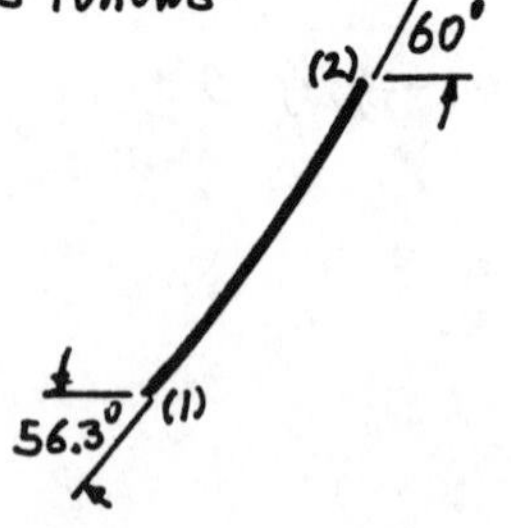

and

$w_{shaft} = U_2 V_{\theta 2} = \left(30\frac{ft}{s}\right)\left(-1.229\frac{ft}{s}\right) = -36.9\frac{ft^2}{s^2}$

12.3R

12.3R (Centrifugal pump) Shown in Fig. P12.3R are front and side views of a centrifugal pump rotor or impeller. If the pump delivers 200 liters/s of water and the blade exit angle is 35° from the tangential direction, determine the power requirement associated with flow leaving at the blade angle. The flow entering the rotor blade row is essentially radial as viewed from a stationary frame.

(ANS: 348 kW)

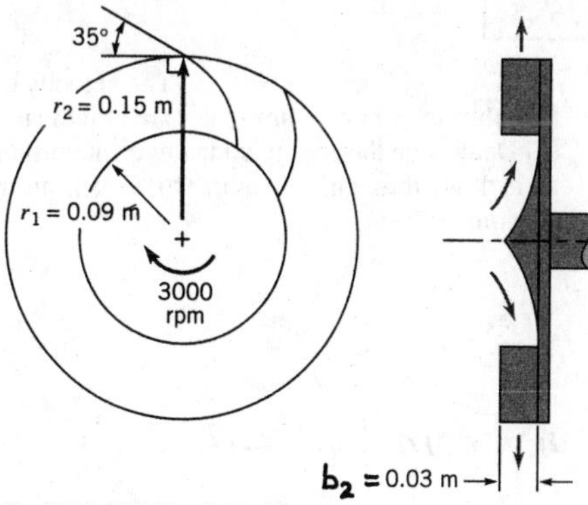

■ FIGURE P12.3R

From Eq. 12.11 with $V_{\theta 1} = 0$ we have

$$\dot{W}_{shaft} = \rho Q (U_2 V_{\theta 2} - U_1 V_{\theta 1}) = \rho Q U_2 V_{\theta 2} \qquad (1)$$

where

$$U_2 = r_2 \omega = (0.15\,m)\left(3000 \frac{rev}{min}\right)\left(2\pi \frac{rad}{rev}\right)\left(\frac{1}{60}\frac{min}{s}\right) = 47.1 \frac{m}{s}$$

To determine $V_{\theta 2}$ we use the exit velocity triangle shown below.

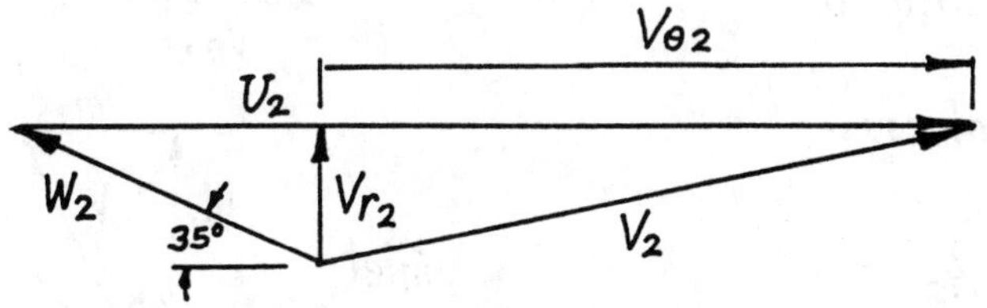

Thus,

$$V_{\theta 2} = U_2 - \frac{V_{r2}}{\tan 35°}$$

with

$$V_{r2} = \frac{Q}{2\pi r_2 b_2} = \frac{\left(200 \frac{liters}{s}\right)\left(\frac{1\,m^3}{1000\,liters}\right)}{2\pi (0.15\,m)(0.03\,m)} = 7.07 \frac{m}{s}$$

Hence,

$$V_{\theta 2} = 47.1 \frac{m}{s} - \frac{7.07}{\tan 35°} = 37.0 \frac{m}{s}$$

From Eq. (1):

$$\dot{W}_{shaft} = \left(999 \frac{kg}{m^3}\right)\left(200 \frac{liters}{s}\right)\left(\frac{1\,m^3}{1000\,liters}\right)\left(47.1 \frac{m}{s}\right)\left(37.0 \frac{m}{s}\right)$$

$$= 3.48 \times 10^5 \frac{N \cdot m}{s} = \underline{\underline{348\,kW}}$$

12.4R

12.4R (Centrifugal pump) The velocity triangles for water flow through a radial pump rotor are as indicated in Fig. P12.4R. **(a)** Determine the energy added to each unit mass (kg) of water as it flows through the rotor. **(b)** Sketch an appropriate blade section.

(ANS: 404 N·m/kg)

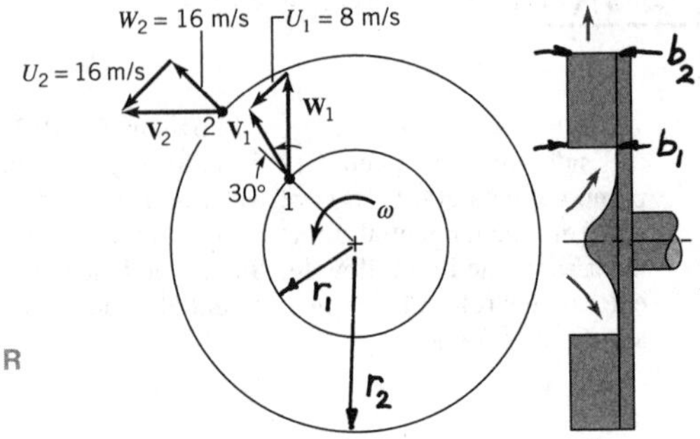

■ FIGURE P12.4R

a) From Eq. 12.12,

$w_{shaft} = U_2 V_{\theta 2} - U_1 V_{\theta 1}$, where since the relative velocity at the exit is in the radial direction (see the figure below), $V_{\theta 2} = U_2 = 16 \frac{m}{s}$ (1)

Also, from the inlet conditions,

$V_{\theta 1} = -V_{r1} \tan 30°$, where the minus sign means that $V_{\theta 1}$ is in the opposite direction of U_1

exit

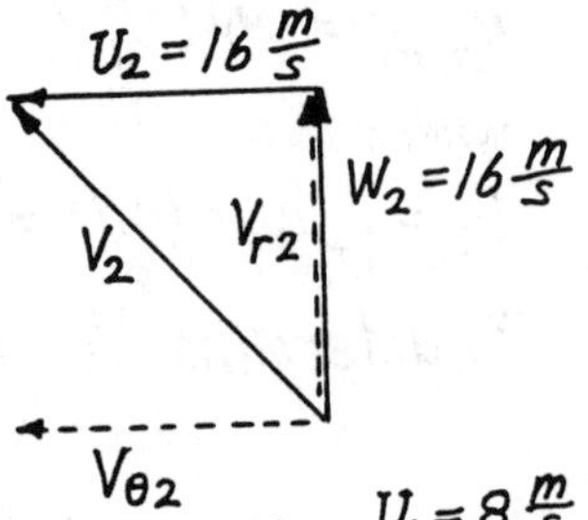

From conservation of mass (with $\rho_1 = \rho_2$) we have

$V_{r1} A_1 = V_{r2} A_2 = W_2 A_2$

inlet

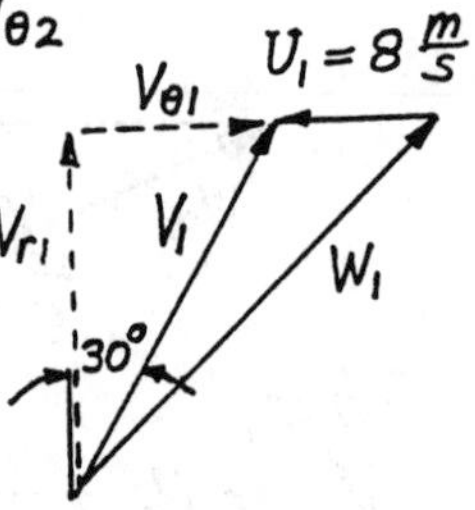

Thus,

$V_{r1} = W_2 \frac{A_2}{A_1} = W_2 \frac{2\pi r_2 b_2}{2\pi r_1 b_1} = W_2 \frac{r_2}{r_1}$ since $b_1 = b_2$

Also,

$U_1 = r_1 \omega$ and $U_2 = r_2 \omega$, or $\frac{r_1}{r_2} = \frac{U_1}{U_2} = \frac{8 \frac{m}{s}}{16 \frac{m}{s}} = 0.5$

Thus, $V_{r1} = (16 \frac{m}{s})/(0.5) = 32 \frac{m}{s}$ so that,

$V_{\theta 1} = -V_{r1} \tan 30° = -(32 \frac{m}{s}) \tan 30° = -18.5 \frac{m}{s}$

From Eq. (1):

$w_{shaft} = (16 \frac{m}{s})(16 \frac{m}{s}) - (8 \frac{m}{s})(-18.5 \frac{m}{s}) = 404 \frac{m^2}{s^2} = 404 \frac{N \cdot m}{kg}$

b) An appropriate blade section would be tangent to W_1 and W_2 at the inlet and exit. Thus, at the exit the blade would be radial. At the inlet,

$\beta_1 = \tan^{-1}\left[\frac{U_1 - V_{\theta 1}}{V_{r1}}\right] = \tan^{-1}\left[\frac{8 \frac{m}{s} - (-18.5 \frac{m}{s})}{32 \frac{m}{s}}\right] = 39.6°$

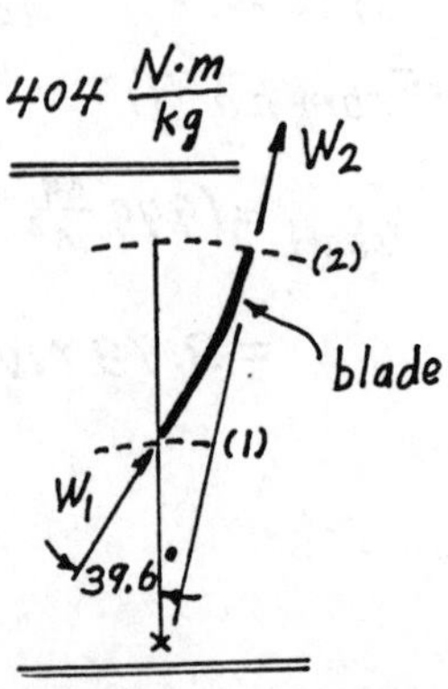

12.5 R

12.5R (Similarity laws) When the shaft horsepower supplied to a certain centrifugal pump is 25 hp, the pump discharges 700 gpm of water while operating at 1800 rpm with a head rise of 90 ft. **(a)** If the pump speed is reduced to 1200 rpm, determine the new head rise and shaft horsepower. Assume the efficiency remains the same. **(b)** What is the specific speed, N_{sd}, for this pump?

(ANS: 40 ft, 7.41 hp; 1630)

a) From Eq. (12.37), for a given pump operating at different speeds

$$\frac{h_{a1}}{h_{a2}} = \frac{\omega_1^2}{\omega_2^2} \quad \text{so that}$$

$$h_{a2} = \left(\frac{\omega_2}{\omega_1}\right)^2 h_{a1} = \left(\frac{1200\ rpm}{1800\ rpm}\right)^2 (90\ ft) = \underline{\underline{40.0\ ft}}$$

Similarly from Eq. (12.38),

$$\frac{\dot{W}_{shaft1}}{\dot{W}_{shaft2}} = \frac{\omega_1^3}{\omega_2^3} \quad \text{so that}$$

$$\dot{W}_{shaft2} = \left(\frac{\omega_2}{\omega_1}\right)^3 \dot{W}_{shaft1} = \left(\frac{1200\ rpm}{1800\ rpm}\right)^3 (25\ hp) = \underline{\underline{7.41\ hp}}$$

b) From Eq. (12.44),

$$N_{sd} = \frac{\omega(rpm)\sqrt{Q(gpm)}}{[h_a(ft)]^{3/4}}$$

so that

$$N_{sd_1} = \frac{(1800\ rpm)\sqrt{700\ gpm}}{(90\ ft)^{3/4}} = \underline{\underline{1630}}$$

Note that for conditions of state (2), from Eq. (12.36)

$$\frac{Q_1}{Q_2} = \frac{\omega_1}{\omega_2} \quad \text{or} \quad Q_2 = \frac{\omega_2}{\omega_1} Q_1 = \left(\frac{1200\ rpm}{1800\ rpm}\right)(700\ gpm) = 467\ gpm$$

Thus,

$$N_{sd_2} = \frac{(1200\ rpm)\sqrt{467\ gpm}}{(40.0\ ft)^{3/4}} = 1630$$

That is, $N_{sd_1} = N_{sd_2}$. The specific speed is constant for the pump regardless of its speed when operating at a constant efficiency.

12.6R

12.6R (Specific speed) An axial-flow turbine develops 10,000 hp when operating with a head of 40 ft. Determine the rotational speed if the efficiency is 88%.

(ANS: 65.4 rpm)

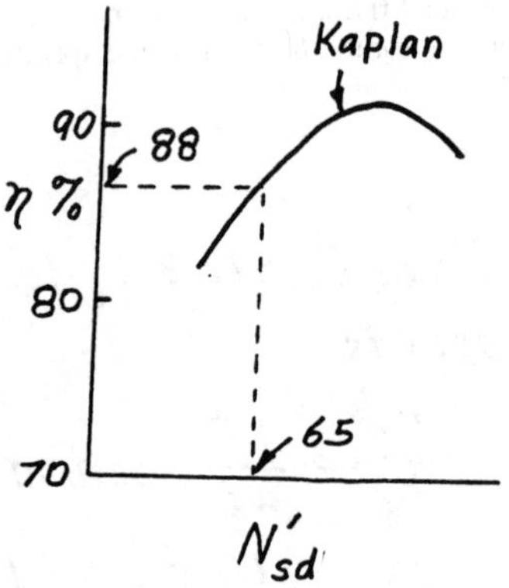

$\dot{W}_{shaft} = 10{,}000\ hp$, $h_T = 40\ ft$, $\eta = 0.88$

From Fig. 12.32, $N'_{sd} \approx 65$

Thus,

$$N'_{sd} = 65 = \frac{\omega \sqrt{\dot{W}_{shaft}}}{h_T^{5/4}} = \frac{\omega \sqrt{10^4}}{40^{5/4}} \quad \text{or} \quad \omega = 65.4\ rpm$$

12.7R

12.7R (Turbine) A water turbine with radial flow has the dimensions shown in Fig. P12.7R. The absolute entering velocity is 50 ft/s, and it makes an angle of 30° with the tangent to the rotor. The absolute exit velocity is directed radially inward. The angular speed of the rotor is 120 rpm. Find the power delivered to the shaft of the turbine.

(ANS: − 1200 hp)

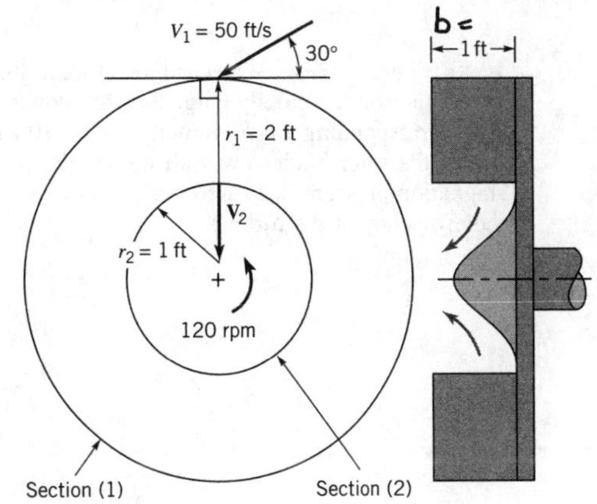

■ FIGURE P12.7R

$$\dot{W}_{shaft} = \rho Q(U_2 V_{\theta 2} - U_1 V_{\theta 1}) \quad \text{where } V_{\theta 2} = 0 \text{ and} \qquad (1)$$

$$V_{\theta 1} = (50 \tfrac{ft}{s}) \cos 30° = 43.3 \tfrac{ft}{s}$$

Also,

$$Q = 2\pi r_1 b V_1 \sin 30° = 2\pi (2 ft)(1 ft)(50 \tfrac{ft}{s}) \sin 30° = 314 \tfrac{ft^3}{s}$$

and

$$U_1 = \omega r_1 = (120 \tfrac{rev}{min})(\tfrac{1 min}{60 s})(\tfrac{2\pi \, rad}{rev})(2 ft) = 25.1 \tfrac{ft}{s}$$

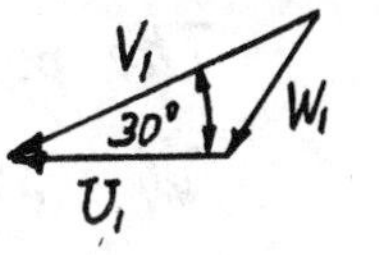

Thus, Eq. (1) gives

$$\dot{W}_{shaft} = (1.94 \tfrac{slugs}{ft^3})(314 \tfrac{ft^3}{s})(-25.1 \tfrac{ft}{s})(50 \cos 30° \tfrac{ft}{s}) = -6.62 \times 10^5 \tfrac{ft \cdot lb}{s}$$

$$\text{or } \dot{W}_{shaft} = (6.62 \times 10^5 \tfrac{ft \cdot lb}{s}) \left(\frac{1 \, hp}{550 \tfrac{ft \cdot lb}{s}} \right) = \underline{\underline{-1,200 \text{ hp}}}$$

12.8R

12.8R (Turbine) Water enters an axial-flow turbine rotor with an absolute velocity tangential component, V_θ, of 30 ft/s. The corresponding blade velocity, U, is 100 ft/s. The water leaves the rotor blade row with no angular momentum. If the stagnation pressure drop across the turbine is 45 psi, determine the efficiency of the turbine.

(ANS: 0.898)

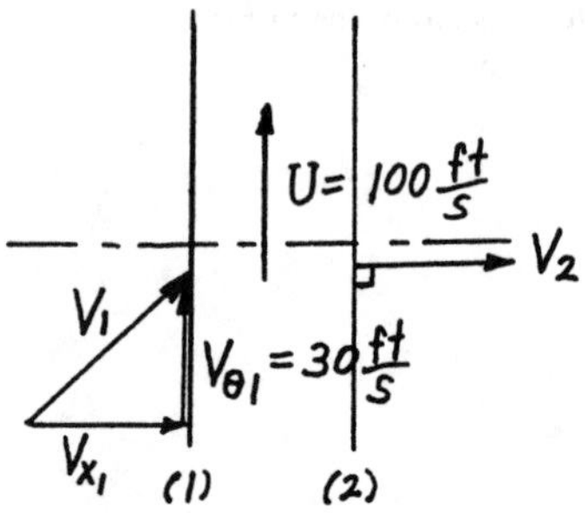

The power removed from the fluid is $\dot{W}_r$, where

$\dot{W}_r = \dot{m}g\left(\frac{\Delta p_s}{\gamma}\right)$, where $\frac{\Delta p_s}{\gamma}$ is the stagnation pressure head drop.

Thus,

$$\dot{W}_r = \rho Q\, g\left(\frac{\Delta p_s}{\rho g}\right) = Q\,\Delta p_s$$

The power removed by the turbine is $\dot{W}_t$, where

$$\dot{W}_t = \rho Q U (V_{\theta 2} - V_{\theta 1}) = -\rho Q U V_{\theta 1} \quad \text{since } V_{\theta 2} = 0$$

Hence,

$$\eta = \text{efficiency} = \frac{|\dot{W}_t|}{\dot{W}_r} = \frac{\rho Q U V_{\theta 1}}{Q\,\Delta p_s} = \frac{\rho U V_{\theta 1}}{\Delta p_s}$$

or

$$\eta = \frac{\left(1.94\,\frac{slugs}{ft^3}\right)\left(100\,\frac{ft}{s}\right)\left(30\,\frac{ft}{s}\right)}{\left(45\,\frac{lb}{in.^2}\right)\left(144\,\frac{in.^2}{ft^2}\right)} = \underline{\underline{0.898}}$$